All-Russian Scientific-Practical Conference with International Participation "Actual Issues of Transport in the Forest Sector"

IOP Conference Series: Materials Science and Engineering
Volume 817

St. Petersburg, Russia
28 - 29 November 2019

ISBN: 978-1-7138-2804-4
ISSN: 1755-8981

Printed from e-media with permission by:

Curran Associates, Inc.
57 Morehouse Lane
Red Hook, NY 12571

Some format issues inherent in the e-media version may also appear in this print version.

No changes have been made to the content of these proceedings. There may be changes to pagination and minor adjustments for aesthetics.

Printed with permission by Curran Associates, Inc. (2021)

Institute of Physics
Dirac House, Temple Back
Bristol BS1 6BE UK

Phone: 44 1 17 929 7481
Fax: 44 1 17 920 0979

techtracking@iop.org

Additional copies of this publication are available from:

Curran Associates, Inc.
57 Morehouse Lane
Red Hook, NY 12571 USA
Phone: 845-758-0400
Fax: 845-758-2633
Email: curran@proceedings.com
Web: www.proceedings.com

TABLE OF CONTENTS

The volume contains a collection of articles on pressing issues of logging, forest transport, the use of forest logging complexes, transport logistics and integrated use of forest products.

The research results presented in the volume were discussed at the All-Russian scientific-practical conference with international participation "Actual issues of transport in the forest sector" held at St. Petersburg State Forest Technical University on November 28-29, 2019.

The articles discuss innovative technologies for the use of logging complexes in forest areas with different environmental conditions, new materials for the construction of forest roads in various conditions, technologies for collecting and processing harvesting residuals and methods for their rational use.

Several methods have been proposed to increase technological efficiency of logging operations and minimize environmental risks associated with the harvesting and transport of forest products. The information on the soil conditions for the use of forest equipment has been analysed. The developments in the area of processing of waste from logging operations have been presented. Original results on the interaction of the moving parts of forest machines with forest soils and surfaces of forest roads have been presented. A model of the consequences of the passage of forestry equipment on a dry road surface has been presented. Methods have been proposed for the construction of forest roads in difficult soil and hydrological conditions and on soils with low bearing capacity.

The articles published in the volume will be of interest to specialists working in the transport, logging and timber processing industries.

The organizing committee of the conference thanks the staff of the Department of Industrial Transport of SPbSFTU who participated in the organization of the conference, as well as a leading specialist of the Congress Department of SPbSFTU, Vadim S. Pavlov, and an assistant professor of the Department of Technological Processes and Machines of the Forestry Complex of SPbSFTU, Ivan V. Bacherikov, who made a valuable contribution to the volume by handling and editing the submitted papers.

Associate Professor of the Department of Industrial Transport, St. Petersburg State Technical University, Ph.D. Gleb V. Grigoriev

IOP Conf. Series: Materials Science and Engineering **817** (2020) 011002 doi:10.1088/1757-899X/817/1/011002

Peer review statement

All papers published in this volume of *IOP Conference Series: Materials Science and Engineering* have been peer reviewed through processes administered by the proceedings Editors. Reviews were conducted by expert referees to the professional and scientific standards expected of a proceedings journal published by IOP Publishing.

IOP Conf. Series: Materials Science and Engineering **817** (2020) 012001 doi:10.1088/1757-899X/817/1/012001

Models for the determination of basic parameters of tracked feller buncher machines

A V Andronov[1*], G S Taradin[1], I A Zverev[2]

[1] Department of Forestry Engineering, Service and Repair, *St. Petersburg State Forest Technical University*, 5 Institutskiy Lane, St. Petersburg 194021, Russian Federation
[2] Institute of Technological Machines and Forest Transportation, *St. Petersburg State Forest Technical University*, 5 Institutskiy Lane, St. Petersburg 194021, Russian Federation

[*]Corresponding e-mail: andronovalexandr@gmail.com

Abstract. Models for the determination of basic parameters of tracked feller bunchers are presented. A statistical analysis has been carried out and the normal law of distribution of the main technical characteristics of the most common models of machinery has been established. Both univariate and multivariate regression equations of the main parameters of the machines have been obtained, depending on the main parameter characterizing the load transmitted from the object of labour, which is cut and packaged by the machine - load capacity.

1. Introduction

The whole tree logging technology currently maintains a strong position both in Russia and around the world. In 2013, in the Russian forest complex about 26% of the total volume of logging was harvested using the whole tree technology [1-10]. The most common system of machines for the whole tree harvesting is: a feller buncher (FB), a skidder and a delimbing machine. The main machine in this system is the FB equipped with a hydraulic manipulator with a gripping-cutting device, a rotary platform with an operator's cabin and a power generating unit mounted on it, and a tracked undercarriage.

The main parameters of the FB are engine power, vehicle mass, and ground pressure. In order to establish the optimal values of the main parameters of any FB, one has to know the nature and strength of influence of the load factor on them.

The purpose of the study was, by means of statistical analysis, to establish the optimal parameters of a FB, an indicator reflecting the load transmitted from a tree to the metal structure of a manipulator and to the whole machine is used as a load factor - load capacity.

2. Methods and Materials

2.1. Object of the Study

Generally speaking, a tracked feller buncher is an excavator type machine. The undercarriage is of a track type, consisting of a track belt equipped with lugs and support rollers. Most modern FBs are equipped with a full-rotary platform, which makes it possible to pack trees behind the machine, thus protecting the undergrowth. Also, some models of machines are equipped with a platform equalizer which allows the machine to work on slopes or uneven ground. An operator's cabin, a power unit compartment and an articulated hydraulic manipulator with a gripping-cutting device, which, in modern machines, is usually equipped with an accumulating device and uses a saw blade as a working body, are mounted on a rotary platform.

FBs are used for cutting standing trees and their further packaging, i.e. for transporting the cut trees using the manipulator behind the machine or at an angle to it. Given a heavy mass of modern FBs (20.000 kg and above), as well as engine power (184 – 246 kW), it can be said that modern models of tracked FBs are designed for clear-cutting, both with and without preservation of the undergrowth. The use of such machines in thinnings, where the ability of the machine to work under the canopy of the forest is required, can be very problematic, due to rather large overall dimensions of the machines in height (about 7 m). It is also possible to use such machines in selective cuttings with a share of the cut component around 0.7, clear fellings, and other types of felling (deforestation for power lines, building construction, etc.). Technical characteristics of the most common FBs are given in table 1.

Table 1. Technical characteristics of the most common feller buncher models.

Company	Model	Power, kW	Mass, kg	Energy saturation, kW/t	Ground pressure, kPa	Load-carrying capacity, kg	Max. manipulator reach, m
John Deere	803M	224	28250	7.93	57.7	5520	8.49
	853M	224	30170	7.42	57.7	6860	8.49
	859M	224	35450	6.32	68.0	6860	8.49
	903M	224	31590	7.09	60.4	7680	8.88
	909M	224	36660	6.11	70.4	7680	8.88
	953M	246	33210	7.41	63.5	10260	8.88
	959M	246	37740	6.52	72.4	10260	8.88
Tigercat	845E	210	26760	7.85	57.3	5760	8.53
	L845E	210	34925	6.01	74.8	5760	8.53
	855E	212	27600	7.68	59.1	8440	8.46
	L855E	212	35600	5.95	75.1	8440	8.46
	860C	224	29260	7.65	62.6	6080	8.46
	870C	224	31300	7.16	66.1	8435	8.46
	L870C	224	35600	6.29	75.1	8435	8.46
	LX870C	224	35600	6.29	75.1	8435	8.46
Caterpillar	511	184	24363	7.55	38.6	7100	8.1
	521	212	27084	7.83	40.8	7100	8.1
	522	212	30410	6.97	45.8	7100	8.1
	532	212	31620	6.71	46.8	7100	8.1
	541	227	30191	7.52	38.4	9400	8.6
	551	227	31057	7.31	39.4	9400	8.6
	552	227	35680	6.36	45.3	9400	8.6
	541-2	226	30826	7.33	39.2	9400	8.6
	552-2	226	35816	6.31	45.4	9400	8.6
	521B	226	27501	8.22	41.1	6100	8.1
	522B	226	32528	6.95	48.5	6100	8.1
Komatsu	XT450L2	224	31545	7.10	64.0	5216	8.18
	XT445L3	224	30926	7.24	61.4	5216	8.18
	XT430L3	224	29239	7.66	57.9	5216	8.18
	XT460L3	224	33710	6.64	68.0	6121	8.18
Valmet	475FXL	246	35951	6.84	73.1	5988	9.49

IOP Conf. Series: Materials Science and Engineering **817** (2020) 012001 doi:10.1088/1757-899X/817/1/012001

The main parameters of an FB are in the following ranges of values:
- Power N_e is from 184 to 246 kW;
- Mass M_m from 24.363 to 35.951 kg;
- Energy saturation N_e/M_m from 5.95 to 8.22 kW/t
- Ground pressure P from 38.4 to 75.1 kPa;
- Load capacity M_g from 5.216 to 10.260 kg;
- Maximum reach L from 8.1 to 9.49 m.

Caterpillar machines produce the lowest pressure on the ground (from 38.4 to 48.5 kPa), despite the standard mass (from 24.363 to 35.680 kg) of modern FBs; this is due to the fact that they have the longest track length compared to models of other companies. The greatest ground pressure is produced by machines with a larger mass.

A wide range of load capacities can be explained by the design features of machines and technological equipment, such as wall thickness and the type of material of the metal structure of the manipulator, the use of a reinforced frame of machines, etc.

The values of power, mass, energy saturation and maximum reach for all machines differ slightly from one to another.

2.2. Research Methods

To assess the reliability of the considered values of the characteristics of feller bunchers, we carried out a statistical analysis. The minimum sample size was 112. It was obtained with a Student criterion of 1.96, a coefficient of variation of 11%, and an accuracy of 2%.

3. Results and Discussion

The results of the statistical analysis are summarized in table 2. The statistical analysis was performed using Statgraphics 18 software.

Table 2. The values of the key statistics of the main characteristics of tracked feller bunchers.

Key Statistics	Key Statistics Values					
	Power N_e, kW	Mass M_m, kg	Energy saturation N_e/M_m, kW/t	Ground pressure P, kPa	Load capacity M_g, kg	Maximum reach L, m
Mathematical Expectation m(X)	223.4	31904.9	7.04	57.4	6958.3	8.47
Standard deviation, σ	11.56	3858.2	0.56	13.02	1597.05	0.33
Distribution law				Normal		
Coefficient of variation C_v, %	5.2	12.1	7.9	22.7	22.9	3.9
Criterion χ^2 (Chi-Square) and its significance level p	12.8 0.23	14.4 0.42	14.8 0.19	11.4 0.25	12.4 0.26	9.4 0.49

According to the data presented in table 2, the values of χ^2 for all parameters $p = 0.19 - 0.49$ were above 0.05. Consequently, the distribution of parameter samples does not differ from the normal distribution. Figure 1 is an example of a graphical representation of the normal energy saturation distribution of a modern tracked FB.

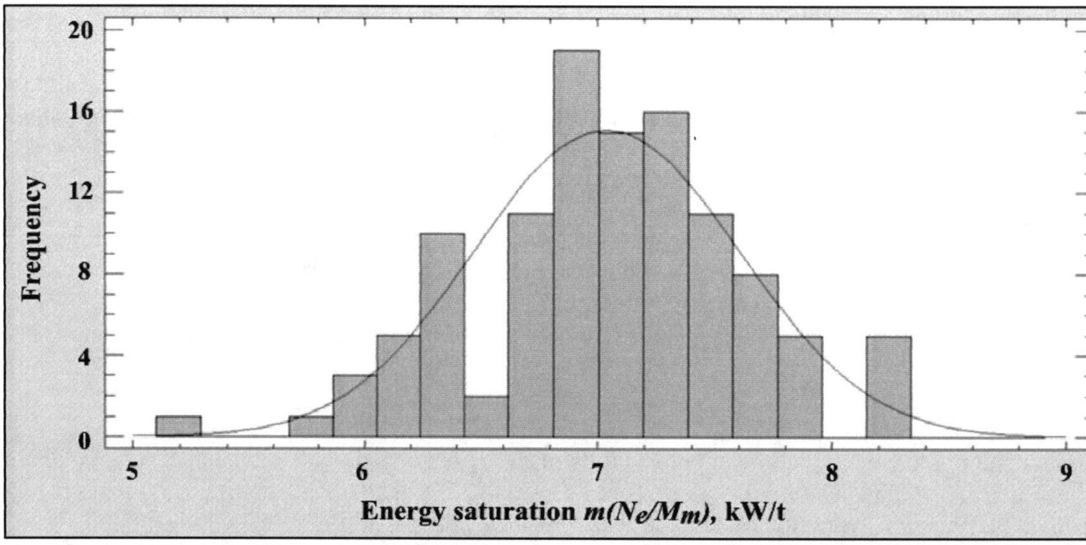

Figure 1. Normal distribution of energy saturation values of modern tracked feller bunchers.

Given the values of the coefficient of variation C_v = 22.7 – 22.9%, the values of mathematical expectation of ground pressure $m(P)$ and load capacity $m(M_g)$ have noticeable fluctuations. Mathematical fluctuations in power $m(N_e)$, energy saturation $m(N_e/M_m)$, mass $m(M_m)$ and maximum reach $m(L)$ vary only slightly. Noticeable fluctuations in ground pressure and load capacity can be explained by the design features of the machines.

One can evaluate the effect of a machine's load capacity M_g on its power N_e, mass M_m and ground pressure P using figures 2 – 4, respectively. These relationships are estimated by linear regressions. The values of the coefficient of determination R^2 of the presented relationships are 0.83, 0.93 and 0.96; therefore, the linear regressions explain 83%, 93% and 96% of variation in power, mass and ground pressure, respectively. The significance levels of the *t*-test and *F*-test statistic for all three equations are less than 0.05, i.e. the coefficients of the equations and the regressions themselves are 95% reliable. The correlation coefficient between the arrays amounted to $r = 0.91; 0.96; 0.98$.

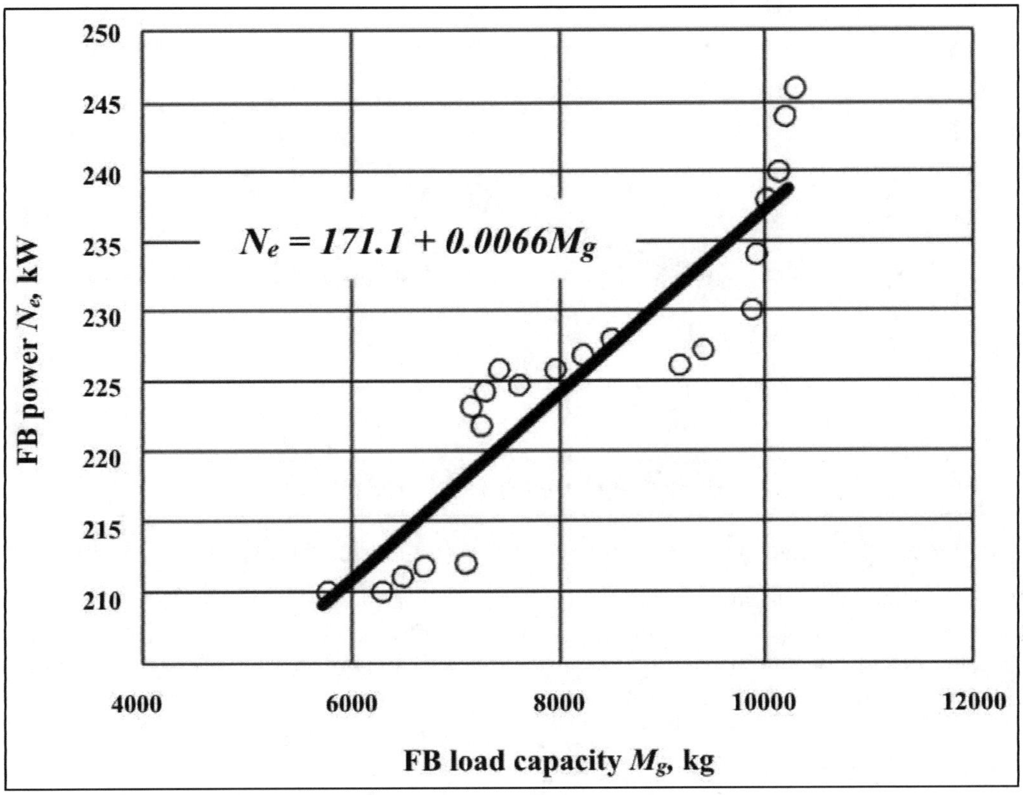

Figure 2. Change in the power of an FB depending on its load capacity.

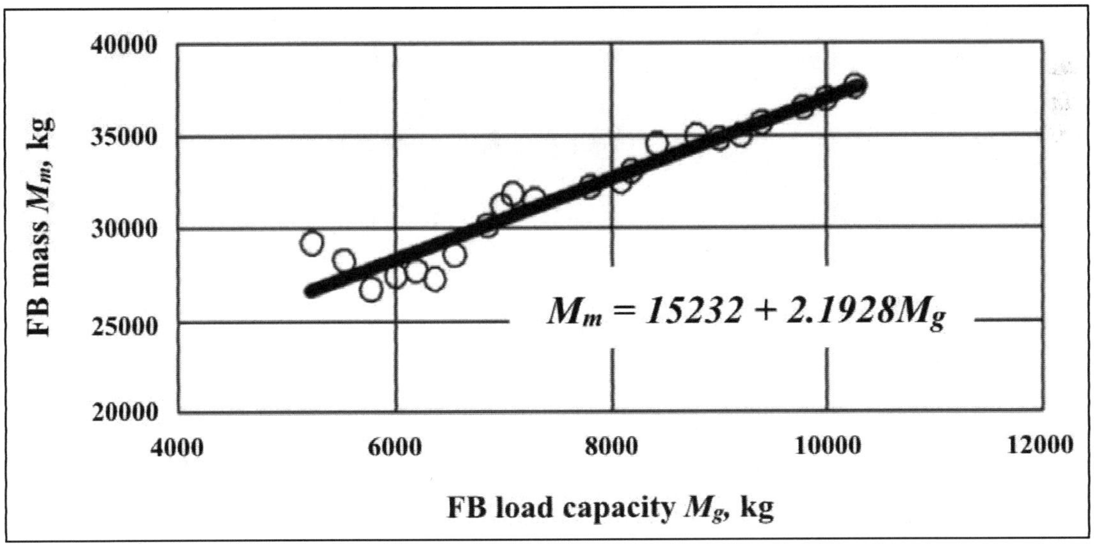

Figure 3. Change in the mass of an FB depending on its load capacity.

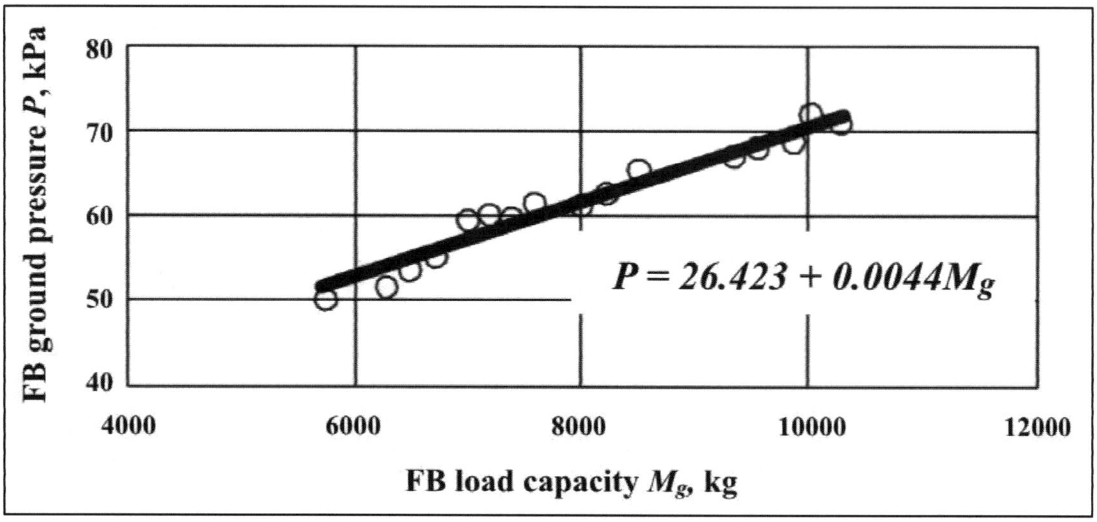

Figure 4. Change in the ground pressure of an FB depending on its load capacity.

Thus, the obtained regression equations show that with an increase in carrying capacity M_g by 1.000 kg, power N_e, mass M_m, and ground pressure P will increase by 7 kW, 2200 kg and 5 kPa, respectively. The given regression equations make it possible to evaluate the main parameters of an FB at the design stage based on the required carrying capacity. Also, these regression equations $N_e = f(M_g)$, $M_m = f(M_g)$ and $P = f(M_g)$ make it possible to determine the values of these parameters for operation depending on the load capacity.

Together with the analysis of univariate regressions, an analysis of multiple linear regressions $N_e = f(M_g, M_m)$, $M_m = f(M_g, N_e)$ and $P = f(M_g, N_e)$ was also carried out; the equations of multiple regressions have the form:

$$N_e = 133.7 + 0.00036M_g + 0.0027M_m, \quad (1)$$

$$M_m = 5508.1 + 49.89N_e + 2M_g, \quad (2)$$

$$P = 10.35 + 0.093N_e + 0.0038M_g. \quad (3)$$

The values of the coefficient of determination R^2 for expressions 1 – 3 are equal to 0.84, 0.97, and 0.96, respectively. The significance levels of the *t*-criterion for the coefficients of all three equations are less than 0.05, therefore, the coefficients of the multiple regressions obtained are 95% reliable. Significance levels of the *F*-test statistics were less than 0.05 for the three equations; therefore, the regression equations themselves are 95% reliable.

4. Conclusion

1. The obtained regression equations made it possible to describe the nature and strength of the main acting factors. Using the obtained equations, it is possible to approximately determine the values of the main parameters of tracked FBs, i.e. power, mass and ground pressure, both at the design stage and when choosing a machine for operation in various natural and industrial conditions, depending on the requirements for load capacity.

2. The resulting multiple regression equations $N_e = f(M_g, M_m)$, $M_m = f(M_g, N_e)$, $P = f(M_g, N_e)$ allow us to estimate the combined effect of the load capacity and mass on the FB power; the effect of the load capacity and power on the mass of the machine; and the effect of the carrying capacity and power on the ground pressure. Thus, these equations make it possible to comprehensively evaluate and select the main parameters of tracked feller bunchers.

IOP Conf. Series: Materials Science and Engineering **817** (2020) 012001 doi:10.1088/1757-899X/817/1/012001

References

[1] Bolshakov B M 2015 Actual issues of the development of forestry machinery and equipment. *Construction and Road Building Machnery* **2** pp 21-24
[2] Deere & Company 2019 Tracked feller bunchers/harvesters John Deere 800M/MH-Series Brochure p 24
[3] Deere & Company 2019 Tracked feller bunchers/harvesters John Deere 900M/MH-Series Brochure p 28
[4] Tigercat Inc. 2019 845E/L845E Feller Buncher Brochure p 6
[5] Tigercat Inc. 2018 855E/L855E Feller Buncher Brochure p 6
[6] Tigercat Inc. 2016 860C/870C/L870C/LX870C Feller Buncher Brochure p 4
[7] Caterpillar Inc. 2009 541/551/552 Tracked feller bunchers and harvesters brochure p 20
[8] Caterpillar Inc. 2009 511/521/522/532 Tracked feller bunchers and harvesters brochure p 20
[9] Caterpillar Inc. 2013 521B/522B Tracked feller bunchers and harvesters brochure p 20
[10] Komatsu America Corp. 2015 XT460L-3 brochure p 16

IOP Conf. Series: Materials Science and Engineering **817** (2020) 012002 doi:10.1088/1757-899X/817/1/012002

Physical and mechanical properties of forest soils: a review

S M Bazarov*, V A Alexsandrov, Y I Belenky, G S Gasimov, A V Kalistratov

Institute of Technological Machines and Forest Transportation, *St. Petersburg State Forest Technical University*, 5 Institutskiy Lane, St. Petersburg 194021, Russian Federation

*Corresponding email: s.bazarow@yandex.ru

Abstract. The purpose of our article is to analyze literary sources and supplement classification of soils with missing information that simplifies implementation of mathematical models of the interaction of movers of forest machines with soils. Our research methods include analysis, compilation and systematization of reference and experimental data, and approximation of calculated data. The classification of forest soils by strength categories has been clarified, based on the identification of the characteristic combinations of the physical and mechanical properties of soils — the general deformation modulus, specific adhesion, internal friction angle, specific gravity, and the thickness of the deformed layer. As a result of the analysis of literary sources and calculations, the classification of forest soils has been supplemented with missing information that simplifies the implementation of mathematical models of the interaction of the movers of forest machines with soils. An updated version of the classification includes data on the rheological properties of the soil, the shear modulus, bearing capacity, and the cone index of the soil. Based on the information obtained, linear relationships have been developed that reveal the interconnections of the physical and mechanical properties of soils.

1. Introduction

Modeling the interaction of forest machines with supporting surfaces such as forest soils is an important area of research in the field of forest industry science [1-5]. Based on the results of mathematical modeling, recommendations have been formulated to increase the environmental friendliness of forest harvesting and transport, reduce damage to the forest ecosystems caused by rutting and soil compaction, and the technological efficiency of skidding associated with the ability of vehicles to move across the terrain [1-5]. In forest engineering, mathematical models based on the Russian theory of the movement of road vehicles in off-road conditions developed by Prof. Y S Ageykin [1-5] have attracted considerable attention. The models use several characteristics of the soil, namely, the elastic modulus E_0, Poisson coefficient ν, general deformation modulus E, long-term deformation modulus E_∂, soil viscosity η, specific adhesion C, angle of internal friction of the particles φ, shear modulus G, thickness of the deformed soil layer H, resistance of soil to sensing CI (cone index), and load bearing capacity p_s [1-5]. A classification of forest soils by strength categories [6] was previously proposed, based on the identification of the characteristic combinations of the physical and mechanical properties of soils: E, C, φ, γ, H. Unfortunately, not all the characteristics of soils were included in the classical version of the classification.

The purpose of our article is to analyze literary sources, carry out calculations, and supplement the classification of soils with missing information that simplifies the implementation of mathematical models of the interaction of the movers of forest machines with soils.

Research methods consist of analysis, compilation and systematization of reference and experimental data, and approximation of calculated data.

2. Results and Discussion

The long-term deformation modulus E_∂ and viscosity η are introduced when rheological models of soils are used to describe the development of deformations, taking into account the time of impact of the mover. This approach allows us to take into account the influence of the speed of the machine on the rutting and compaction of the soil [7]. The construction of complex combinations of rheological models of soil and their use were considered in [8]; it was shown [7] that the physical phenomena accompanying the formation of a rut can be well described using rheological models of bodies of Maxwell – Thompson and Burgers [7].

It was established [7, 8] that the total and long-term deformation moduli*E and* E_∂ are approximately equal:

$$E_\partial \approx E \tag{1}$$

In addition, it was experimentally established that the viscosity of forest soils η is closely related to the total deformation modulus E [7]:

$$\eta = (0.78 \pm 0.06)E^{1.92 \pm 0.024} \tag{2}$$

To determine the shear modulus of loose soils, such as sand, the following relationship is used [8]:

$$G = \frac{E_0}{2 \cdot (1+\nu)}, \tag{3}$$

moreover, the elastic modulus of forest soils exceeds the total deformation modulus by approximately five times [8]:

$$E_0 \approx 5E. \tag{4}$$

However, forest soils are compact; in this case, an estimate of the shear modulus G by the formula [8] is more accurate:

$$G = V_S^2 \frac{\gamma}{g} \approx 290\gamma \tag{5}$$

where, V_S is the shear strain propagation velocity in the medium (≈ 53.34 m/s), g is the gravity acceleration.

To calculate the soil resistance to sensing, the following equation is given in [6, 9-12]:

$$\begin{cases} CI = -C\cot\varphi + \Theta \cdot \dfrac{24G^m(\tan\alpha + \tan\varphi)(1+\sin\varphi)\tan\alpha}{d^2\gamma^2(m-2)(m-3)(3-\sin\varphi)\tan^3\varphi} \\ \Theta = [C + (Z+l)\cdot\gamma\tan\varphi]^{3-m} - (C + Z\cdot\gamma\tan\varphi)^{2-m} \cdot [C + (Z+3l-lm)\cdot\gamma\tan\varphi] \\ m = \dfrac{4\sin\varphi}{3(1+\sin\varphi)} \end{cases} \tag{6}$$

where, d is the diameter of the base of a cone pressed into the soil (0.0363 m), α is the angle at the apexof the cone (30°), L is the length of the conical part of an indenter (0.0376 m), Z is the distance from the surface of the soil to the conical part of the indenter ($10L$).

To calculate the bearing capacity of the soil the following equation have been obtained [13-16]:

$$\begin{cases} p_s = p_{s0}\alpha_Z \\ p_{s0} = 0.5K_1N_1\gamma b + N_2\gamma h + K_3N_3C \\ K_1 = \dfrac{l}{l+0.4b}; K_3 = \dfrac{l+b}{l+0.5b} \\ N_1 = \dfrac{1-S^4}{S^5}; N_2 = \dfrac{1}{S^2}; N_3 = \dfrac{2(1+S^2)}{S^3}; S = \mathrm{tg}\left(\dfrac{\pi}{4} - \dfrac{\varphi}{2}\right) \\ \alpha_Z = 1 + \dfrac{H^*h}{2H\cdot(H-h-0{,}25H^*)}; H^* = \dfrac{\sqrt{2}}{2}\exp\left[\left(\dfrac{\pi}{4}+\dfrac{3\varphi}{4}\right)\tan\dfrac{3\varphi}{4}\right]b\cos\dfrac{3\varphi}{4}\,\mathrm{tg}\,\varphi \end{cases} \quad (7)$$

where,p_{s0} is the bearing capacity with unlimited thickness of the soil layer, α_z is the correction coefficient for taking into account the thickness of the deformed soil layer, N are the coefficients for taking into account the frictional properties of the soil, S, H^* are supplementary designations, K are the coefficients for taking into account the geometry of the contact area of the mover with the soil, b is the width of the area of contact of the mover with the soil, and l is the length of the area of contact of the mover with the soil.

The characteristic values of the physical and mechanical properties of forest soils [6], supplemented by the results of calculations using formulas (1), (2), (4) – (7), are presented in table 1. When calculating the bearing capacity, it is accepted: b = 0.7 m, l = 0.8 m, h = 0.2 m for soil categories II and III, and h = 0.05 m for category I.

Table 1. Physical and mechanical properties of forest soils of various strength categories.

Parameter	Abbreviation	Unit	III (weak)	II (medium strength)	I (strong)
General deformation modulus	E	MPa	0.4	1	3
Long-term deformation modulus	E_∂	MPa	0.4	1	3
Elasticity modulus	E_0	MPa	2.0	5.0	15.0
Shear modulus	G	MPa	2.192	2.44	2.77
Poisson coefficient	ν	-	0.35	0.25	0.15
Viscosity	η	MPa s	1.343	7.8	64.29
Specific cohesion	C	MPa	0.005	0.012	0.024
Internal friction angle	φ	o	11	15	16
Specific gravity	γ	MN/m^3	0.0075	0.0085	0.0095
Cone index	CI	MPa	0.243	0.464	1.003
Load bearing capacity	p_s	MPa	0.0504	0.117	0.287
Deformed layer thickness	H	m	0.8	0.4	0.3

The data in table 1 related to the cone index and the bearing capacity of forest soils are illustrated by the graphs in figure 1.

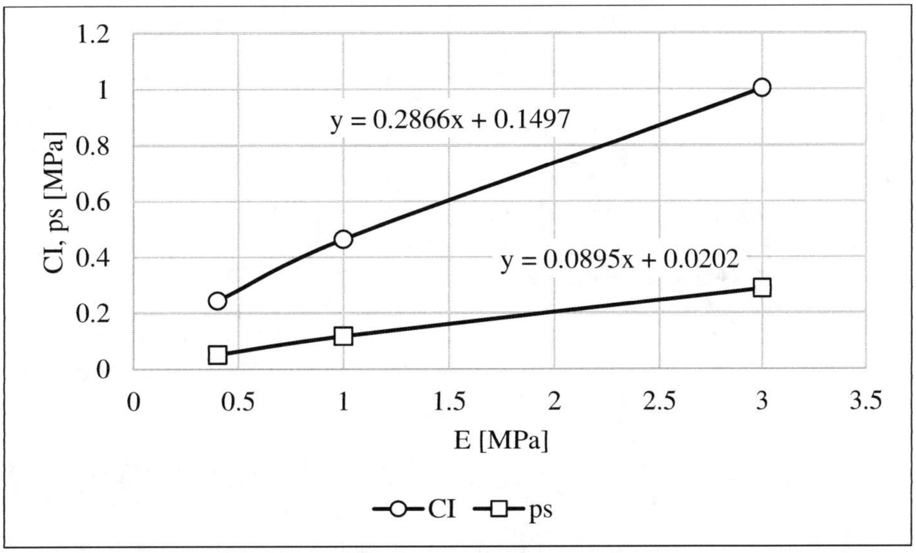

Figure 1. Relationship between the bearing capacity and cone index of forest soils and the deformation modulus.

Based on the information obtained, linear relationships have been developed that reveal the interconnections of the physical and mechanical properties of soils — the module of general deformation, load bearing capacity, and cone index. Processing the calculation results shows that the cone index *CI* [MPa] and the bearing capacity of forest soils p_s [MPa] can be expressed with high accuracy by the linear functions of the general deformation modulus *E* [MPa]:

$$CI = 0.2866E + 0.1497 \quad (8)$$

$$p_s = 0.0895E + 0.0202 \quad (9)$$

3. Conclusion

Earlier, a classification of forest soils by strength categories was proposed, based on the identification of the characteristic combinations of the physical and mechanical properties of soils — the general deformation modulus, specific adhesion, internal friction angle, specific gravity, and the thickness of the deformed layer. As a result of the analysis of literary sources and calculations, the classification of forest soils has been supplemented with missing information that simplifies the implementation of mathematical models of the interaction of the movers of forest machines with soils. An updated version of the classification includes data on the rheological properties of soils, the shear modulus, load bearing capacity, and the cone index of the soil. Based on the information obtained, linear relationships have been developed that reveal the interconnections of the physical and mechanical properties of soils — the module of general deformation, bearing capacity, and cone index.

References

[1] Khitrov E, Andronov A, Iliushenko D and Kotenev E 2019 Comparing approaches of calculating soil pressure of forestry machines Int. Multidisciplinary Scientific GeoConference Surveying Geology and Mining Ecology Management, SGEM 19 (vol 19) issue 3.2 pp 649–655 DOI: 10.5593/sgem2019/3.2/S14.084

[2] Khitrov E, Andronov A, Bogatova E and Kotenev E 2019 Development of recommendations on environmental certification of forestry machinery drives Int. Multidisciplinary Scientific GeoConference Surveying Geology and Mining Ecology Management, SGEM 19 (vol 19) issue 3.2 pp 689–696 DOI: 10.5593/sgem2019/3.2/S14.089

[3] Khitrov E G and Andronov A V 2019 Mathematical model of interaction between forest

machine mover and consolidating soil IOP Conf. Series: J. Phys. 1177 012030 DOI: 10.1088/1742-6596/1177/1/012030

[4] Ivanov V, Stepanishcheva M, Khitrov E and Iliushenko D 2018 Theoretical model for evaluation of tractive performance of forestry machine's wheel Int. Multidisciplinary Scientific GeoConference Surveying Geology and Mining Ecology Management, SGEM 18 pp 997–1004 DOI: 10.5593/sgem2018/3.2/S14.127

[5] Kochnev A and Khitrov E 2018 Theoretical models for rut depth evaluation after a forestry machine's wheel passover Int. Multidisciplinary Scientific GeoConference Surveying Geology and Mining Ecology Management, SGEM 18 pp 1005–12 DOI: 10.5593/sgem2018/3.2/S14.128

[6] Khitrov E, Kochnev A, Ivanov V and Stepanishcheva M 2018 Linking the deformation moduli and cone indices of forest and peatland soils Int. Multidisciplinary Scientific GeoConference Surveying Geology and Mining Ecology Management, SGEM 18 pp 297–304 DOI: 10.5593/sgem2018/3.2/S13.039

[7] Peskov V B 2018 Improving models for assessing rutting and compaction of soil under the influence of movers of wheeled forest machines, diss. cand. tech. sciences (Arkhangelsk: S(A)FU)

[8] Lukhminsky V A 2018 Improving models and methods for predicting the ability of tracked forest machines to negotiate the terrain, diss. cand. tech. sciences (Arkhangelsk: S(A)FU)

[9] Khitrov E G, Grigoriev G V, Dmitrieva I N and Ilyushenko D A 2014 Calculation of the cone index by the magnitude of the deformation modulus of the forest soil Systems. Methods. Technologies 4 127–131

[10] Khitrov E G, Dmitrieva M N, Lukhminsky V A, Khakhina A M, Grigoryev I V and Ivanov V A 2017 Theoretical estimation of wetland soil cone index [In Russian – Teoreticheskij rasch et konusnogo indeksa zabolochennogo grunta] Systems. Methods. Technologies 4 152–156 DOI: 10.18324/2077-5415-2017-4-152-156

[11] Khitrov E G, Khakhina A M, Lukhminsky V A and Kazakov D P 2017 Study of relationship between cone index and deformation modulus of various soil types [In Russian - Issledovanie svyazi konusnogo indeksa i modulya deformacii razlichnyh tipov gruntov] Resources and Technology 14 1–16 DOI: 10.15393/j2.art.2017.3941

[12] Rohani B and Baladi G Y 1981 Correlation of mobility cone index with fundamental engineering properties of soil (US Army Engineer Waterways Experiment Station) p 41

[13] Larin V V 2007 Methods for predicting the support ability of multi-axle wheeled vehicles in the terrain, DSc. Thesis: 05.05.03 (Moscow)

[14] Khitrov E G, Bozhbov V E and Ilyushenko D A 2014 Calculation of load-bearing capacity for forest potting soils influenced by the wheeled running gears [In Russian - Raschet nesushhej sposobnosti lesnyh pochvogruntov pod vozdejstviem kolesnyh dvizhitelej] Systems. Methods. Technologies 4 122–126

[15] Khitrov E G, Kotenev E V, Andronov A V, Taradin G S and Bozhbov V E 2019 Theoretical calculation of bearing capacity of cohesive soil on the cone index and mechanical properties [In Russian - Teoreticheskij raschet nesushchej sposobnosti svyaznogo grunta po konusnomu indeksu i mekhanicheskim svojstvam] Izvestia Sankt-Peterburgskoj lesotehniceskoj akademii 226 111–123 DOI: 10.21266/2079-4304.2019.226.111-123

[16] Khitrov E G, Peskov V B, Kazakov D P, Bozhbov V E and Stepanishcheva M V 2018 Method of solving the problem of stamp indention into inhomogeneous soil massif [In Russian - Metod resheniya zadachi o vdavlivanii shtampa-dvizhitelya v neodnorodnyj massiv grunta] Systems. Methods. Technologies 2 116–120 DOI: 10.18324/2077-5415-2018-2-116-120

IOP Conf. Series: Materials Science and Engineering **817** (2020) 012003 doi:10.1088/1757-899X/817/1/012003

System analysis of forest harvesting logistics

S M Bazarov*, V A Alexsandrov, Y I Belenky, G S Gasimov, A N Soloviev

Institute of Technological Machines and Forest Transportation, *St. Petersburg State Forest Technical University*, 5 Institutskiy Lane, St. Petersburg 194021, Russian Federation

*Corresponding email: s.bazarow@yandex.ru

Abstract. From the general position of logistics, forest harvesting technological complexes are multi-factor logistic flows formed by target operations of a complex production process. System analysis is included in forest harvesting logistics by constructing the functional time of the integrated connectivity of the operations performed. Minimum time of storage operations and maximum synchronization of operations in functional time are necessary conditions for optimizing logging production technologies.

1. Introduction

Under the market conditions, a sustainable development of forest industry enterprises largely depends on how much the basics of logistics, systems theory, analysis of multivariate operations, optimal management, mathematical economics, ecology, etc., are used in the formation and management of forest harvesting and transport taking into account the specifics of forest technologies.

Rational planning, organization, management and control of the processes of production and movement of material, energy, financial, information and other flows in their deep integrated connectivity is formulated by logistics. In view of the widespread use of the general principles of logistics in various areas of national economic activity, one distinguishes between the following types of logistics: military, industrial, forestry, business, economic, transport, customs, etc. [1-5]. In the logistic structure, the key element is production in its cohesive unity [6].The aim of the study is to adapt the basic tools of logistics: time, optimality, minimum energy and financial costs, etc., to the production of specific logging complexes, which are important components of forest industry logistics [2].

2. Methods and Materials

In the mathematical modeling of logistic flows as deeply integrated systems, their presentation as algebraic structures is of decisive importance: the set of elements endowed with a property (objective function) and the mutual relation between them [7]. An important algebraic system is a group, when the binary relations between elements are of a group nature, which ensures integrity and harmony [8, 9]. With a systematic approach, the connectivity of logistics flows is investigated in the functional time of production associated with the dynamics of the production process [10].

In forest harvesting, as a coherent system of technological operations, material, energy and financial flows are characterized by dynamic parameters: capacity, technical power, cost and unit cost. With a systematic approach, the coherence of operations occurs in the functional time of production of a unit of the subject of labor, and of investment of a unit of energy and a monetary unit [10].

IOP Conf. Series: Materials Science and Engineering **817** (2020) 012003 doi:10.1088/1757-899X/817/1/012003

3. Results and Discussion

In the North-West of the Russian Federation, two characteristic processes for the production of assortments are applied:

- based on machine technology: harvester - forwarder - timber truck - sawmill;
- based on mechanized technology: chainsaw - forwarder - timber truck –sawmill.

In the functional time of the coherence of production operations, their effective capacity *P*, power *N*, cost *C*, and unit cost are respectively equal to [10]

$$P = n / \sum_{i=1}^{n} P^{-1}_{i}, \qquad (1)$$

$$N = n^2 / \sum_{i=1}^{n} N_i^{-1}, \qquad (2)$$

$$C = n^2 / \sum_{i=1}^{n} C_i^{-1}, \qquad (3)$$

$$c = C/P, \qquad (4)$$

where, n is the number of sequentially performed technological operations by a complex of forestry equipment.

Based on the information provided by a forestry enterprise (FE), we perform a systematic analysis of these technologies.

Table 1 provides three-factor information of the FE on the production of assortments based on machine technology.

Table 1. Three-factor information on the productionof assortments based on machine technology.

	Capacity, m^3/h	Power, kW	Cost, Rub/h
Harvester	15	205	1614
Forwarder	15	205	1899
Forest truck	16	134	1225
Sawmill	4	28	480

System analysis is feasible for three options of continuous operation:

- without storage of assortments (table 2);

Table 2. Three-factor information on the productionof assortments based on machine technology without storage of assortments.

	Capacity, m^3/h	Power, kW	Cost, Rub/h
Harvester	15	205	1614
Forwarder	15	205	1899
Forest truck	16	134	1225
4 sawmills	16	112	1920

- two hour storage of assortments (table 3);

Table 3. Three-factor information on the production of assortments based on machine technology with two hour storage.

	Capacity, m^3/h	Power, kW	Cost, Rub/h
Harvester	15	205	1614
Forwarder	15	205	1899
Forest truck	16	134	1225
2 sawmills	8	56	960

- four hour storage of assortments (table 4);

Table 4. Three-factor information on the production of assortments based on machine technology with four hour storage.

	Capacity, m^3/h	Power, kW	Cost, Rub/h
Harvester	15	205	1614
Forwarder	15	205	1899
Forest truck	16	134	1225
2 sawmills	4	28	480

The first option (table 2). The productivity of the harvester-forwarder-timber truck system is 15.2 m^3/hour, with continuous synchronized operation of the entire system four frames are needed, in this case in the system: the functional capacity is 15.5 m^3/hour; functional power of 147.8.0 kW; functional cost of 1666.7 Rub/hour; accordingly, the efficiency coefficient of technological capacity is 0.25; coefficient of efficiency of technological power is 0.23; coefficient of efficiency of technological costs is 0.25; and the unit cost is 107.5 Rub/m^3.

The second option (table 3). With two-hour storage of assortments, two frames are necessary; in this case in the system: the functional capacity is 7.8 m^3/hour; functional power is 108.0 kW; functional cost is 1333.3 Rub/hour; accordingly, the efficiency coefficient of technological capacity is 0.14; efficiency coefficient of technological power is 0.19; efficiency coefficient of technological costs is 0.23; and the unit cost is 171.0 Rub/m^3.

The third option (table 4). With four-hour storage of assortments, one frame works: the functional capacity is 4.0 m^3/hour; functional power is 70.2 kW; functional cost is 1000.0 Rub/hour; accordingly, the efficiency coefficient of technological capacity is 0.08; efficiency coefficient of technological power is 0.12; efficiency coefficient of technological costs is 0.19; and the unit cost is 250 Rub/m^3.

The calculation results of the three options for machine technology are summarized in table 5.

Table 5. Three factor information on the technical and economic efficiency of assortment production based on machine technology.

Options	1	2	3
Capacity, m^3/hour // efficiency coefficient	15.5 // 0.25	7.8 // 0.14	4.0 // 0.08
Power, kW // efficiency coefficient	147.8 //0.23	108.0 // 0.19	70.2 // 0.12
Cost, Rub/hour // efficiency coefficient	1666.7 // 0.25	1333.3 // 0.23	1000.0 // 0.19
Unit cost, Rub/m^3	107	171.0	250

Based on the information provided by the FE (table 6), we perform a system analysis of the mechanized technology of production of assortments.

Table 6. Three-factor information on the procurement production of assortments based on mechanized technology.

	Capacity, m^3/hour	Power, kW	Cost, Rub/h
Chainsaw	3	3.4	588
Forwarder	15	205	1899
Forest truck	16	134	1225
Sawmill	4	28	480

System analysis is feasible for three options for continuous operation:
- without storage of assortments (table 7);

Table 7. Three-factor information on the production of assortments based on machine technology without storage of assortments.

	Capacity, m^3/hour	Power, kW	Cost, Rub/h
5 chainsaws	15	17	2940
Forwarder	15	205	1899
Forest truck	16	134	1225
4 sawmills	16	112	1920

- two hour storage of assortments (table 8);

Table 8. Three-factor information on the production of assortments based on mechanized technology with two hour storage of assortments.

	Capacity, m^3/hour	Power, kW	Cost, Rub/h
5 chainsaws	15	17	2940
Forwarder	15	205	1899
Forest truck	16	134	1225
2 sawmills	8	56	960

- four hour storage of assortments (table 9);

Table 9. Three-factor information on the production of assortments based on mechanized technology with four hour storage of assortments.

	Capacity, m^3/hour	Power, kW	Cost, Rub/h
5 chainsaws	15	17	2940
Forwarder	15	205	1899
Forest truck	16	134	1225
Sawmill	4	28	480

The first option (table 7). The performance of the system five chainsaws - forwarder - timber truck is 15.5 m^3/hour, with continuous synchronized operation of the entire system, four frames are needed, in this case in the system: the functional capacity is 15.5 m^3/hour; functional power is 49.4 kW; functional cost is 1904.8 Rub/hour; accordingly, the efficiency coefficient of technological capacity is

0.25; efficiency coefficient of technological power is 0.23; efficiency coefficient of technological costs is 0.25; and the unit cost is 38.6 Rub/m^3.

The second option (table 8). With two hour storage of assortments, two frames are needed, in this case in the system: the functional capacity is 7.8 m^3/hour; functional power is 43.5 kW; functional cost is 1538.5 Rub/hour; accordingly, the efficiency coefficient of technological capacity is 0.14; the efficiency coefficient of technological power is 0.11; efficiency coefficient of technological costs is 0.22; and the unit cost is 35.4 Rub/m^3.

The third option (table 9). With four hour storage of assortments, one frame works: the functional capacity is 4.0 m^3/hour; functional power is 35.7 kW; functional cost is 1111.2 Rub/hour; accordingly, the efficiency coefficient of technological capacity is 0.08; efficiency coefficient of technological power is 0.09; efficiency coefficient of technological costs is 0.17; and the unit cost is 31.1 Rub/m^3.

The results of the calculations of mechanized technology are summarized in table 10.

Table 10. Three factor information on the technical and economic efficiency of assortment production based on mechanised technology.

Options	1	2	3
Capacity, m^3/h // efficiency coefficient	15.5 // 0.25	7.8 // 0.14	4.0 // 0.08
Power, kW // efficiency coefficient	49.4 // 0.23	43.5 // 0.11	35.7 // 0.09
Cost, Rub/h // efficiency coefficient	1904.8 // 0.25	1538.5 // 0.22	1111.2 // 0.17
Unit cost, Rub/m^3	38.6	35.4	31.1

4. Conclusion

The technical and economic efficiency of the three-factor flows of logging production (material, energy and financial), as a deeply integrated dynamic structure, is formed in the functional time of the flow of related operations of the technological process. The operation of storing the subject of labor reduces the productivity of the complexes, the optimization of technology is achieved by maximizing the synchronization of the dynamic parameters of production flows in the functional time of the operations performed.

References

[1] Omel`chenko *et al. Industrial Logistics* 1997 [in Russian – Promyshlennaya logistika] ed A A Kolobov (Moscow: Moscow State University) pp 3-4
[2] Salminen E O, Borozna A A and Tyurin NA 2001 *Forest industry logistics* [in Russian – Lesopromyshlennaya logistika] (Saint Petersburg: SPbFTU) pp 3-4
[3] Semenenko A I 1997 *Entrepreneurial logistics* [in Russian – Predprinimatel'skaya logistika] (Saint Petersburg: Polytechnica) pp 24-29
[4] Smekhov A A 1995 *Basics of transport logistics* [in Russian – Osnovy transportnoj logistiki] (Moscow: Transport) pp 8-11
[5] Stakhanov D V and Stakhanov V N 2000 *Customs logistics* [in Russian – Tamozhennaya logistika] (Moscow: PRIOR) pp 3-4
[6] Protsenko O P and Protsenko I O 2012 Logistics and supply chain management - a look into the future [in Russian – Logistika i upravlenie cepyami postavok – vzglyad v budushchee] (Moscow: DELO) pp 6-10
[7] Gurov S V 2008 *Theory of systems analysis and decision making* [in Russian – Teoriya sistemnogo analiza i prinyatiya resheniya] (Saint Petersburg: SPbFTU) pp 7-9
[8] Faure R, Kaufmann A and Denis-Papin M 1964 *Mathematiques nouvelles* (Paris) pp 68-69
[9] Fried E 1972 *Absztrakt algebra* (Budapest) pp 27-28
[10] Bazarov S M, Belenky Y I and Soloviev A N 2018 *Fundamentals of a systems analysis of production processes* [in Russian – Osnovy sistemnogo analiza proizvodstvennyh processov] (Saint Petersburg: SPbFTU) pp 5-8

IOP Conf. Series: Materials Science and Engineering **817** (2020) 012004 doi:10.1088/1757-899X/817/1/012004

Management of transport and logistic infrastructure of the territory: methodological tools and their improvement

T L Bezrukova[1*], T R Tereshkina[2], S Y Nesterov[1], I V Kuksova[1], O A Pecherskaya[1]

[1]Department of Economics and Finance, *Voronezh State University of Forestry and Technologies named after G F Morozov,* 8 Timiryazeva Street, Voronezh 394087, Russian Federation
[2]*Saint-Petersburg State University of Industrial Technologies and Design*, 4 Ivana Chernykh Street , St. Petersburg, 198095, Russian Federation

[*]Corresponding: e-mail: bezrukova_t_l@mail.ru

Abstract. This article discusses the main problems of transport and logistics infrastructure management, which consist in the absence of one criterion for assessing the effectiveness of its operation, in the presence of institutional barriers for the development of transport and logistics complexes of territories, in the need for a new scheme and management system due to changing economic conditions dictated by globalization. The priority directions for the development of constituent entity of the Russian Federation - Voronezh region - are presented. Voronezh region has an advantageous geographical position and important transport corridors pass through its territory. A multi-level mathematical model has been built by its example for projects of regional transport infrastructure development, given by the initial conditions using multi-criteria optimization of the regional transport system that solves the problem in a dual formulation, based on a hypergraph superimposed on spatial planning scheme of this area.

1. Introduction

Business organizations, realizing their products, while meeting the needs of population in a certain territory and beyond its borders, need a developed transport and logistics infrastructure. This is a necessary criterion in the operation of economic entities, due to the limited number of commodity markets in the regions in terms of capacity and in the import of marketable products from other territories. This implies that a condition for the development of entrepreneurship and trade depends on the quality of transport infrastructure management, which predetermines the region's ability to participate and benefit from this during national and international division of labor and production specialization [1, 2].

Business environment of entrepreneurial structures changed when the Russian Federation switched to the market type of economic development; therefore it is necessary to change the existing management system of territorial transport and logistics infrastructure [3]. Due to the fact that flow of goods transported across the territory of Voronezh Region has significantly increased, greater importance is given to the transport and logistics infrastructure from the standpoint of business activity of business structures, and the role of private sector in its formation has increased significantly [4]. Therefore, in current conditions, changes in the management system of transport and logistics infrastructure of the territory and development of methodological approaches and methods to improve

IOP Conf. Series: Materials Science and Engineering **817** (2020) 012004 doi:10.1088/1757-899X/817/1/012004

the tools for its optimal management are needed.

2. Methods and Materials

Transport and logistics infrastructure is a tool for the development of economy of separate regions. One of the conditions for its development is competition in transportation, storage, processing, quality of provided services; the second condition is employment of local population, not the population invited by network operators.

Dynamics of changes in the main indicators of transport in Voronezh region is presented in table 1. Quantitative indicators on the example of a single territory (Voronezh Region) clearly show that passenger traffic through railway transport has decreased by 2.5%, and bus transport - by 13.66%. Freight traffic by rail has increased by 18.26%, while highway transport has decreased by 30.19%.The length of roads has increased by 0.13%, and the length of paved roads has increased by 2.69%.

Table 1. Main indicators of transport in Voronezh region.

Indicators	Years				
	2013	2014	2015	2016	2017
Mileage (km)					
Public railway tracks	1708	1684	1673	1671	1671
Highways - total:	26132	29410	29556	29905	29943
Including hard surface	16810	17563	17777	17972	18456
Cargo transportation (mln.t.)					
Transport - total	17.42	17.66	46.5	48.6	39.5
Railway	11.32	10.92	10.6	11.5	13.6
Highway	5.58	6.23	35.9	37.1	25.9
Air	0.03	0.01	-	-	-
Inland water	0.49	0.5	-	-	-
Cargo turnover (mln.t./km)					
Transport - total	27472.4	25897.5	27426.1	28420	30372
Railway	26859.2	24994.5	25472.1	26240	27440
Highway	499.7	846.8	1954.0	2180	2932
Air	108.3	50.1	-	-	-
Inland water	5.2	6.1	-	-	-
Passenger transportation (mln. people)					
Public transport – total	283.0	294.1	263.6	260.6	225.9
Railway	9.8	9.8	8.9	8.0	7.8
Bus	266.3	276.7	254.7	252.6	218.1
Surface rail transit	-	-	-	-	-
Trollebus	6.6	7.3	-	-	-
Air	0.3	0.3	-	-	-

Quantitative indicators on the example of a single territory (Voronezh Region) clearly show that passenger traffic through railway transport has decreased by 2.5%, and bus transport - by 13.66%.

Freight traffic by rail has increased by 18.26%, while highway transport has decreased by 30.19%.The length of roads has increased by 0.13%, and the length of paved roads has increased by 2.69%.

At present, Russian transport and logistics infrastructure is being adapted to the European development model. One of the effective ways of social and economic development of Russia in general and its regions, in particular, is development and implementation of regional transport and logistics systems (TLS) [5].

The Central Black Earth region (CBER) acts as an important link in the overall development strategy of the world and Russian transport and logistics systems, since over 30% of the population of

IOP Conf. Series: Materials Science and Engineering **817** (2020) 012004 doi:10.1088/1757-899X/817/1/012004

the Russian Federation lives within a radius of 600 km with a circle center in Voronezh. CBER transport system is represented by federal highways (rail, road, aviation), which are in the format of European transport corridors.

Voronezh region has good geographical position, which allows expanding the service area by increasing flow of goods in the CBER and in transit cargo transportation from the regions of Central Federal District in the directions to the Ukraine and the regions of Southern Federal District along transport corridors.

Objects of transport and logistics infrastructure of Voronezh region are divided into PL providers depending on the level of comprehensiveness of provided services and the depth of integration [6]. Transport nodes are located in Voronezh, Liski, Borisoglebsk and Rossosh.

Criteria for assessing transport and logistics infrastructure are presented by two main groups [7]:

– Criteria directly reflecting characteristics of transportation process itself (grouping, cost, time of delivery, etc.). They play an important role in operational management of the economic entity;

– Criteria reflecting activities of transport and logistics infrastructure, or for each individual economic entity as a whole (that is, they are characterized by qualitative indicators of work, share of services rendered in a certain market segment, etc.). They have great importance in the strategic management of a business entity [8].

Classification of indicators of functioning of transport and logistics infrastructure enterprises is presented in table 2.

In a market economy, it is impossible to evaluate the efficiency of transport and logistics infrastructure by one criterion, a system of criteria with expert assessment approaches is needed, depending on the functions performed by transport and logistics infrastructure and the impact on economic and social development of a country's subject [9].

These include, for example, meeting the needs of territory's population with transport services, solving social problems, etc. However, one should not forget that transport and logistics infrastructure is assessed not only by internal indicators, but also by effective indicators of those industries and industries that serve it, such as: meeting transport and logistics needs of material production and population of a specific subject of the Russian Federation and country as a whole.

Table 2. Classification of performance indicators of enterprises of transport and logistics infrastructure.

Indicator classification criteria	Indicators
By initial values	Past statistics, forecast indicators and engineering calculations
By transport group	Sectoral, general transport, specific, work of a particular type of transport in the subject of the Russian Federation is evaluated with their help, revealing the possibilities and reserves of resources to increase the efficiency of their activities
By performance	Characterizing the work of transport and transportation processes in physical terms
By economic indicators	Absolute and specific indicators that make it possible to evaluate the effectiveness of transport process in value terms

The main factors that provide cumulative effect of transport and logistics infrastructure functioning in the region [10] are:

- increasing inter-sectoral and inter-regional coordination;
- use of modern logistics technologies for the organization of transport process;
- creation of appropriate institutional environment that forms a basis of a public-private partnership.

Of course, it is necessary to form and implement integrated schemes for the development of transport and logistics infrastructure in each region, which must cover the issues of creation of newly created or development of existing infrastructure facilities of transport types, available in a given subject, development of transport corridors, creation of multimodal logistics centers and terminal complexes, organization of multimodal transport, formation of single information space, usage of integrated information management system, establishment of appropriate legal and regulatory framework, compliance with higher environmental standards, etc.[11].

Institutional barriers to the development of transport and logistics complex of subjects of the Russian Federation include:

- Large number of public authorities responsible for the development of transport and logistics complex in the region, resulting in difficulties with coordination of various authorities and increasing frequency of their authority crossing;
- Lack of communication and coordination of actions of various authorities, due to unclear organizational structure and uncertain responsibilities, authorities and accountability, as well as lack of dialogue between various authorities in solving transport and logistics problems;
- Lack of power to make management decisions for most of the authorities, in connection with which there are difficulties in implementation of regional policy in the development of transport and logistics complex, as well as lack of autonomy of authorities responsible for certain activities;
- Lack of resources for development of transport and logistics complex in the region and lack of mechanisms for monitoring and evaluating the effectiveness of policy implementation [12];
- Lack of clear framework for the development of transport and logistics complex of the region, due to the lack of proper principles, clear objectives and consistency of its development policies.

Development of new approaches to the long-term regulation of activities of transport and logistics complex contributes to building a model of its functioning and development for a long period of time.

Improving the management system for the development of transport and logistics infrastructure in the federal government system in the context of globalization significantly improves the efficiency of production, financial, managerial and human resources. This fact is a significant aspect for entrepreneurial structures with such organizational and economic form while increasing their competitiveness in the market conditions, as well as other restrictions that result from this and affect their independence from financial and economic point of view.

3. Results and Discussion

The priority directions of transport and logistics system of the territory are maintenance of:

1) Import cargo traffic that comes from Moscow, St. Petersburg and the seaport of Novorossiysk for consumption in the regions of the Central Federal District;
2) Export freight of the regions of the Central Federal District, in which goods are transported from the port of Novorossiysk by sea transport;
3) Import-export cargo traffic from the Ukraine, and in future from Kazakhstan and China, which follows the route of the international transport corridor "West - East": the Ukraine - Kursk - Voronezh - Borisoglebsk - Saratov - Kazakhstan - China;
4) Interregional transportation.

Increased interest in the theory of spatial economics gives a rise to an actualized search for points of growth in the regions of the Russian Federation.

A common feature is inherent in those studies in which formulation and solution of problems of transport and logistics are performed at the regional level. In the study this fact is closely related to the construction of the graph which informs and displays transport structure of the region (figure 1). In order to solve this series of problems, construction of a graph or hypergraph is carried out with the display of all necessary informative databases (key factor) in it [13].

The methodology which implies the development of projects for the development of regional transport infrastructure is as follows: a multilevel mathematical model is set by initial conditions; the use of multi-criteria optimization of the regional transport system solves the problem in a dual formulation.

IOP Conf. Series: Materials Science and Engineering **817** (2020) 012004 doi:10.1088/1757-899X/817/1/012004

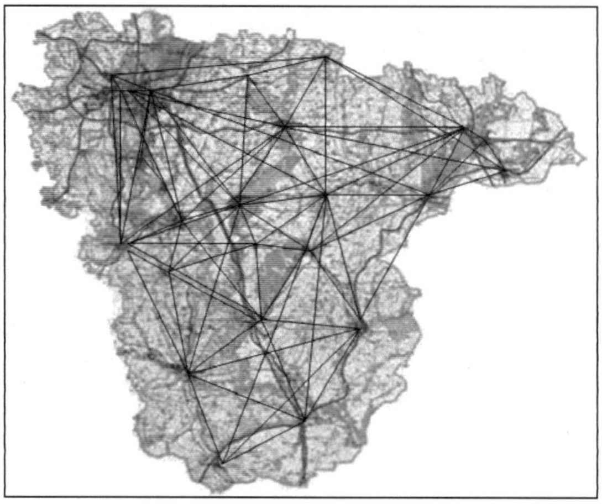

Figure 1. Hypergraph of the transport system of Voronezh region with the main inter-regional transport links.

The construction of the model is based on the analysis of transport infrastructure of the territory of the Russian Federation. Morphology of economic space of the subject of the Russian Federation is determined the graph of transport network, represented by the main railways and road transport, because there is an anthropogenic factor in the process of its formation. Existing and project (for the development of the resources of the territory) main transport lines can become a locomotive or a brake in the development of the territory. Transport infrastructure is accessed through technical and technological parameters of the infrastructure and from the perspective of probability factor of reliability of transport subsystems.

At creation of model the parameters presented to table 3 which list can be continued are fixed.

Table 3. Parameters of model transport level.

Lines	Vertices
Distance Number of ways, lanes Maximum axial load Speed limit Average speed on the site Travel time on the site Freight traffic density	Loading unloading Freight transit Passenger traffic by type of transport Passenger traffic in long-distance trains Passenger traffic in suburban and city traffic Population traffic mobility
Traffic capacity Traffic-carrying capacity Number of traffic accidents	Number of transfer passengers at the connections of international flights (air) and domestic flights
Length of public roads complying with regulatory requirements for transport and operational indicators	Level of public satisfaction with measures taken by the executive authorities to ensure the safety of population in transport
Increase in transit traffic Level of supervisory authority equipment by technical means Share of roads maintained with long-term contracts Share of the length of sections of the railway network where there are restrictions on throughput and carrying capacity Probability of uptime	

At the next stage, the resources, available in this territory, are determined; they are adjacent to the vertices of the graph in building the model. Favorable natural resources play an important, and in most cases crucial, role in spatial economics and transport optimization.

In the nearest future, development of the territories of our country will be uneven, on the basis of this it is possible to identify areas with favorable geographical position in which made investments achieve the greatest effect.

The territory having the most favorable natural conditions should be covered by a well-developed dense network of transport infrastructure.

When forming the model determining criterion values, data on the parameters of the level "Favorable natural conditions" in Voronezh region are determined (table 4).

The average score for each factor of utility of natural conditions of the territory has been determined by expert means, for which 20 independent experts have been involved.

The principle must be fulfilled in order to realize the model: the available data, which in different planes reflect the parameters of the model and are recorded only once, and the possibility of their use can be achieved many times depending on the functions when solving various tasks.

Vector optimization is the basis for received data processing, provided by the method and ensuring conditions fulfillment. In other words, vectors display optimization criterion indicators that allow fixing values that act as elements of vectors for an array of data in the Python programming language using "list".

Table 4. Parameters of the level "Favorable natural conditions" in Voronezh region.

Favorable factors of the natural conditions of the territory	Points
Recreational assessment	6.7
Seismic hazard level	0
Drinking water supply	8.9
Landscape and Recreational Assessment favored for industrial and civil construction	8.5
Assessment favored for agriculture	9.2
Average temperature	9.1
Accessibility and sufficient length of roads	7.2
The level of ecological cleanliness of the environment along the hard-surfaced roads	6.7

Using this technique, one can carry out a multi-criteria analysis of the transport network and optimize it in regional economic development projects, get an assessment and compare various data obtained by related branches of science about the studied subject. Moreover, in this context, transport network acts as an infrastructure framework that links social and economic processes, as well as spatial configuration and flow that can pass through it. Using this methodology, it is possible to improve the quality of management of the development of territorial transport infrastructure through the implementation of original approaches by computer programs.

4. Conclusion

Enterprises of transport and logistics infrastructure, as well as business structures of other industries, can develop in two directions:

- expanding the range of services provided in the existing market segment;
- expanding the segment of management, based on consumer preferences and development programs developed by the authorities of other regions.

Global economy dictates its own conditions, for example, it requires an assessment of the effectiveness of economic entities activities on a national economy scale and in comparison with global competitors.

Profitability, cost effectiveness of the company for the period, the dynamics of revenue growth, the number of applications, capital productivity, output - these are the main indicators that characterize economic and managerial performance of the organization for a certain period, in other words, characterize the efficiency of motor transport logistics enterprise (MTLE).

IOP Conf. Series: Materials Science and Engineering **817** (2020) 012004 doi:10.1088/1757-899X/817/1/012004

Financial responsibility of the company, its turnover rates and growth rates of staff salaries, as well as the share of regular customers in the total traffic volume, characterize the stability of the company's development.

During the study, the authors established the fact that it is impossible to determine the dynamics of MTLE development using only a standard approach to the calculation of performance indicators. This aspect is explained as follows:

- Due to the fact that the share of goods transported by owns MTLE vehicles with large volumes of order fulfillment is very small, this makes it impossible to use it for an objective assessment of the efficiency of using fixed assets;
- Since the planning of MTLE activity is connected with the execution of all applications received from customers, but not with obtaining maximum profit, i.e. it is based on a customer-oriented approach to the company's activity, which does not allow planning the traffic volumes in the future period, and, consequently, coefficient of agreement of actual traffic volume with the planned one loses its economic sense;
- MTLE activities are carried out in the form of intermediation - implementation of the supply chain management process of various companies through the organization of cargo transportation by owners of vehicles which are not owned by the company.

References

[1] Akhmetov L A 2010 Improving Business Activity in Motor Transport in Russia, *Scientific Notes: Role and Place of Civilized Entrepreneurship in the Russian Economy*: Collection of scientific papers. Issue XXIII, (Moscow: Russian Academy of Entrepreneurship; Press Agency "Science and Education")

[2] Bezrukova T L, Nesterov S Yu, Kostenko Yu A and Pecherskaya O A 2013 Background of the formation of new types of freight road transport and logistics enterprises, *Problems, trends and prospects for the development of modern logistics science*, (Moscow: ITKOR Institute), pp 73–96

[3] Bezrukova T L, Pecherskaya O A and Nesterov S Yu 2015 Improving the management of transport and logistics infrastructure: organizational and methodological foundations, effective tools, optimization methods and models, (Moscow: KnoRus), 110

[4] Bezrukova T L, Yerknapeshyan E Zh 2011 Development of transport management systems: methodological foundations, foreign and Russian practice, (Moscow: KnoRus)

[5] Kizim A A, Isaulova S S 2004 Virtual transport and logistics system. Logistics centers as points of virtual economic analysis, *Finance and credit*, Vol 1, Issue 139, pp 63–67

[6] Nesterov S Yu, Tereshkina T R and Pecherskaya O A 2012 Systems approach to the management of motor transport logistics company, *Prospects for the innovative development of the modern world community: economic, legal and social aspects*: materials of the International anniversary scientifi and practical conference VEPI-VGLTA 2012, Voronezh, pp 166-169

[7] Kupavykh S S and Bezrukova T L 2015 Background of improving the management system of transport complex development in the context of globalization, *Alternative energy sources in transport and technological complex: problems and prospects for rational use*, Vol 2, Issue 2, 779

[8] Kupavykh S S, Bezrukova T L and Pecherskaya O A 2016 Optimization of transport and logistics infrastructure of the regions as a factor in improving the competitiveness of the Russian economy, *Alternative energy sources in transport and technological complex: problems and prospects for rational use*, Vol 1, Issue 3, pp 430-432

[9] Kupavykh S S 2016 Transport and logistics infrastructure: assessment criteria, problems and development prospects, *Alternative energy sources in transport and technological complex: problems and prospects for rational use*, 2016, Vol 1, Issue 3, pp 427-429

[10] Prokofieva T A and Adamov N A 2011 Strategy for the development of logistics infrastructure

in the transport complex of Russia, Moscow: Ekonomicheskaya Gazeta, p 302
[11] Pecherskaya O A 2014 Priority directions for the development of transport and logistics complexes *Alternative energy sources in transport and technological complex: problems and prospects for rational use*, Vol 1, pp 415-420
[12] Pecherskaya O A, Bezrukova T L and Nesterov S Yu 2015 Method for assessing the effectiveness of organizing the interaction of participants in the region's transport and logistics infrastructure, *Forestry Engineering Journal*, Vol 3, pp 45-52
[13] Nesterov S Yu 2012 Development of regional logistics in the Central Black Earth region, *Logistics*, pp 32–34

IOP Conf. Series: Materials Science and Engineering **817** (2020) 012005 doi:10.1088/1757-899X/817/1/012005

Impregnation of lumber due to the pressure of the impregnating liquid

A Birman*, F Svoykin, V Neslukhov, A Borozna, M Taraban

St. Petersburg State Forest Technical University, 5 Institutskiy Lane, St. Petersburg 194021, Russian Federation

*Corresponding email: birman1947@mail.ru

Abstracts. Storage of the container is one of the most serious problems faced by the manufacturer. At temperatures above 5 °C, the fungus and blue stain affect the container board. Because of this, enterprises suffer losses, which most often occur due to the fact that customers reject wooden containers, since it is important to use wooden pallets without mold and mildew in food and packaging industries. The article describes the methodology and results of experimental studies on the impregnation of wood due to the hydrostatic pressure of the impregnating liquid. Graphic dependencies are made to determine the main technological parameters of the impregnation process - its duration. We studied wood samples of spruce, birch, alder and aspen, the main species from which container sets and pallets are made. A description is given of the use of a densitometer of a fundamentally new design, which makes it possible to determine the density of samples with positive buoyancy.

1. Introduction

Enterprises involved in sawing round wood are aimed at increasing the volume of products, one of which is the container board, which is widely in demand in the production of pallets, crates, flooring. Most often, such lumber is sawn from slabs, low-grade boards, or when finalizing unclaimed lumber. Moreover, some factories focused on the production of pallets are cutting only the container board.

Storage of the container is one of the most serious problems faced by the manufacturer. At temperatures above 5 °C, the fungus and blue stain affects the container board [1]. Because of this, enterprises suffer losses, which most often arise due to the fact that customers reject wooden containers, since it is important to use wooden pallets without mold and mildew in food production and packaging production.

The requirements for wooden pallets are regulated by GOST 33757-2016 [2], which states that only mushroom sapwood colors of wood that are not resulting from storage of wood are allowed. Also, large recipient plants of pallets compose their own technical specifications, which indicate that fungus and any manifestations of mold are unacceptable. Therefore, manufacturers of pallets need to responsibly approach the storage of raw materials for the production of pallets.

The main measures that are used when storing a container board cut from freshly cut wood can be attributed [3]:

- stacking boards crosswise so that the bulk of the surface of the lumber is ventilated. The method is the most common, but ineffective, because with prolonged storage in the places of contact, the juice forms with each other.

- surface treatment with antiseptics by spraying them on the surface of the board. This technique is the most time-consuming, but at the same time the risk of mold is reduced. The disadvantage of this method is that the antiseptic solution is not deeply absorbed, and during long-term storage the container board is also covered with mold.
- heat treatment of the packaging board, providing a reduction in humidity up to 25-30%. This processing method is the most reliable, provided that the board is no longer damaged by the fungus. The downside of the drying process is its energy consumption. Therefore, the production of dry wooden containers often becomes economically inexpedient.
- impregnation of the packaging board by immersion in an antiseptic solution. This is a method in which the ability to store a container increases to several months.

The main objective of the impregnation process is to achieve an optimal temporary impregnation mode, during which the raw materials for the production of pallets can be stored for a long time.

2. Methods and Materials

The experimental technique. To solve this problem, an experimental setup was used, figure 1.

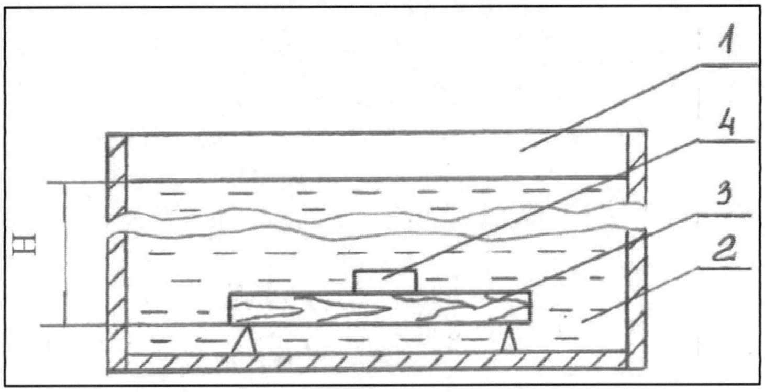

Figure 1. Scheme of the experimental setup. H - height of the column of impregnating liquid, H = 4 000 mm; 1 - reservoir, 2 - impregnating liquid, 3 - sample, 4 - cargo.

Prototypes (10 pieces each) were made from the sapwood of freshly cut wood species most commonly used in the manufacture of pallet boards, table 1.

Table 1. Characteristics of the samples.

Wood species	Aspen	Birch	Alder	Spruce
Size, mm	22×100×1 200	22×100×1 200	22×100×1 200	22×100×1 200
Average humidity, %	80	70	70	90

Samples were immersed in an aqueous solution of methylene blue for temporary periods that were multiples of 5 minutes. The amount of liquid absorbed due to hydrostatic pressure was determined by the increase in the mass of the liquid when extracting samples from the installation, tables 2-5.

The mass gain Δm is determined by the equation [4]:

$$\Delta m = \frac{m_1 - m_0}{m_0} 100\ \%, \quad (1)$$

where, m_0 – weight before impregnation, g; m_1 – weight after impregnation, g.

Table 2. Impregnation of aspen wood samples.

Sample number	Duration, min	Initial mass, g	Mass after impregnation, g	Weight gain, %
1	5	2 100	2 140	2%
2	10	2 070	2 170	5%
3	15	2 065	2 200	7%
4	20	2 140	2 270	6%
5	25	2 120	2 330	10%
6	35	2 120	2 370	12%
7	40	2 100	2 420	15%
8	45	2 090	2 445	17%
9	50	2 080	2 480	19%
10	55	2 110	2 570	22%

The data in Table 2 is illustrated in figure 2.

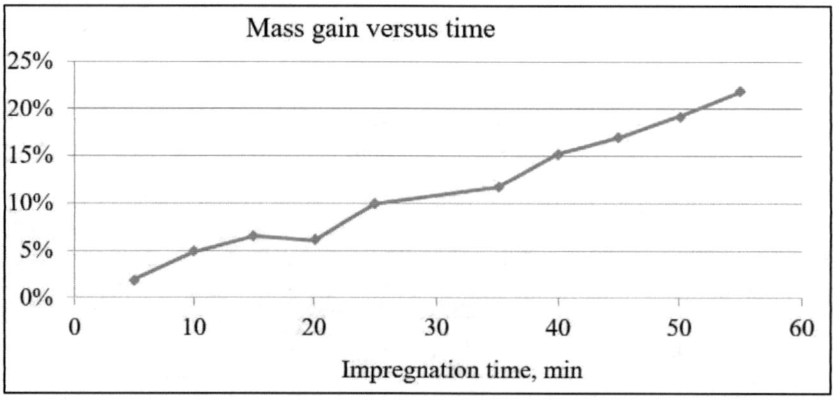

Figure 2. The increase in mass of the impregnating fluid over time. Aspen.

Table 3. Impregnation of birch wood samples.

Sample number	Duration, min	Initial mass, g	Mass after impregnation, g	Weight gain, %
1	5	2 510	2 570	2%
2	10	2 530	2 630	4%
3	15	2 490	2 630	6%
4	20	2 460	2 705	10%
5	25	2 580	2 965	15%
6	35	2 520	2 970	18%
7	40	2 390	2 890	21%
8	45	2 540	3 125	23%
9	50	2 560	3 225	26%
10	55	2 700	3 450	28%

The data in table 3 is illustrated in figure 3.

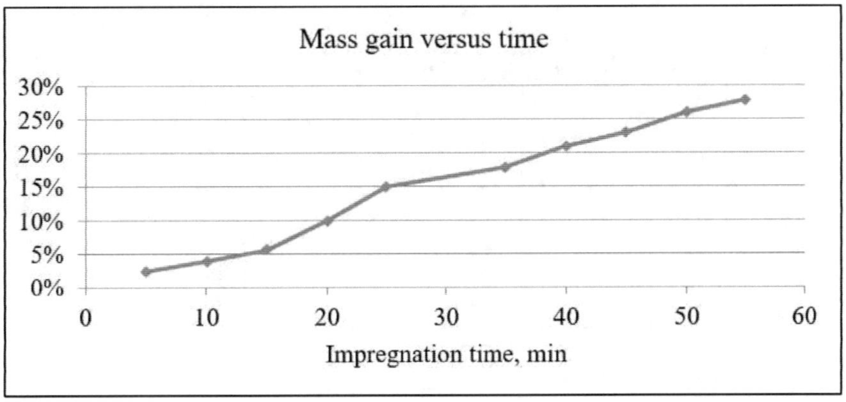

Figure 3. The increase in mass of the impregnating fluid over time. Birch.

Table 4. Impregnation of alder wood samples.

Sample number	Duration, min	Initial mass, g	Mass after impregnation, g	Weight gain, %
1	5	1 980	2 010	2%
2	10	1 920	2 010	5%
3	15	2 050	2 150	5%
4	20	2 045	2 145	5%
5	25	1 990	2 110	6%
6	35	1 985	2 145	8%
7	40	2 000	2 170	9%
8	45	2 000	2 210	11%
9	50	2 055	2 300	12%
10	55	2 010	2 290	14%

The data in table 4 is illustrated in figure 4.

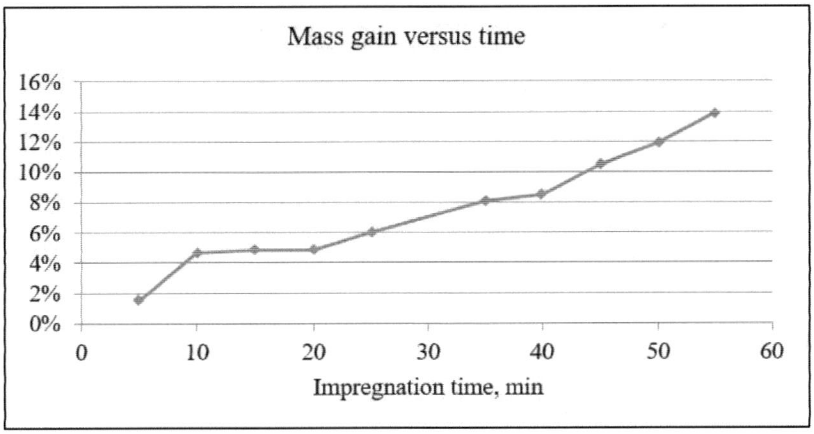

Figure 4. The increase in mass of the impregnating fluid over time. Alder.

IOP Conf. Series: Materials Science and Engineering **817** (2020) 012005 doi:10.1088/1757-899X/817/1/012005

Table 5. Impregnation of spruce wood samples.

Sample number	Duration, min	Initial mass, g	Mass after impregnation, g	Weight gain, %
1	5	1 700	1 730	2%
2	10	1 670	1 720	3%
3	15	1 650	1 730	5%
4	20	1 695	1 780	5%
5	25	1 725	1 845	7%
6	35	1 715	1 850	8%
7	40	1 700	1 860	9%
8	45	1 695	1 890	12%
9	50	1 705	1 910	12%
10	55	1 690	1 960	16%

The data in table 5 is illustrated in figure 5.

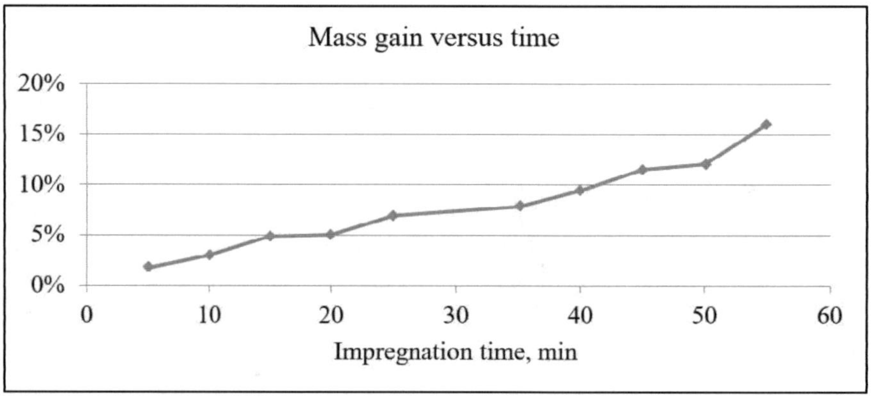

Figure 5. The increase in mass of the impregnating fluid over time. Spruce.

After the experiments, the samples were split along the fibers and the depth of the zone dyed with methylene blue was measured [5, 6]. Based on the graphs in Figures 2-5, the technological parameters of the impregnation process were determined for each of the tested wood species.

The second stage of the experiments was their conduct for samples of 3 × 3 × 3 cm in size using a new construction densitometer designed and manufactured at St. Petersburg State Forest Technical University (SPbFTU).

The objective of the experiment was to determine the density of the samples before and after impregnation. The determined density value was an indicator of the amount of impregnating liquid in the sample after completion of the impregnation process.

The density of a substance is determined by the well-known formula:

$$\rho = \frac{m}{V}, \tag{2}$$

where, m – mass of the sample, V – volume of the sample.

The mass of the sample is easily determined by weighing it. The volume of the sample in the case of its irregular shape (after impregnation and swelling, the shape of the samples is incorrect) is determined by known devices, based on measuring the volume of liquid displaced by the sample when it is immersed in a reservoir with liquid. So, for example, a pycnometric method of measuring density is carried out, which consists in the following [7, 8].

A pycnometer (a glass vessel, the volume of which is known with great accuracy), is filled with distilled water (up to the mark) and weighed. The mass of the pycnometer with water M_0 is equal to:

$$M_0 = \rho_0 V_p + m_p, \tag{3}$$

where, ρ_0 – density of water, V_p – volume of the vessel, m_p – mass of the vessel.

Then weighed the test sample. Its mass m is equal to:

$$m = \rho V \tag{4}$$

Where ρ – desired density of the sample, V – volume of the sample.

Then the weighed sample is immersed in a pycnometer with water and the excess water is removed so that its level again coincides with the mark on the picnometer vessel.

Then determine the mass M of the pycnometer with water and a sample:

$$M = \rho V + \rho_0 (V_p - V) + m_p = m + \rho_0 (V_p - V) + m_p, \tag{5}$$

Then, subtracting (4) from (2) and adding (3), the mass of displaced water is determined and, having determined the volume V from this, we obtain the expression for the desired density:

$$\rho = \frac{m}{V} = \frac{m\rho_0}{M_0} - M + m, \tag{6}$$

The technique for carrying out such measurements is simple; the results obtained are accurate. But all this applies to the studied samples with negative buoyancy [9, 10]. Samples with positive buoyancy (in our case, wood) are not completely immersed in the liquid of the measuring instruments, that is, they displace a liquid volume smaller than its own volume. And their forced complete immersion with the help of additional devices immersed in the liquid of the device together with the sample sharply reduces the measurement accuracy [7, 11].

The density meter design is shown in figure 6.

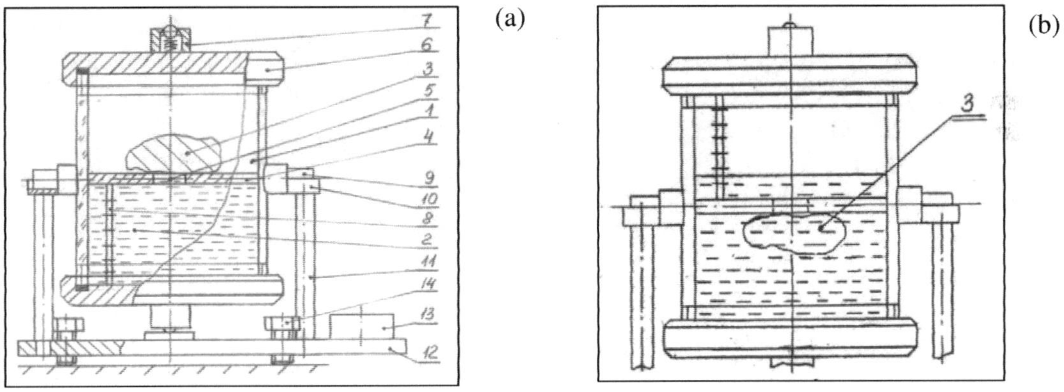

Figure 6. Density meter, *a* - view of the density meter with the sample placed in it before determining the volume of the sample, *b* - the same when determining the volume of the sample.

The use of a new universal densitometer provides the ability to measure volumes of solids with both negative and positive buoyancy relative to the liquid used in the device without the use of additional devices designed to ensure their complete immersion in the liquid, as well as simplifying the measurement process and achieving opportunities to work in the field (outside the laboratory) conditions with the involvement of unskilled personnel.

Conducting experiments to determine the density of wood samples before and after impregnation is as follows.

Having installed the densitometer on a horizontal surface, they achieve a horizontal position in the space of its bed «12» using a standard horizontal level «13» and height-adjustable legs «14».

Liquid «2» is poured into the vessel «1», vertically fixed by the retainer «7», with the threaded upper sealed stopper 6 removed, in an amount equal to half the capacity of the vessel «1». A test sample «3» is placed on the rigid partition «4» and the vessel «1» is sealed with the stopper «6», Figure 6, (a).

Turning the vessel «1» in a vertical plane by 1 800 mm on the cylindrical fingers «9» located on the supports of rotation «10» of the uprights «11», the vertical position of the vessel «1» is fixed with the help of the retainer «7» of the upper tube «6» turned upside down.

Upon completion of the rotation of the vessel «1», the liquid «2» through the hole «5» in the rigid partition «4» flows into the cavity of the vessel «1», where the test sample is located.

In this position, figure 6, (b), in the upper cavity of the vessel «1» there is a volume of liquid displaced by sample «3» from the opposite cavity of the vessel, and, obviously, equal to the volume of the sample. The value of this volume is determined visually by the reading of the scale of the vertical ruler «8», graduated in units of volume. Sample «3» floats in the fluid volume of the lower cavity of the vessel «1», or rather, is pressed by the buoyancy force to the lower plane of the rigid partition «4».

The density of the substance ρ of a pre-weighed sample «3» with a mass m is determined by the Formula (2), where V is the reading of the scale of line «8».

At the end of the process of examining the sample, the densitometer, turning the vessel «1» to 1 800, is returned to its original position, remove the plug «7», remove the sample and proceed to the next study.

The condition for the axis of the fingers «9» to pass through the center of gravity of the vessel «1» must be satisfied to ensure reliable operation of the latches «7». The lower sealed plug «6» is removable to allow periodic cleaning of the vessel «1».

The fluid used in the densitometer can be opaque or tinted in order to more accurately visually fix its level in relation to the risks of graduation of the ruler «8». The fluid was replaced when it was contaminated by particles that separated from the sample during the study period.

The frame «12» of the device can be mounted on a scale, which allows measurement of the mass of the samples simultaneously with the determination of their volume.

The capacity of the vessel «1» with the installed partition «4» was determined before its first operation with the help of the measured capacity.

The densitometer used by us is a fairly accurate instrument, easy to manufacture and convenient to operate.

3. Results and Discussion

It is possible to carry out experimental studies of the degree of impregnation of wood samples of the impregnating liquid on the basis of an experimental setup simulating the impregnation process due to the hydrostatic pressure of the impregnating liquid. The degree of impregnation of wood, as a body with positive buoyancy, is proposed to be carried out using a new construction densitometer.

References

[1] Lomakin A 1990 *Protection of wood and wood materials* [in Russian - Zashhita drevesiny` i drevesny`x materialov] (Moscow: Forest industry) p 153

[2] GOST RF 33757-2016 *Flat wooden pallets. Specifications* 2016 (Moscow: FGUP Standartinform)

[3] Birman A and Krivinogova A 2015 *The use of methods for fedding long assortments of raw materials* [in Russian - Ispol`zovanie metodov propitki dlinnomerny`x sortimentov sy`r`ya] Tomsk State University Journal 1 (Nizhnevartovsk: NizhSU) pp 45-48

[4] Shamaev V, Nikulina N and Medvedev I 2013 *Wood modification* [in Russian - Modificirovanie drevesiny`] (Moscow: Science) p 456

[5] Ermolin V 1999 *The basics of increasing the liquid fluid permeability of coniferous species: monograph* [in Russian - Osnovy` povy`sheniya proniczaemosti zhidkostyami dre-vesiny` xvojny`x porod] (Krasnoyarsk: SibGTU) p 100

[6] Patyakin V, Tishin Yu and Bazarov S 1990 *Technical hydro-dynamics of wood* [in Russian - Texnicheskaya gidrodina-mika drevesiny`] (Moscow: Forest Industry) p 304

[7] Kivilis S 1980 *Density meters* [in Russian - Plotnomery`] (Moscow: Energy. Literature) p 279

[8] Sergovsky P and Rassev A 1987 *Hydrothermal processing and preservation of wood* [in Russian - Gidrotermicheskaya obrabotka i kon-servirovanie drevesiny`] (Moscow: Forest Industry) p 360

[9] Birman A, Lokshtanov B, Krivinogova A and Toan Nguyen Van 2016 *A critical analysis of the use of impregnation methods in order to improve the quality characteristics of wood raw materials* [in Russian - Kriticheskij analiz ispol`zovaniya sposobov propitki s cel`yu povy`she-niya kachestvenny`x xarakteristik drevesnogo sy`r`ya] Actual problems of the development of the forest complex (SPb: SPbFTU) pp 58-60

[10] Krivonogova A and Birman A 2013 *Using the effectiveness of the method of hydrostatic pressure of an impregnation liquid with an increase in the qualitative characteristics of timber materials* [in Russian - Ispol`zovanie e`ffektivnosti meto-da gidrostaticheskogo davleniya propitochnoj zhidkosti pri povy`shenii kachestvenny`x xarakteristik lesomaterialov] Modern Scientific Bulletin **9(1)** p 310

[11] Kunitskaya O, Burmistrova S and Goncharov Yu 2015 *Search for new technical solutions to increase the efficiency of impregnation of wood* [in Russian - Poisk novy`x tex-nicheskix reshenij dlya povy`sheniya e`ffektivnosti propitki drevesiny`] Intensification of the formation and protection of intellectual property: materials of the republic. Scientific and practical conferens reg. 75th anniversary of PetrSU (Petrozavodsk: PetrSU) p 16-17

IOP Conf. Series: Materials Science and Engineering **817** (2020) 012006 doi:10.1088/1757-899X/817/1/012006

Industrial approbation of temporary road structures for transport-assisted exploitation of forests

A A Borozna[1]*, D A Skachek[1], M T Naskovets[2], S P Mokhov[2], S A Golyakevich[2]

[1]*St. Petersburg State Forest Technical University*, 5 Institutskiy Lane, St. Petersburg 194021, Russian Federation

[2]*Belarusian State University,* 4 Nezavisimosti Avenue, Minsk 220030, Republic of Belarus

*Corresponding e-mail: 3349280@mail.ru

Abstract. The article discusses topical issues for the forest industry of the Russian Federation and the Republic of Belarus in the field of construction of automobile forest roads in difficult soil and hydrological conditions with low bearing capacity of soils and lack of road-building materials. To solve this problem, the authors of the article proposed road structures with the use of geotextile fabric, which could save up to 30% of road-building materials without reducing the strength characteristics of the structure. Recommendations have been developed for the construction of roads using geotectile fabric based on the results of pilot testing in industrial conditions.

1. Introduction

In recent years, in the Russian Federation and the Republic of Belarus there has been an increase in the volume of construction of roads in forested areas. At the same time, the composition of transport networks has been replenished not only with various categories of roads of year-round operation, but also with seasonal and temporary roads. Moreover, constructive solutions for temporary roads should take into account both the regional soil and hydrological conditions of the area where they are planned to be built and the negative impact of weather and climatic factors on the operation of the temporary transport and technological routes. In this case, the construction of temporary forest roads is largely constrained by the type of forest stands, as well as the characteristics of the soil and peat deposits on which they grow. After all, it is precisely these circumstances that mainly determine the bearing capacity of various types of soils in terms of the movement of timber transport vehicles, as well as the operation of vehicles involved in the road construction process (table 1).

Table 1. Types of forest stands, and characteristics of weak foundations and their strength parameters.

Types of forests, tree species, vegetation cover	Characteristics of peat massifs and soil types	Load bearing capacity, MPa
Pine-shrub, pine-sphagnum, ridges with small *Eriophorum vaginatum /Sphagnum spp.* agglomerations (upland), lowland wetland, sedge swamp (lowland).	Bogs. The presence of tree trunks is characteristic, and for floodplain bogs, layers of mineral soil and dense peat. The swamp water level is 0.3 – 0.5 m. The freezing depth is 0.5 – 0.7 m.	0.025–0.055
Ridges with medium *Eriophorum vaginatum /Sphagnum spp.* agglomerations (upland), lowland wetland, sage bog (lowland).	Upland and lowland swamps. Low stability, medium waterlogged. Interlayers and lenses of peat of a fluid consistency and liquid swamp formations are found. The swamp water level is 0.1 - 0.3 m. The freezing depth is 0.4 - 0.8 m.	0.025 –0.01
Ridges with large *Eriophorum vaginatum / Sphagnum spp.* agglomerations, lakeside (upland), sphagnum swamp (upland). Bogs.	Unstable, heavily waterlogged. Lenses of water and liquid wetland formations. Lakes and canals. The swamp water level is 0.1 m. The freezing depth is 0.3 - 1 m.	> 0.01
Pine and spruce forests with *Oxalis spp*, secondary aspen and birch forests. Pine forests with moss, lichen, heather, and lingonberry. Spruce forests with lingonberry, and secondary birch and aspen forests.	Sandy and light loamy. Sandy. Sandy and sandy loam.	0.15 – 0.20
Pine forests with blueberry. Spruce forests with blueberry, and secondary aspen and birch forests. Pine forests with grass.	Sandy loam on top. At a depth of 0.30 – 0.50 m, they are underlain by loam or clay. Ordinary sandy loam, lined at a depth of 0.30 – 0.50 m with loam or clay, less often loamy. Loamy.	0.10 – 0.15
Pine and spruce forests with blueberry and tall mosses, and secondary aspen and birch forests. Spruce stands near streams (in gullies), secondary aspen and birch forests. Pine and spruce forests with grass, sedge and sphagnum, secondary aspen and birch forests.	Loamy. In the upper part, there is a layer of humus up to 0.40 m thick. Loam or clay, on top of a layer of humus with a thickness of up to 0.5 m. Peat, more than 50 cm thick.	0.055 – 0.1

2. Methods and Materials

The data presented in the table indicate that, in order to ensure the passage of forestry vehicles and the movement of road-building equipment across territories characterized by low bearing capacity of soil, it is necessary to develop temporary road structures based on reinforcing layers of geosynthetic fabric,

IOP Conf. Series: Materials Science and Engineering **817** (2020) 012006 doi:10.1088/1757-899X/817/1/012006

brushwood and local soil. The positive effect of the use of geosynthetic layers is explained by the fact that, when the underlying soil is deformed, they stretch and absorb a part of the load redistributing it to a significantly larger surface area. Moreover, a weak soil base protected by a geosynthetic layer is less loaded than a soil under the pavement without a protective layer [1].

Brushwood lining is a local and widely available material for forest enterprises and therefore is often used for temporary roads. A sealed cushion of wood harvesting residues creates a dense deformable base. At the same time, in the process of arranging a brushwood cushion, the brunches are laid with a large margin in height which depends on soil and hydrological conditions. The use of brushwood allows us to increase the rigidity of the soil base when exposed to and transferring the load, which reduces the specific pressure of vehicles.

To prevent the mixing of materials of structural layers and allow the redistribution of pressure from vehicle movers to a large area of weak underlying soil, alongside with brushwood cushions layers of various types of geosynthetic materials are used, which, in addition to high strength, are wear-resistant, do not decay in the ground, do not reduce the strength gap, and filter water well. They are also an effective solution that allows us to use all the positive properties of local building materials in order to reduce the cost of construction.

For the purposes of road construction, a relatively large number of geosynthetic materials have been used. One of such promising materials is DuPont's Typar, characterized by:

1. High modulus of elasticity, due to which the material can absorb significant loads and perform the function of reinforcing at relatively small deformations.
2. Large elongations at break: depending on the type of material, Typar has elongations under maximum load up to 45%. Thus, local damage does not lead to the destruction of the material and Typar continues to perform its functions.
3. Universal filtering ability, due to the specific structure of the material, which makes it almost impossible to introduce soil particles into the pores and clog them. This allows Typar to maintain consistently good filtering properties under soil pressure and in conditions of strong vibration.
4. High tear and puncture resistance, which is especially valuable when laying out the material.

Other important properties of Typar material are:

1. Easy to install: the material rolls are small and light, which reduces transport and storage costs, as well as labor costs.
2. Easy to handle: Typar rolls can be sawed directly in place with a chain or hand saw.
3. Typar does not absorb water: when used in wet conditions, the weight of the rolls remains unchanged.
4. Typar is resistant to natural acids and alkaloids, as well as to insects, bacteria and mold.
5. Has the same degree of stretching in the longitudinal and latitudinal directions, which allows it to achieve uniform stretching of the geosynthetic layer [2].

3. Results and Discussion

In the industrial conditions, in order to determine which structural and technological solutions can be used on the forest site selected for road construction, visual inspection of the site has to be carried out. A comprehensive study of the territory where road construction works are to be carried out, as a rule, identifies the waterlogged sites and the degree of waterlogging of the terrain (figure 1), as well as the presence of wetlands and peat deposits.

IOP Conf. Series: Materials Science and Engineering **817** (2020) 012006 doi:10.1088/1757-899X/817/1/012006

Figure 1. General view of the sites designated for the construction of experimental facilities.

The next stage of a pilot-industrial testing is the selection and development of structural and technological solutions at the experimental facilities of temporary automobile roads. First of all, places are determined where it is possible to use either a layer of local soil or arrange a brushwood cushion made of forest harvesting residues. Then, objects are selected on which structures with layers of brushwood, geosynthetic materials (figure 2) or various types of combined layers (figure 3) will be arranged.

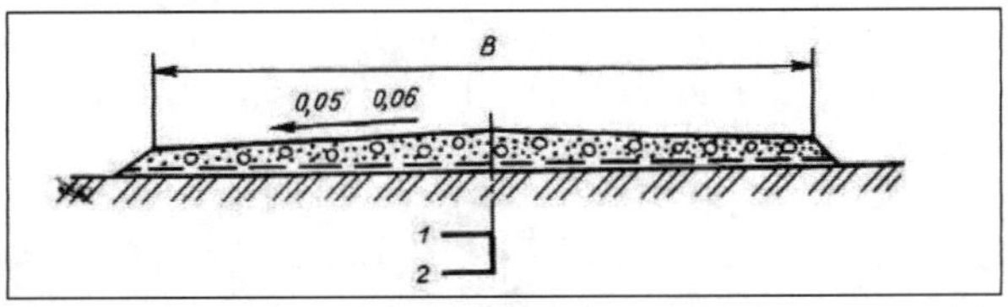

Figure 2. Cross section of the road structure with a layer of synthetic material: 1 - road surface; 2 - geosynthetic fabric, *B* - width.

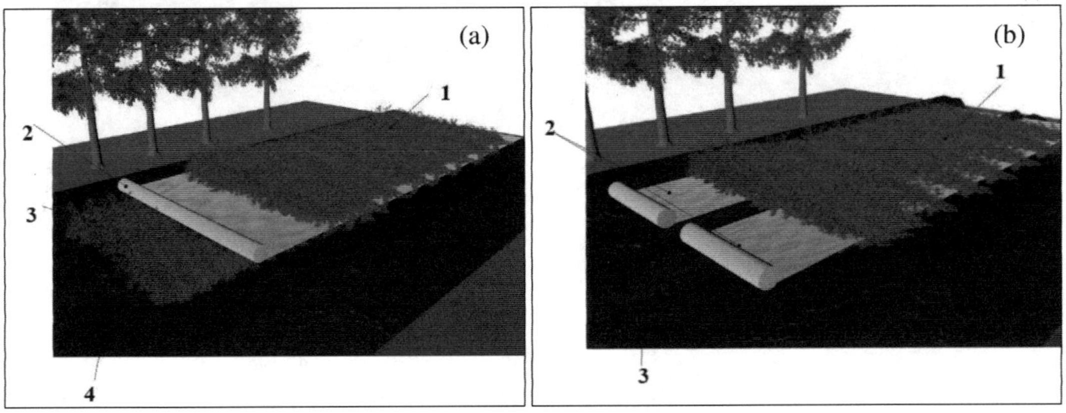

Figure 3. Visualization of the cross sections of temporary roads on weak bases made of a combination of brushwood and layers of geosynthetic material, laid across the entire width of the road surface; (a): 1– upper layer of brushwood; 2 – geosynthetic fabric; 3 – bottom layer of brushwood; 4 – base layer with a low load bearing capacity; and in the wheel drives; (b): 1 – brushwood; 2 – geosynthetic fabric; 3 – base layer with a low load bearing capacity.

As practice shows, the use of interlayers in the construction of temporary roads is the least time-consuming construction approach in which geosynthetic fabric is used on soil bases. In this case, the technological processes are quite simple and are carried out by the road-building equipment available at the road building enterprises. The production technology includes: rolling out the geosynthetic material, filling and leveling the sand coating layer, and its further compaction (figure 4).

Figure 4. Construction of experimental plots with a geosynthetic layer and local soil.

The technology of the construction of a road based on a brushwood lining and geosynthetic material includes leveling, planning and compaction of the base. The process starts with laying out along the entire width of the carriageway forest harvesting residues (figure 5) with their subsequent compaction, then a layer of a non-woven synthetic material is rolled out over the brushwood lining with fastening along the edges of the roadway. Next, on the underlying layer of brushwood lining and interlayers of geosynthetic material, the upper layer of brushwood lining is laid and compacted [3].

Figure 5. Longitudinal tapes and wood harvesting residues.

The design of constructive elements and choice of surface materials for forest transport systems in conditions of excessively moist soils requires the use of effective and at the same time inexpensive materials. As a road-building material, used car tires in combination with local soil can be used. It is the development of new designs of forest transport systems based on used tires that can serve as a fairly rational solution for increasing the carrying capacity of dirt roads on which ruts appear after the movement of timber trains.

In order to more evenly transfer the load from the wheels of moving vehicles along the depth of the road and reduce the specific pressure on a weak base, a road construction based on used tires has been developed. The proposed method for a tire road surface includes the formation of ruts by their preliminary broadening and deepening, as well as the assembly of tires in the form of a tape. Tires are assembled into tapes by first cutting them along the perimeter of the cross section and making two holes symmetrical with respect to the cut line with a subsequent sequential establishment by the method of rotation of each subsequent tire into the holes of the previous one. A distinctive feature of the proposed design is that the tires connected into continuous tapes, to a certain extent, serve as absorbers of pressure from the wheels of vehicles (figure 6).

Figure 6. Use of tires in industrial conditions.

At the end of the processes of pilot construction of road structures, a number of production tests should be carried out in order to assess the operability of a particular technical solution. For these purposes, simplified methods, instruments and equipment should be applied. The operational qualities of temporary roads, in particular the gauge depth after the passage of timber road trains, and the strength of the base should be assessed using a densitometer (figure 7).

Figure 7. Testing the performance of the constructed road surface.

By means of the latter, it has been determined: the indentation resistance of the working tip (E_w, N/cm^2), the modulus of elasticity of the road surface (E_g, MPa), specific adhesion (C_g, MPa), the angle of internal friction (φ_g, deg.), and the working capacity of the experimental section by the number of passes.

Tests carried out at the established experimental facilities in the process of timber transport showed that the developed temporary road structures are quite operational and provide the passage of forestry equipment in difficult terrain conditions.

References

[1] Golyakevich S A and Goronovskii A R 2017 Evaluation of loading dynamics and fatigue life for a forwarder half-frame articulation *Journal of Machinery Manufacture and Reliability* **46** pp 463 – 471
[2] Naskovets M T and Drachilovsky A I 2012 Improvement of road structures to increase the

transport and operational indicators of forest roads [In Russian - Sovershenstvovanie dorozhny`x konstrukcij dlya povy`sheniya transportno-e`kspluatacionny`x pokazatelej lesny`x dorog] *Proceedings of BSTU: Forest and woodworking industry* **2** pp 80–82

[3] Naskovets M T, Korin G S and Drachilovsky A I 2013 Application of Spanbel geotextile material in the construction of pilot objects of forest roads of secondary importance and the results of their tests [In Russian - Primenenie geotekstil`nogo materiala «Spanbel» pri stroitel`stve opy`tny`x ob``ektov lesny`x dorog vtorostepennogo znacheniya i rezul`taty` ix ispy`tanij] *Proceedings of BSTU: Forest and woodworking industry* **2** pp 31–33

IOP Conf. Series: Materials Science and Engineering **817** (2020) 012007 doi:10.1088/1757-899X/817/1/012007

Improving the physical and mechanical properties of fortified soil for road construction in the forest zone

S Chudinov

Department of Transport and Road Construction, *Ural State Forestry University*, 37 Siberian Road, Yekaterinburg 620100, Russian Federation

Corresponding e-mail: serg-chudinov@yandex.ru

Abstract. In order to purposefully control the processes that determine the formation of the structure and properties of cement soils, studies were carried out on the strengthening of soils with Portland cement with a water-soluble additive hydropropylene emulsion (WHE). The purpose of the laboratory studies was to study the change in the ultimate strength in compression, water saturation and frost resistance of cement soils, depending on the particle size distribution of the original soil, the amount of mineral binder - Portland cement and the WHE additive. Based on the results of the research, soil strengthening with Portland cement with WHE additive allows to form a complex crystallization-coagulation structure of materials with the required values of compressive strength, water saturation and frost resistance, depending on the dosages of binding components for given climatic and transport-operating conditions. The effectiveness of the WHE additive increases with increasing dispersion of soils, so the technology of strengthening of clay soils in conditions of wet and acidic soils of the forest zone for the construction of forest roads is promising.

1. Introduction

Soil reinforcement technology for the construction of pavement layers of construction has proven itself in road construction, especially in regions with an insufficient amount of stone materials. Reinforced soils can be used in any climatic and transport-operational conditions in the construction of public roads and industrial, including forest roads.

As a binder to strengthen the soil can be used mineral and organic substances. One of the most common mineral binders for strengthening the soil is Portland cement. However, the effectiveness of the structure formation of Portland cement in cement soils depends largely on the degree of activity of sorption and ion exchange processes on the surface of fine soil particles, especially clay particles. When soil is strengthened with portland cement in the forest zone, the acidification of clay soils, which are characterized by incomplete saturation of the exchange complex with hydrogen and calcium ions, also results in crystal formation and, as a result, absorb calcium hydroxide from hydrating cement [1].

Thus, to strengthen the soil with Portland cement, especially in the forest zone, an improvement in the process of crystal formation of the mineral binder is required. One of the effective technologies for improving the quality of fortified soils in the forest zone is the use of polymer additives in the composition of cement-ground mixtures. Conducted theoretical studies have allowed to determine that WHE allows to improve the conditions of hardening of Portland cement while strengthening clay soils and improve the physical and mechanical properties of cement grounds.

To confirm this hypothesis were conducted laboratory studies. The purpose of the laboratory studies was to study the change in the ultimate strength in compression, water saturation and frost resistance of cement soils, depending on the particle size distribution of the original soil, the amount of mineral binder - Portland cement and the WHE additive.

1.1. Research tasks

• Sampling of natural soils of various types: sand, sandy loam, heavy loam, clay. Characterization of initial soil samples.

• Production of soil samples reinforced with Portland cement and WHE.

• Determination of the compressive strength of water-saturated soil samples reinforced with Portland cement and WHE.

• Determination of water saturation of soil samples reinforced with Portland cement and WHE additive.

• Determination of the frost resistance of soil samples reinforced with Portland cement and WHE additive.

• Analysis of the results of the ultimate strength in compression, water saturation and frost resistance of cement soils, depending on the particle size distribution of the original soil, the amount of mineral binder - Portland cement and the WHE additive.

2. Methods and Materials

For laboratory studies were selected natural soils from forest roads in the Sverdlovsk region of the Russian Federation. Soil tests were carried out on the basis of relevant regulatory documents. Types of tests are shown in table 1.

Table 1. Types of tests and relevant regulatory documents.

Name of soil characteristics	Regulatory Document
Dry soil density	GOST 5180 [2]
Plasticity limits	GOST 5180 [2]
Granulometric composition	GOST 12536 [3]
Optimum humidity and maximum density with standard compaction	GOST 22733 [4]
Acidity of the water extract	GOST 26423 [5]
Organic Content	GOST 23740 [6]

2.1. Characteristic of portland cement

For laboratory research, Portland cement M400 brand produced by Brozex Dry Mixes Plant LLC was used in accordance with GOST 10178 [7].

2.2. Characteristics supplements WHE

The stabilizing additive WHE is a water-soluble hydropropylene emulsion with an agglomerative effect on clay soil particles. Physico-chemical characteristics of the WHE supplements are presented in table 2.

IOP Conf. Series: Materials Science and Engineering **817** (2020) 012007 doi:10.1088/1757-899X/817/1/012007

Table 2. Physical and chemical indicators of the WHE supplement.

Name of the indicator	Method of Definition	Magnitude	Unit
pH	DIN 38 404-C5	11.0	-
Temperature (for pH)	DIN 38 404-C4	25.0	°C
Chloride	DIN EN ISO 10304	1.0	mg / l
DOC	DIN EN 1484	1.9	mg / l
Zinc	DIN EN ISO 11885	0.03	mg / l

2.3. Determination of the minimum required number of observations in laboratory studies

When studying the issues of soil reinforcement, it is considered that sufficient reliability of the experiment is ensured with a measurement accuracy of 5% and a confidence level of 0.95. The required number of samples, when determining any indicator of the strength or deformation properties of the soil, was established by the formula (1):

$$n = \frac{u^2 \sigma N}{(N-1)\Delta^2 + u^2 \sigma^2} \quad (1)$$

where, n – the required number of observations; σ – the sample variance; N – the number of preliminary tests performed; Δ – the permissible deviation from the average value; u – the deviation value depending on the specified confidence level ($p = 0.95$; $u = 1.96$).

On the basis of preliminary data obtained experimentally, the minimum number of observations was determined when studying such properties of cement soil as the compressive strength of water-saturated samples. The calculation of the average value of the experiment from n measurements was performed by the formula (2):

$$\bar{y}_n = \frac{1}{n}\sum_{i=1}^{n} y_i \quad (2)$$

where, $\bar{y}_n$ – the average value of the experience; n – the number of samples in the experiment; y_i – the single result of observation in experience.

The estimate of the variance of the sample of n values was determined by the formula (3):

$$\sigma = \frac{1}{n-1}\sum_{i=1}^{n} (y_i - \bar{y}_n)^2 \quad (3)$$

The results of statistical processing of indicators of preliminary tests are presented in table 3

Table 3. Statistical processing of preliminary test results.

Observation results, MPa			Number of measurements, N	Average value, MPa	Estimation of variance, σ	Number of observations required	The deviation value, U	Tolerance, Δ, MPa
2.8	2.64	2.65	24	2.715	0.0114	2.57	1.96	0.135
2.69	2.75	2.54						
2.74	2.9	2.5						
2.83	2.66	2.75						
2.72	2.82	2.73						
2.06	1.94	1.78						
1.80	1.88	1.85						
1.95	2.12	2.02						

Thus, as a result of statistical processing of data from preliminary tests of the compressive strength of water-saturated samples of cement grounds with a base area of 20 cm^2, it can be concluded that the compressive strength should be defined as the arithmetic average of the results of testing three samples.

2.4. The composition of the mixtures of soils reinforced with Portland cement with the addition of WHE

In order to study the changes in compressive strength, water saturation and frost resistance of cement ground depending on the particle size distribution of the soil, the amount of mineral binder - Portland cement and WHE additives, various compositions of cement – ground mixtures were prepared and tested.

For laboratory studies, various types of soil were used: coarse sand, sandy sandy loam, heavy silt loam, light silt clay. The amount of Portland cement introduced into the soil, was: 0%, 3%, 5%, 7%, 9% by weight of dry soil. The amount of WHE additive introduced into the soil was 0%, 0.03%, 0.06%, 0.2% by weight of dry soil.

The list of the compositions of soil mixtures: coarse sand, sandy sandy loam, heavy silty loam, light silt clay strengthened with Portland cement with WHE additive for laboratory research are presented in table 4.

Table 4. The list of soil mixes: coarse sand, sandy sandy loam, heavy silt loam, light silt clay, fortified with Portland cement with WHE additive.

Amount of portland cement % by weight of dry soil	*The amount of additive WHE,% by weight of dry soil*			
	0	**0.03**	**0.06**	**0.20**
0	+	+	+	+
3	+	+	+	+
5	+	+	+	+
7	+	+	+	+
9	+	+	+	+

IOP Conf. Series: Materials Science and Engineering **817** (2020) 012007 doi:10.1088/1757-899X/817/1/012007

3. Results and Discussion

3.1. The results of laboratory studies to determine the ultimate strength in compression of water-saturated samples of fortified soil

The results of determining the compressive strength (Z, MPa) of water-saturated cement-ground samples depending on the particle size distribution of the soil and the dosages of Portland cement (X,%) and the WHE additive (Y,%) are presented in Figure 1-4.

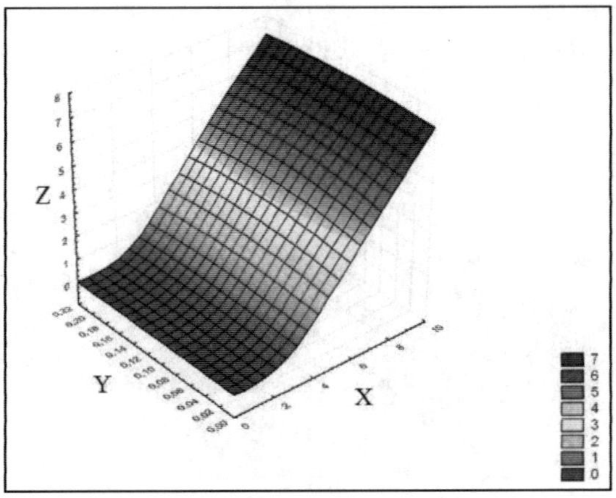

Figure 1. The surface values of the ultimate compressive strength for coarse sand, depending on the content of binders.

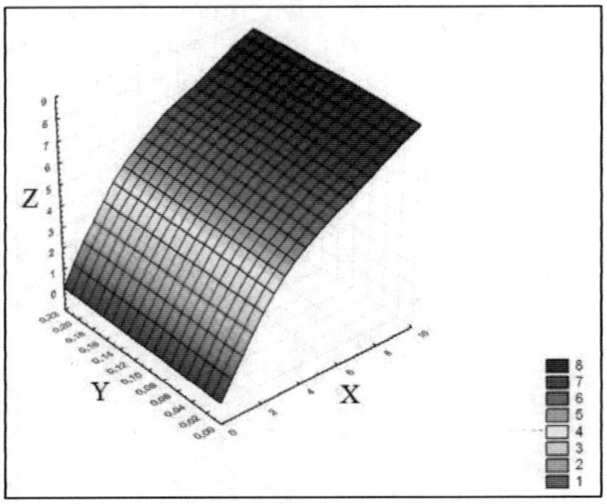

Figure 2. The surface values of compressive strength for sandy sandy loam depending on the content of binding materials.

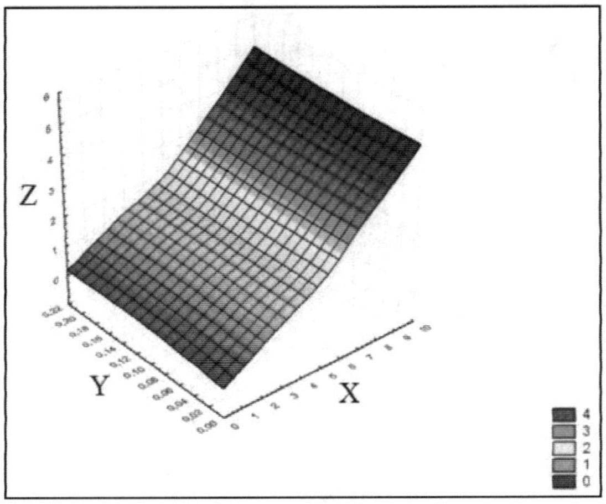

Figure 3. The surface values of the ultimate compressive strength for heavy loam depending on the content of binders.

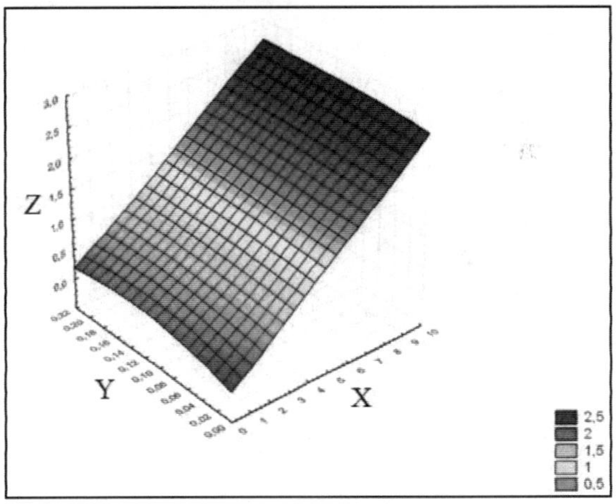

Figure 4. The surface values of the ultimate compressive strength for clay dust depending on the content of binders.

Thus, based on the results of laboratory studies of the dependence of the compressive strength of water-saturated soil samples: coarse sand, sandy sandy loam, heavy loamy silty clay, light silt clay strengthened with Portland cement with the WHE additive, the following conclusions can be made:

1. The use of the WHE additive in strengthening the soil without a mineral binder - Portland cement leads to a slight increase in the strength of the soil.

IOP Conf. Series: Materials Science and Engineering **817** (2020) 012007 doi:10.1088/1757-899X/817/1/012007

2. The effectiveness of the application of the WHE additive in strengthening the soil without a mineral binder - Portland cement increases with increasing dispersion of soils in the direction: sand - sandy loam - loam - clay.

3. The use of the WHE additive in conjunction with Portland cement allows, depending on the dosages of the binding components, to achieve the required values of the compressive strength of the cement for the given climatic and transport and operating conditions.

3.2. The results of laboratory studies to determine the water saturation of samples of fortified soils

The results of determination of water saturation (Z,%) of samples of fortified soils depending on the particle size distribution of the soil and dosages of Portland cement (X,%) and the WHE additive (Y,%) are presented in Figure 5-8.

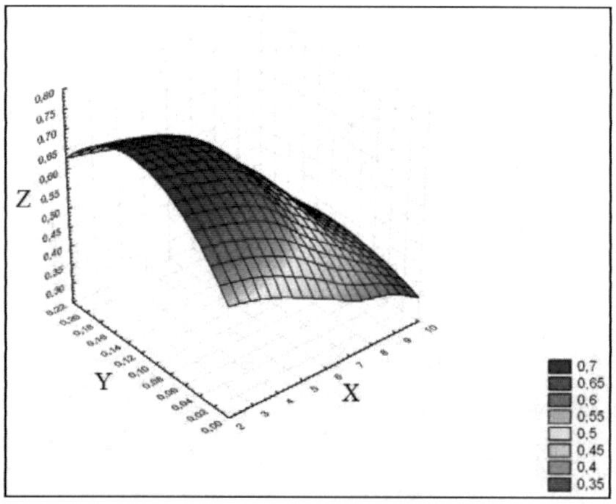

Figure 5. The surface water saturation values for coarse sand depending on the content of binders.

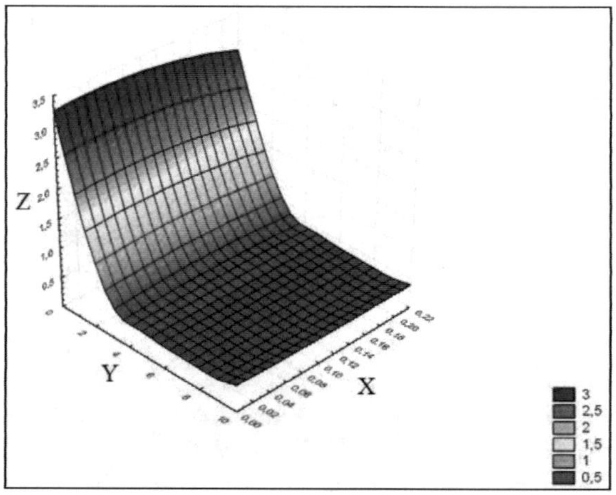

Figure 6. The surface water saturation values for sandy sandy loam depending on the content of binding materials.

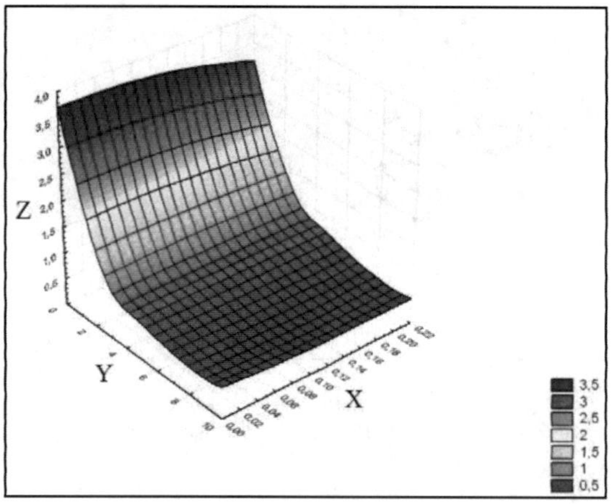

Figure 7. The surface water saturation values for heavy loam depending on the content of binders.

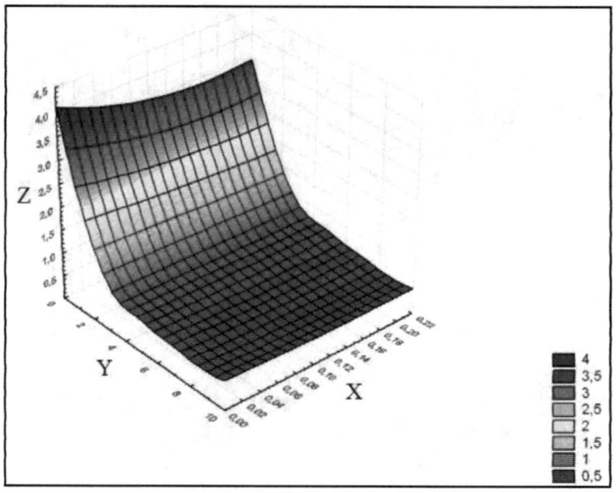

Figure 8. The surface water saturation values for clay dust depending on the content of binders.

Thus, based on the results of laboratory studies of the dependence of water saturation of soil samples: coarse sand, sandy sandy loam, heavy silty clay loam, light silt clay strengthened with Portland cement with WHE additive, we can draw the following conclusions:

1. The use of the WHE additive in strengthening the soil without a mineral binder - Portland cement leads to a slight decrease in the water saturation of the stabilized soil.
2. The effectiveness of the application of the WHE additive in strengthening the soil without a mineral binder - Portland cement increases with increasing dispersion of soils in the direction: sand - sandy loam - loam - clay.
3. The use of the WHE additive together with Portland cement allows, depending on the dosages of the binding components, to achieve the required values of water saturation for given climatic and transport and operating conditions.

3.3. The results of laboratory studies to determine the frost resistance of samples of fortified soil

The results of determining the frost resistance of samples of fortified soils depending on the particle size distribution of the soil and dosages of Portland cement and the WHE additive are presented in Table 5-8.

Table 5. The results of studies of frost resistance of coarse sand, reinforced with Portland cement with WHE additive

Brand on frost resistance, F		Amount of additive WHE, % by weight of dry soil			
		0	**0.03**	**0.06**	**0.20**
The amount of Portland cement, % by weight of dry soil	0	–	–	–	–
	3	F0	F0	F0	F0
	5	F5	F5	F5	F5
	7	F5	F5	F5	F5
	9	F10	F10	F10	F10

Table 6. The results of studies of frost resistance of sandy, reinforced with Portland cement with WHE additive

Brand on frost resistance, F		Amount of additive WHE, % by weight of dry soil			
		0	**0.03**	**0.06**	**0.20**
The amount of Portland cement, % by weight of dry soil	0	–	–	–	F0
	3	F0	F0	F0	F0
	5	F0	F0	F0	F0
	7	F5	F5	F5	F5
	9	F10	F10	F10	F10

Table 7. The results of studies of frost resistance of heavy loam hardened with Portland cement with WHE additive

Brand on frost resistance, F		Amount of additive WHE, % by weight of dry soil			
		0	**0.03**	**0.06**	**0.20**
The amount of Portland cement, % by weight of dry soil	0	–	–	F0	F0
	3	F0	F0	F0	F0
	5	F0	F0	F0	F0
	7	F0	F0	F0	F0
	9	F5	F5	F5	F5

Table 8. The results of studies of the frost resistance of clay light silt, reinforced with Portland cement with the additive WHE

Brand on frost resistance, F		Amount of additive WHE, % by weight of dry soil			
		0	0.03	0.06	0.20
The amount of Portland cement, % by weight of dry soil	0	–	F0	F0	F0
	3	F0	F0	F0	F0
	5	F0	F0	F0	F0
	7	F0	F0	F0	F0
	9	F5	F5	F5	F5

Thus, based on the results of laboratory studies of the dependence of frost resistance of soil samples: coarse sand, sandy sandy loam, heavy silty clay loam, light silt clay, fortified with Portland cement with WHE additive, we can draw the following conclusions:

1. Frost resistance of cement soils increases with decreasing dispersion of the original soil in the direction: clay - loam - sandy loam - sand.
2. The effectiveness of the application of the WHE additive in strengthening the soil without a mineral binder - Portland cement increases with increasing dispersion of soils in the direction: sand - sandy loam - loam - clay.
3. The use of the WHE additive together with Portland cement allows, depending on the dosages of the binding components, to achieve the required values of frost resistance for given climatic and transport and operating conditions.

4. Conclusions

1. As shown by the results of studies of water saturation, ultimate compressive strength of water-saturated samples and frost resistance of soils, different granulometric composition: coarse sand, sandy sandy loam, heavy clay loam and clay light dusty, reinforced with portland cement with the WHE additive, depending on the number of binding components , cement grounds have high strength characteristics. The use of the WHE additive in soil consolidation with Portland cement allows the formation of a complex structure of cement ground - crystallization-coagulation, characterized not only by high strength, but also by plasticity and crack resistance [8].

2. The use of the WHE additive without a mineral binder - Portland cement allows the formation of a coagulation structure of fortified soils with elastic-viscous-plastic properties, low strength and water saturation [9]. The effectiveness of the application of the WHE additive in strengthening the soil without a mineral binder - Portland cement increases with increasing dispersion of soils in the direction: sand - sandy loam - loam - clay.

3. Strengthening of soils with Portland cement with the WHE additive allows forming a complex crystallization-coagulation structure of materials, possessing the required values of compressive strength, water saturation and frost resistance, depending on dosages of binding components for given climatic and transport and operating conditions [10].

4. Due to the fact that the effectiveness of the WHE additive increases with increasing dispersion of soils, the technology of strengthening clay soils in conditions of wet and acidic soils of the forest zone for the construction of forest roads is promising.

References

[1] Nevolin D G 2015 *Innovative technologies for the design and construction of roads* [in Russian - Innovatsionnyie tehnologii proektirovaniya i stroitelstva avtomobilnyih dorog] (Ekaterinburg: Ural State University of Railway Transport) p 291
[2] GOST 5180 *Soils. Laboratory methods for determination of physical characteristics* 2015 p 19
[3] GOST 12536 *Soils. Methods of laboratory granulometric (grain-size) and microaggregate*

distribution 2014 p 18
[4] GOST 22733 *Soils. Laboratory method for determining of maximum density* 2016 p 12
[5] GOST 26423 *Soils. Methods for determination of specific electric conductivity, pH and solid residue of water extract* 2015 p 7
[6] GOST 23740 *Soils. Methods of laboratory determination of organic composition* 2016 p 10
[7] GOST 10178 *Portland cement and portland blastfurnace slag cement. Specifications* 2015 p 7
[8] Bezruk V M 1982 *Fortified soil (Properties and application in road and airfield construction)* [in Russian - Ukreplennyie gruntyi] (Moscow: Transport) p 231
[9] Chudinov S A and Buldakov S I 2010 Theoretical studies of the processes of structure formation of *clay soils, reinforced with Portland cement with the addition of polyelectrolyte* [in Russian - Teoreticheskie issledovaniya protsessov strukturoobrazovaniya glinistyih gruntov, ukreplennyih portlandtsementom s dobavkoy polielektrolita] (Arkhangelsk: News of higher educational institutions. Forest Journal. № 5) pp 82–88
[10] Chudinov S A 2019 *Improving the quality of cement-soil coatings of forestry road* [in Russian - Povyishenie kachestva tsementogruntovyih pokryitiy lesovoznyih dorog] (Omsk: Education. Transport. Innovation. Building. Collection of materials of the II National Scientific and Practical Conference) pp 459–462

IOP Conf. Series: Materials Science and Engineering **817** (2020) 012008 doi:10.1088/1757-899X/817/1/012008

Method for reducing the hourly fuel consumption in the transport mode of operation of a forestry machine and tractor unit

M Y Durmanov*, **B G Martynov, O A Mikhaylov, S V Spiridonov**

St. Petersburg State Forest Technical University, 5 Institutskiy Lane, St. Petersburg 194021, Russian Federation

*Corresponding e-mail: DurmanovMJ@yandex.ru

Abstract. By improving energy consumption in technological operations, the amount of fuel required and the operating costs of a machine-tractor unit (MTU) can also be reduced, which ultimately lowers the costs of forestry operations and reforestation. Inefficient fuel consumption significantly reduces the performance indicators of an MTU, which is also characteristic of the transport mode, although its duration is short. The article investigates the influence of a corrective device in the diesel fuel supply system on the hourly fuel consumption of a forestry MTU. For the transport mode, the diesel fuel consumption per hour is determined for dynamic loads by component: from resistance on the rise; from the MTU inertial forces; from oscillations of the MTU suspension in the longitudinal-vertical plane; from mechanical losses in the friction pairs of the engine, as well as its total value. The maximum values of the total dynamic component of the hourly fuel consumption in the transport mode are observed at frequencies close to 2.5 s^{-1}, i.e. its own, energy-consuming frequency of oscillations of the rotational speed of the crankshaft of an SMD-20T.04 engine. To improve the fuel efficiency of an MTU diesel engine, it is necessary to introduce a corrective device into the design of the high-pressure fuel pump (HPFP). As a corrective device, it is advisable to use an oscillation damper of the rail of the fuel pump (FPR).

1. Introduction

Minimizing the amount of energy required to carry out technological operations reduces the amount of fuel required and the operating costs of a machine-tractor unit (MTU) [1, 2], which ultimately reduces the costs of forestry operations and reforestation. Inefficient fuel consumption significantly reduces the performance indicators of an MTU, which is also characteristic of the transport regime, although its duration is short [3-6].

It is known [1, 2, 7] that each type of engine and its system for controlling the speed and fuel supply have their own frequencies. For the diesel engines SMD-18BN and SMD-20T.04 at values of natural frequencies close to 2.51 s^{-1}, the greatest deviations of the fuel pump rail (FPR) from the average set values have been observed.

Studies of the effect of the high pressure fuel pump (HPFP) of the LSTN-49010 type of the SMD-18BN diesel engine on the quality of the fuel supply showed [8] that, when the fuel is cut off, a hydrodynamic force arises with an amplitude at a frequency of 2.51 s^{-1}, which turns the plunger and associated FPR in the direction of reducing the magnitude of the cyclic feed and turns off the speed controller (SC). The speed controller, trying to restore the given mode, returns the FPR to its original position with an overshoot, accompanied by an increased cyclic feed, and as a result, by an increase in the amplitude of the oscillations of the crankshaft rotation speed (figure 1) and a decrease in the constant component of the rotational speed.

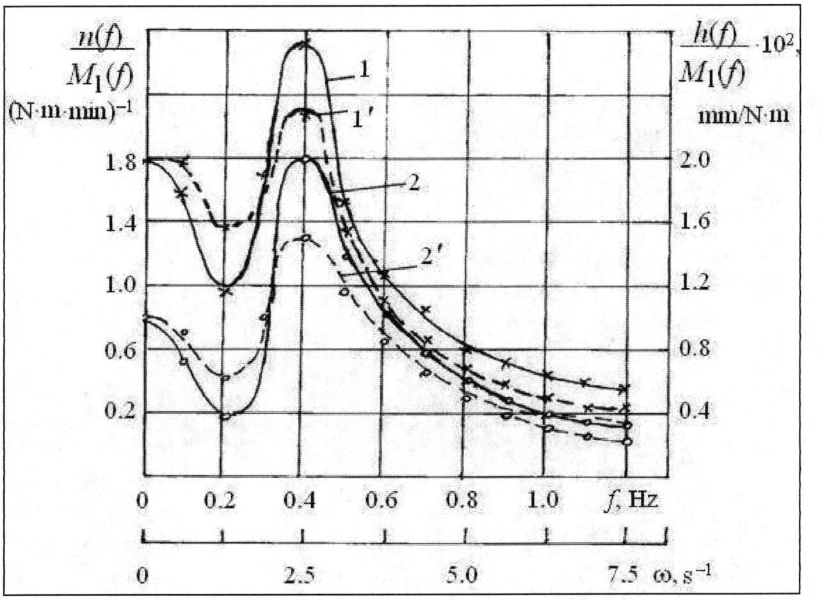

Figure 1. AFC of the rotation frequency (1, 1′) and the oscillations of the rail of the fuel pump (2, 2′) of the SMD-18BN diesel engine: 1, 2 - without an FPR oscillation damper; 1′, 2′ - with an FPR oscillation damper [2].

Of the four natural frequencies of the amplitude-frequency characteristic (AFC) of the hydrodynamic force of the HPFP, only one ($\omega_{1e} = 2.5$ s^{-1}) coincides with the amplitude-frequency characteristic of the SC [8], being an additional source of excitation, enhancing the oscillatory process in the system "HPFP - speed regulator", and creating conditions under which precise dosing of the cyclic feed and the beginning of fuel injection are impossible.

In order to reduce the amplitude of the rail vibrations and the negative effect of the hydrodynamic force on the quality of the speed control process, an oscillation damper has been developed [9]. Figure 1 shows the frequency characteristics of the FPR and the rotational speed of the SMD-18BN drive with and without an FPR oscillation damper. Figure 2 shows an HPFP of the LSTN-49010 type with an FPR oscillation damper mounted on a test bench [8].

The purpose of the work is to confirm the efficiency of using an FPR oscillation damper to reduce the hourly fuel consumption of the SMD-20T.04 diesel during the transport [10] and arable [11] modes of operation of a forestry MTU. The proposed method for increasing the operational efficiency of an MTU, especially relevant at the stage of tractor design, serves to improve the quality of design and engineering work, and to improve the design of a forestry MTU [12, 13].

2. Methods and Materials

A model of a forestry MTU functioning in the transport mode was implemented in the MathCAD mathematical environment, and the Microsoft Office Excel application was used to process the results of calculations and for graph-building. The calculation method is described below.

The main external influences in the model were: the surface profile of the felled area, defined by the correlation function and spectral density, frequency of load oscillations on the tractor's sprockets, and speed of the MTU. The unsteady nature of forces of resistance to the motion of the MTU, described by an ergodic stationary random process, is modelled for the entire frequency spectrum of the input load moment. The article discusses the transport mode of operation of an LHT-100 tractor with a PKL-70 plow. For the MTU under consideration, a sufficient amount of operational indicators has been accumulated, which are necessary for estimating modelling errors, since it has been widely used in forestry and reforestation operations.

Figure 2. The high-pressure fuel pump (HPFP) of type LSTN-49010 with an FPR oscillation damper and sensors mounted on a test bench: 1 - the FPR oscillation damper; 2 – the FPR displacement sensor; 3 - the camshaft speed sensor of the HPFP.

2.1. Analytical expressions for determining the hourly fuel consumption of an MTU diesel engine in the transport mode

The expression for the hourly fuel consumption of an MTU diesel engine in the transport mode has two components - regular B_0^t and variable $B^t(\omega,\upsilon_0)$, written in the operator form (at $s = j\omega$) [2, 10]:

$$B_0^t = E_1 n_0 \Big[\frac{R}{i_t \eta_m} A_1^t + D_1(a_1 + b_1 n_0)\Big]; \quad (1)$$

$$B^t(\omega,\upsilon_0) = E_1 M_n^a \omega \cdot |U_{11}(j\omega,\upsilon_0)| \cdot \Big\{\frac{R}{i_t \eta_m}\Big[A_1^t + 2A_4\omega \cdot |U_{11}(j\omega,\upsilon_0)| + +2f_{rf}|\Theta_v^t(j\omega,\upsilon_0)|\Big]$$
$$+ D_1 \cdot \Big[\frac{a_1}{P_{mn}}\left(\frac{4\omega^2+\gamma^2}{\omega^2+\gamma^2}\right)^{0.5} \cdot |G_{61}(j\omega,\upsilon_0)| + 2b_1\omega \cdot |U_{11}(j\omega,\upsilon_0)|\Big]\Big\}; \quad (2)$$

where, $E_1 = \frac{3.6\pi}{30\eta_i H_u}$; $A_1^t = mg(\sin\alpha + f_{rf}\cos\alpha)$; $D_1 = \frac{V_c i_e}{\pi\tau_e}$; $A_4 = \frac{\pi R}{30 i_t} m$,

where, $n_0 = 0.8n_n$; n_n - the nominal speed of the diesel crankshaft; R - the radius of the leading sprockets; i_t - the transmission ratio ; η_m - the mechanical efficiency, $\eta_m = \eta_t \cdot \eta_{mov}$; η_t, η_{mov} - the efficiency of the transmission and propulsion, respectively; a_1, b_1 - the coefficients obtained experimentally for each type of engine; M_{n}^{a} - the amplitude value of the input disturbance: $M_{\text{n}}^{a} = 0.15M_{\text{en}}$; M_{en} - the nominal effective torque; ω - the angular speed of rotation of the crankshaft; $|U_{11}(j\omega)|$ - the transfer function of the rotational speed of the crankshaft of the MTU power unit; υ_0 - the MTU movement speed; m – the mass of the MTU, $m = m_{\text{t}} + m_{\text{pl}}$; m_{t}, – the mass of the tractor; m_{pl} – the mass of the plow; α - the angle of the elevation of the track; f_{rf} - the coefficient of friction of the rolling propulsion; $|\Theta_{\text{v}}^{\text{t}}(j\omega)|$ - the transfer function of the oscillations of the MTU frame in the longitudinal-vertical plane in the transport mode; P_{mn} - the oil pressure in the main oil line (MOL) at nominal rotation speed; $\gamma = 10^{-3}\omega_{\text{n}}$; ω_{n} - the nominal angular speed of rotation of the crankshaft. $|G_{61}(j\omega)|$ - the transfer function of oil pressure in the MOL; η_i - the indicator engine efficiency; H_u - the lowest calorific value of fuel; V_{c} - the working volume of one engine cylinder; i_{e}, τ_{e} - the number of cylinders and the engine cycle, respectively.

The transfer function of the rotational speed of the crankshaft of the MTU power unit in a disturbing action (with a fixed position of the body $h(t)$ that sets the value of the cyclic fuel supply: $h_0 = \text{const}$) [1, 2]:

$$|U_{11}(j\omega)| = \frac{k_{\text{e1}}(T_2^2 s^2 + 2T_2\xi_1 s + 1)}{(T_1 s + 1)(T_3^2 s^2 + 2T_3\xi_2 s + 1)(T_4 s + 1)}, \tag{3}$$

where, k_{e1} - the transmission speed ratio; T_1, T_2, T_3, T_4 - the time constants; ξ_1, ξ_2 - the attenuation coefficients.

For the engine SMD-20T.04: k_{e1} = 1.80; T_1 = 1.073 s; T_2 = 0.531 s; T_3 = 0.398 s; T_4 = 0.354 s; ξ_1 = 0.300; ξ_2 = 0.250 [1, 2].

The input impacts for the MTU from the side of the felling are unevenness of the micro-relief and heterogeneity of the soil which are the cause of fluctuations $F_r(t)$ in the resistance forces on the leading sprockets.

Note that for the MTU, taking into account its mass m, the moment of inertia given to the crankshaft I_{a} and the time constant T_1 are

$$I_{\text{a}} = I_{\text{e}} + \frac{mR^2}{i_{\text{t}}^2}; \; T_1 = \frac{\pi}{30} I_{\text{a}} \frac{n_{\text{n}}^2}{N_{\text{n}}}, \tag{4}$$

where, I_{e} - the moment of inertia of the engine; N_{n} - rated engine power.

The transfer function of the oil pressure in the engine MOL [1, 2]:

$$|G_{61}(j\omega)| = k_{\text{m1}} s \cdot U_{11}(j\omega), \tag{5}$$

where, k_{m1} - the transfer rate of oil pressure in MOL: k_{m1} = 1.20.

The transfer function of the oscillations of the MTU frame in the longitudinal-vertical plane in the transport mode was obtained theoretically [2, 6, 10]:

$$|\Theta_{\text{v}}^{\text{t}}(j\omega)| = \frac{c_1 z_{\text{mr}}\omega^2 \cdot \left(m_\Sigma''^2 (c_1 - m_\Sigma'\omega^2)^2 + (m_\Sigma''\beta_1\omega)^2\right)^{0.5}}{\left([m_\Delta^2\omega^4 - (c_1 m_\Sigma + \beta_1^2)\omega^2 + c_1^2]^2 + [\beta_1\omega(2c_1 - m_\Sigma\omega^2)]^2\right)^{0.5}}, \tag{6}$$

where, z_{mr} - the height of a micro-roughness; c_1 - stiffness of the suspension springs;

$$m'_{\Sigma} = m_{10} + m_{20}\,;\;\; m''_{\Sigma} = m_0 k_1 + m_2 k_2\,;\;\; m_{\Sigma} = m_1 + m_2\,;\;\; m_{10} = m_1 + m_0\,;\;\; m_{20} = m_2 + m_0\,;$$

$$k_1 = \left(\left(\frac{m_1}{m_0}\right)^2 + 1 - \frac{2m_1}{m_0}\cos 0.39\right)^{0.5};\;\; k_2 = \left(\left(\frac{m_0}{m_2}\right)^2 + 1 - \frac{2m_0}{m_2}\cos 0.39\right)^{0.5}$$

$$\beta_1 = 2\nu_n\sqrt{c_1 m}\,;\;\; m_1 = \frac{I_s + (m_{ts} + m_{pl})l_2^2}{l^2}\,;\;\; m_2 = \frac{I_s + (m_{ts} + m_{pl})l_1^2}{l^2}\,;$$

$$m_0 = \frac{I_s - (m_{ts} + m_{pl})l_1 l_2}{l^2}\,;\;\; m_{\Delta}^2 = m_1 m_2 - m_0^2\mu_{\Delta}\,;\;\; \mu_{\Delta} = 1 - \frac{m_0^2}{m_1 m_2}\,;$$

$\upsilon_n = 0.1$; m_{ts} - the sprung mass of the tractor with a plow, $m_{ts} = 8400$ kg; $l_1 = 0.65$ m; $l_2 = 1.10$ m; l - distance between the axes of the carriages of the tracked mover, $l = l_1 + l_2 = 1.75$ m; I_S - the moment of inertia of the system: $I_S = 26154$ kg·m^2 [2, 6, 10].

By substituting in (2) the values of $U_{11}(j\omega)$ from (3), $G_{61}(j\omega)$ from (5), and $\Theta_v^t(j\omega)$ from (6), we determine the surface of the state of the frequency characteristics (FC) of the hourly fuel consumption of the MTU diesel engine in the transport mode (figure 3). The numerical values of the hourly fuel consumption are determined by substituting the following source data in (1) and (2): V_c =1.575 l; $i_e = 4$; $\tau_e = 2$; $a_1 = 0.45$; $b_1 = 0.97 \cdot 10^{-3}$; $\eta_i = 0.40$; $H_u = 41300$ kJ/kg; $P_{mn} = 0.6$ MPa; n_n =1900 min^{-1}; $N_{en} = 88$ кW; $I_e = 2.40$ kg·m^2; $M_{en} = 480$ N·m; $\gamma = 0.188$ s^{-1}; $F_r^a = 0.15\,M_{en}$; $\eta_m = 0.68$; $R = 0.238$ m; $m_t = 10400$ kg; $m_{pl} = 520$ kg; $f_{rf} = 0.15$; $z_{mr} = 0.06$ m; $\alpha = 5°$ [2, 6, 10].

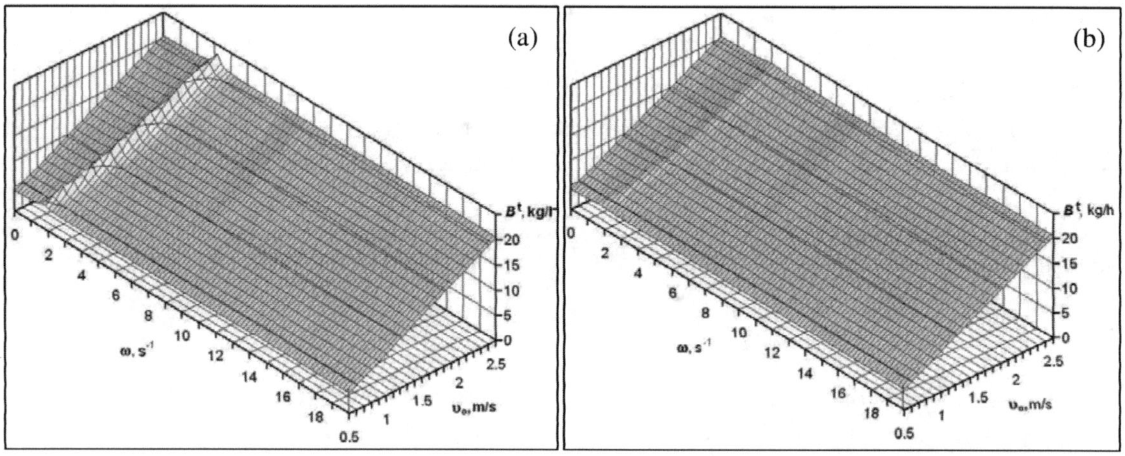

Figure 3. The state surfaces of the FC of the hourly fuel consumption of the MTU diesel engine based on the LHT-100 tractor with the PKL-70 plow in the transport mode depending on the speed of movement, with the standard suspension stiffness $c_1 = 1400$ kN /m: (a) - with a standard diesel engine; (b) - with an oscillation damper of the FPR.

Performing sections in a longitudinal-vertical plane (figures 4 and 5) for any fixed value of the regular speed υ_0 of movement of the MTU and over the entire range of frequencies ω of oscillations of the load on the leading sprockets, one can calculate the regular component B_0^t and the dynamic component of the hourly fuel consumption $B^t(\omega, \upsilon_0)$ in the transport mode.

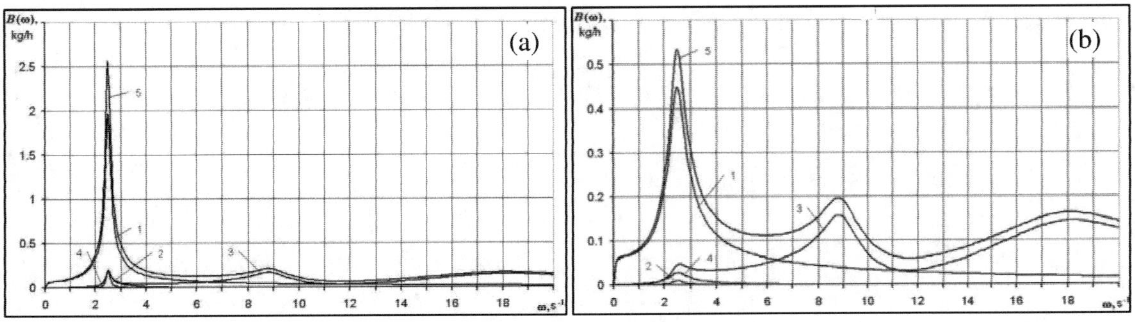

Figure 4. Frequency response of the hourly fuel consumption by components in the MTU transport mode based on the LHT-100 tractor with the PKL-70 plow with the suspension stiffness $c_1 = 1400$ kN/m and travel speed $\upsilon_0 = 1.15$ m/s (*III* gear): (a) - with a standard diesel engine ; (b) - with an oscillation damper of the FPR; 1 - resistance to the movement of the MTU on the rise; 2 - MTU inertial forces; 3 - tractor oscillations in the longitudinal-vertical plane; 4 - mechanical losses in the engine friction pairs; 5 – the total consumption for dynamic loads.

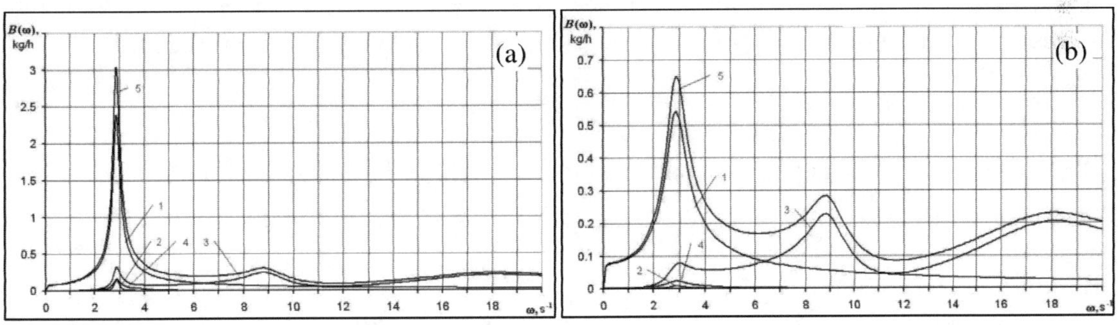

Figure 5. Frequency response of the hourly fuel consumption by components in the MTU transport mode based on the LHT-100 tractor with the PKL-70 plow with suspension stiffness $c_1 = 1400$ kN / m and travel speed $\upsilon_0 = 1.78$ m/s (*IV* gear): (a) - with a standard diesel engine ; (b) - with an oscillation damper of the FPR; the legend as in figure 4.

2.2. Analysis of the hourly fuel consumption components of the MTU diesel engine in the transport mode

Using formula (2), the losses of the hourly fuel consumption per dynamic load for each component in the frequency spectrum 0 ... 20 s^{-1} were calculated, as well as the total for different values $c_1 = 1800$; 1400; 1000 kN/m:

1 - from the resistance on the rise at an angle α=5°

$$B_1^{t}(\omega) = \frac{E_1 R \cdot M_n^a}{i_t \eta_m \omega_{le}} A_1^{t} \cdot \int_{\omega=0}^{\omega=20} \omega \cdot U_{11}(j\omega)\, d\omega ; \qquad (7)$$

2 - from the inertial forces of the MTU

$$B_2^{t}(\omega) = \frac{2E_1 R \cdot M_n^a}{i_t \eta_m \omega_{le}} A_4 \cdot \int_{\omega=0}^{\omega=20} \omega^2\, U_{11}^2(j\omega)\, d\omega ; \qquad (8)$$

3 - from oscillations of the suspension of the MTU in the longitudinal-vertical plane

$$B_3^{\mathrm{t}}(\omega) = \frac{2E_1 R \cdot f_{\mathrm{rf}} \cdot M_{\mathrm{n}}^{a}}{i_{\mathrm{t}}\eta_{\mathrm{m}}\omega_{\mathrm{le}}} \cdot \int_{\omega=0}^{\omega=20} \omega \cdot \left|U_{11}(j\omega)\right| \cdot \left|\Theta_{\mathrm{v}}^{\mathrm{t}}(j\omega)\right| \mathrm{d}\omega\,; \tag{9}$$

4 - from mechanical losses in the engine friction pairs

$$B_4^{\mathrm{t}}(\omega) = \frac{E_1 D_1 \cdot M_{\mathrm{n}}^{a}}{\omega_{\mathrm{le}}} \cdot \int_{\omega=0}^{\omega=20} \omega \cdot \left|U_{11}(j\omega)\right| \left[\frac{a_1}{P_{\mathrm{mn}}} \left(\frac{4\omega^2 + \gamma^2}{\omega^2 + \gamma^2} \right)^{0.5} \cdot \left|G_{61}(j\omega)\right| + 2b_1\omega \cdot \left|U_{11}(j\omega)\right| \right] \mathrm{d}\omega \tag{10}$$

Total losses of hourly fuel consumption for dynamic loads

$$B^{\mathrm{t}}(\omega) = \frac{M_{\mathrm{n}}^{a}}{\omega_{\mathrm{le}}} \cdot \int_{\omega=0}^{\omega=20} \left|G_{31}^{\mathrm{t}}(j\omega)\right| \mathrm{d}\omega\,. \tag{11}$$

where, ω_{le} - the intrinsic lowest frequency of load oscillations; $G_{31}^{\mathrm{t}}(j\omega)$ - the transfer function of the hourly fuel consumption in the transport mode [2].

The actual hourly fuel consumption $B_{\mathrm{f}}^{\mathrm{t}}$ in the transport mode, taking into account the consumption due to dynamic loads, for various values of c_1 = 1800; 1400; 1000 kN/m is determined by the sum

$$B_{\mathrm{f}}^{\mathrm{t}} = B_0^{\mathrm{t}} + B^{\mathrm{t}}(\omega)\,. \tag{12}$$

The results of the calculation of the regular component B_0^{t} and the actual hourly fuel consumption $B_{\mathrm{f}}^{\mathrm{t}}$ are summarized in table 1.

3. Results and Discussion

Despite a short duration of the transport mode of a forestry MTU, it is relatively energy-intensive and is characterized by a high hourly fuel consumption.

In the transport mode, the frequency response of the MTU hourly fuel consumption for dynamic loads is realized with two extremes (figures 4 and 5, curves 5). The highest consumption occurs in the range of natural frequencies of oscillations of the engine rotation speed ω_{le} = 2.23 ... 2.88 s^{-1}. In terms of the structure of the dynamic component, the components of resistance to the movement of the MTU on the rise and oscillations of the tractor in the longitudinal-vertical plane have the highest values of fuel consumption (figures 4 and 5, curves 1 and 3).

The speed of movement υ_0 has a significant impact on the size of the regular component of fuel consumption per hour B_0^{t}, as well as on the elements of the dynamic component $B_i^{\mathrm{t}}(\omega)$ (table 1), therefore the choice of the transmission ratio of the transmission and gearbox at the design stage should provide effective traction and speed modes of the MTU.

To improve the fuel efficiency of the MTU diesel engines, the design of which includes the fuel pump rail (FPR), it is necessary to use a corrective device. As a corrective device, it is advisable to use an FPR oscillation damper [2, 8, 9, 10]. The oscillation damper is used to reduce the amplitude of oscillations of the rail and the negative effect of the hydrodynamic force on the quality of the speed control process. The damping properties of the oscillation damper can reduce the effect of hydrodynamic forces during fuel cut-off, and also provide resistance when the FPR moves towards an increase in the cyclic fuel supply.

Table 1. Calculation results of the hourly MTU fuel consumption based on the LHT-100 tractor with the PKL-70 plow in the transport mode. In the numerator - with a standard diesel engine, in the denominator - with an oscillation damper of the FPR.

MTU speed υ_0, m/s	Regular component B_0^t, kg/h	Total dynamic components $B^t(\omega)$, kg/h with different stiffness c_1, kN/m			Components the dynamic component $B_i^t(\omega)$, kg/h with stiffness c_1 = 1400 kN/m			
		1 000	1400	1800	1[d]	2[d]	3[d]	4[d]
0.86[a]	6.983	1.135 / 0.853	1.174 / 0.891	1.163 / 0.880	0.616 / 0.413	0.025 / 0.004	0.483 / 0.455	0.050 / 0.019
1.15[b]	9.337	1.414 / 1.045	1.461 / 1.092	1.447 / 1.079	0.771 / 0.502	0.023 / 0.004	0.615 / 0.567	0.052 / 0.018
1,78[c]	14.453	1.765 / 1.280	1.821 / 1.337	1.802 / 1.320	0.959 / 0.606	0.020 / 0.004	0.795 / 0.711	0.047 / 0.016

[a] $\omega_{le} = 2.23\ s^{-1}$ – *II* gear (i_t = 44.226);

[b] $\omega_{le} = 2.51\ s^{-1}$ – *III* gear (i_t = 32.854);

[c] $\omega_{le} = 2.88\ s^{-1}$ – *IV* gear (i_t =21.271);

[d] Components: 1 - resistance to the movement of the MTU on the rise; 2 - inertial forces of the MTU; 3 - tractor oscillations in the longitudinal-vertical plane; 4 - mechanical losses in the friction pairs of the engine.

Acknowledgments
The authors express their deep gratitude to Associate Prof. V P Antipin for his invaluable assistance with the manuscript.

References

[1] Antipin V P 2012 *Power consumption of a machine-tractor unit* (St Petersburg: Publishing House of The Polytechnic University)

[2] Antipin V P 2017 *Productivity, energy consumption and resource of the machine-tractor unit* (St Petersburg: Publishing House of The Polytechnic University)

[3] Antipin V P, Durmanov M Y and Mikhaylov O A 2019 Choosing Transmission Gearset for Agricultural Aggregates Based on Energy Consumption *Proc. of the 4th Int. Conf. on Industrial Engineering, ICIE 2018* ed A Radionov *et al.* (Springer, Cham) pp 1591–1600

[4] Durmanov M Y, Martynov B G and Spiridonov S V 2019 Energy and Fuel Consumption of an Agricultural Aggregate *Proc. of the 4th Int. Conf. on Industrial Engineering, ICIE 2018* ed A Radionov *et al.* (Springer, Cham) pp 1601–1612

[5] Vlasov E N, Mikhailov O A, Durmanov M Y and Epifanova A Yu 2015 Estimating the amount of energy spent on the implementation of the tangential power of the tractor LHT-100 in the plowing mode *Izvestia Sankt-Peterburgskoj lesotehniceskoj akademii* (St Petersburg: SPbGLTU) **212** pp 104–112

[6] Vlasov E N, Mikhailov O A, Durmanov M Y and Epifanova A Yu 2015 Estimating the amount of energy spent on the implementation of the tangential power of the tractor LHT-100 in the transport mode *Izvestia Sankt-Peterburgskoj lesotehniceskoj akademii* (St Petersburg: SPbGLTU) **213** pp 138–147

[7] Krutov V I 1978 *Internal combustion engine as an adjustable object* (Moscow: Mechanical Engineering) 472 p

[8] Shevtsov A A 1987 *Improving the fuel economy of skidders on transient operating modes*, Candidate of Tech. Sciences thesis, Leningrad, 182 p

[9] Antipin V P, Gribov S A, Shevtsov A A, Kozlov A V and Crystal M E 1986 *Direct-effect speed controller for internal combustion engines* [In Russian - Regulyator skorosti pryamogo deystviya dvigatelya vnutrennego sgoraniya] Certificate 1276843 (15 December 1986) Bulletin 46

[10] Durmanov M Y, Kulikov V N and Mikhailov O A 2019 The influence of the parameters of a forestry machine and tractor unit on the hourly consumption of diesel fuel in the transport mode *Proc. of Scientific and Technical Conf. of the Institute of Technological Machines and Forest Transport based on the results of research work in 2018* (*Electronic resource*) ed V A Sokolova (St. Petersburg: SPbGLTU) pp 110–120

[11] Martynov B G, Spiridonov S V and Durmanov M Y 2019 Influence of the parameters of a forestry machine-tractor unit on the hourly fuel consumption of a diesel engine *IOP Conf. Series: Earth and Environmental Science* (IOP Publishing) **316** 012035

[12] Spiridonov S V, Durmanov M Y and Mikhailov O A 2017 Energy costs for the implementation of the tangential force of the forestry MTU in the transport mode *2nd Int. Scientific and Technical Conf. "Forests of Russia: politics, industry, science, education"*(St Petersbirg: SPbGLTU) vol 3 pp 43–46

[13] Durmanov M Y, Spiridonov S V and Mikhailov O A 2018 Estimated hourly fuel consumption of a forestry machine-tractor unit at the design stage *3rd Int. Scientific and Technical Conf. "Forests of Russia: politics, industry, science, education"* (St Petersbirg: SPbGLTU) vol 2 pp 164–167

IOP Conf. Series: Materials Science and Engineering **817** (2020) 012009 doi:10.1088/1757-899X/817/1/012009

The influence of transport vibrations on the condition of Russian cultural heritage objects

I Dyachkova*, M Skachkova, V Kovyazin

Department of Engineering Geodesy, *Saint Petersburg Mining University*, 2, 21st Line, St. Petersburg 199106, Russian Federation

*Corresponding email: irinad95@mail.ru

Abstract. This article is devoted to the negative impact of transport vibrations on the condition of historical and cultural monuments. The causes of the occurrence of transport fluctuations and the consequences of their impact were considered. The authors proposed a classification of the negative impact of transport on monuments, ways to reduce of the negative impact were given, as well as the regions of Russian Federation that are the most susceptible to such impact were analyzed.

1. Introduction

Russia has a rich cultural and historical heritage. In 2019 the UNESCO World Heritage List includes 29 objects which located in Russia, and they are 2.6% of the total (1 121 objects in 2019 in total). There are 268 127 real historical and cultural monuments on the Russian territory in 2018 (figure 1) [1].

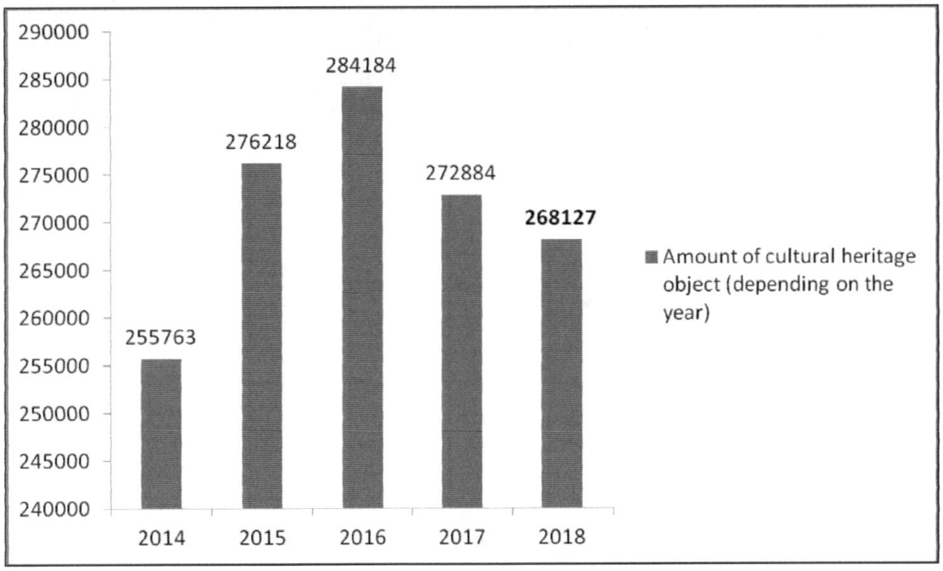

Figure 1. The total amount of Russian cultural heritage objects.

IOP Conf. Series: Materials Science and Engineering **817** (2020) 012009 doi:10.1088/1757-899X/817/1/012009

Despite the obvious high importance of the cultural heritage objects, reality shows disappointing facts: according to the All-Russian Society of the Protection of the History and Culture Monuments (RSPHCM) more than 2.5 thousand historical and cultural monuments have died in the Russian Federation over the last 10 years, and the annual loss is about 150–200 monuments, which is approximately 1.4% of the total number of historical and cultural monuments for 2017 [1].

The reasons of the loss and destruction of cultural heritage objects in Russia are legal instability, abandonment, vandalism, legislative imperfections in the fields of protection of valuable objects, as well as anthropogenic and natural environmental problems [2]. According to RSPHCM, there are about 2.3 thousand cultural heritage objects that are threatened due to the impact of anthropogenic factors.

The most common causes of the anthropogenic physical destruction and loss of cultural heritage objects are disturbances of the geological environment as a result of economic activity during the active development of the adjacent territory, air pollution from industrial emissions, vehicles and public utilities, chemical pollution of soils and earth, visual disturbance of landscapes and unregulated building, as well as noise, vibration and other disturbances of natural physical parameters of the environment, including the movement of various modes of transport: rail, road, tram communications, subways, etc. [3, 4].

The development of transport infrastructure in the cities ensures the social and economic well-being of the population. Modern cities are permeated by a network of transport routes that ensure the movement of passengers and goods, communication between individual segments of urban space. Every year transport infrastructure is actively developing and expanding in connection with the increasing economic needs of society and population growth (table 1) [5, 6].

Table 1. The length of transport routes, thousand km.

Type of transport / Years	2005	2010	2015	2016	2017
Railways	85	86	86	86	87
Car roads	858	1004	1642	1658	1666
Tram rails	2.8	2.6	2.5	2.5	2.5
Trolleybus lines	4.9	4.9	5.3	5.3	5.2
Subway ways	0.436	0.475	0.517	0.532	0.542

According to the Federal State Statistics Service, population growth, the development of cities and transport links result in an increase of the number of cars (in 2017 there was an increase of almost 1.74 times compared with 2005) and the number of metro stations (in 2017 there was an increase of 1.22 times since 2005), passenger turnover and the number of passenger traffic are growing too (figure 2) [5].

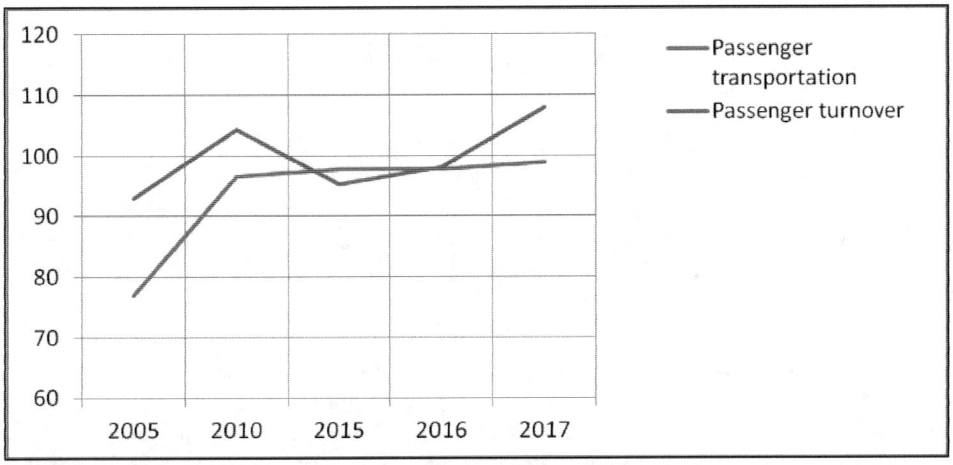

Figure 2. Passenger transportation index and passenger turnover index by types of public transport (as a percentage of the previous year).

In connection with the development of transport infrastructure, an increase of passenger traffic, and, consequently, an increase of the physical load of transport infrastructure on adjacent objects, which is especially acute in the historical centers of cities, the problem of the negative impact of transport vibrations on the physical condition of historical and cultural monuments located near transport routes arises.

The aim of this work is to identify and classify the main negative effects of transport vibrations on the condition of historical and cultural monuments.

2. Methods and Materials

Research methods are analysis, comparison, classification. The problem of the influence of transport vibration on historical and cultural monuments was investigated by such Russian scientists as S G Alimov, A V Lukin, M E Krogius, E K Borisov, A G Usov, I R Machine, Yu A Koryazhmina, E N Sychkina [2, 3, 7], and foreign scientists: I Roselli, V Fioriti, I Bellagamba, M Mongelli, A Tatì, M Barbera, M M Cianetti, G de Canio, X Gao, L Zhang, G Liu, Z Ba [7, 8] and others.

2.1. Research of transport vibrations

All modes of transport transmit vibrational energy to buildings and structures located near transport routes. The resulting vibration negatively affects the technical condition of the buildings and the sanitary and hygienic conditions of use.

If we consider the structure from a physical point of view, then it will be a consumer of vibrational energy. And this energy is transmitted to the structure depending on its nature, the feature of the contact with the energy-bearing medium and the physical and mechanical characteristics of this medium.

Vibrations are the mechanical vibrations which are characterized by a certain repeatability in time, amplitude and frequency. Types of mechanical vibrations are presented in figure 3 [9].

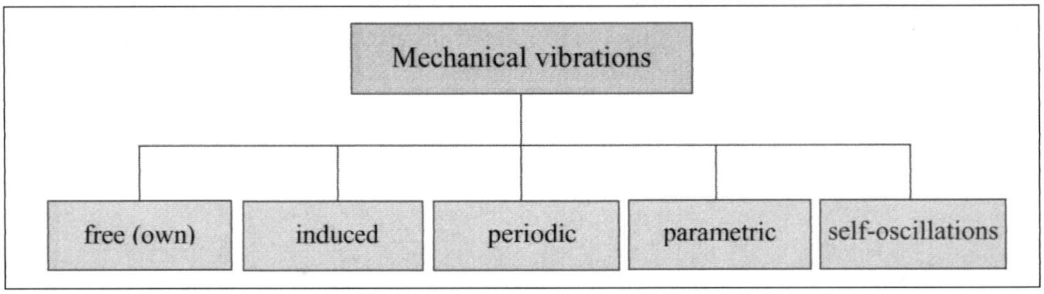

Figure 3. Types of mechanical vibrations.

According to the specificity of the transfer of vibrational energy to the structure, in all cases transport vibration is a kinematic disturbance of the protected object (self-oscillation).

When the natural frequencies of building structures oscillations coincide with the frequency of external influences, resonance phenomena occur. And it leads to complex destruction of the building or its parts, violation of the external appearance, and cracks [3].

In the historical part of any large city there are narrow streets laid during its initial construction, which eventually turned into noisy highways [3]. A large flow creates a vibrational load on a valuable historical building. State standard RF 52892-2007 "Vibration and shock. Vibration of buildings. Measurement of vibration and assessment of its impact on the structure" says that vibration is capable of changing the properties of the soils where the building is located [10]. For example, local compaction of the soil occurs. Due to uneven settlement of the soil base, it can lead to serious structural damage. Soil compaction can occur even with a low level of vibration, if it is long-term.

Another example of changing soil properties is soil liquefaction. It is more dangerous, because under vibration exposure the liquefied soil loses its bearing capacity. This phenomenon is more evident in water-saturated soils and is indirect evidence of the effect of vibration on buildings and structures. Therefore, it is necessary to obtain a comprehensive assessment of the impact of vibration and the involvement of specialists - geotechnicians [3, 11].

Thus, transport vibrations affecting cultural heritage objects can be classified as follows:

1) by source of vibration:
– car,
– tram,
– trolleybus,
– rail,
– subway;

2) by the nature of the impact:
– immediate (causing cracks, destruction of individual parts of the building, etc.);
– indirect (causing deformation of soils and surrounding objects);

3) by period of exposure:
– long-term;
– short-term.

In St. Petersburg Petropavlovsky Cathedral was covered by through cracks and a result was subway vibration, Isaakievsky Cathedral tilted 25 cm from the vertical, construction work of the vestibule of the Teatralnaya metro station led to a serious upsetting of the oldest Mariinsky Theater. Reality knows a lot of such examples, therefore it is necessary to look for methods to reduce the influence of transport vibrations [3], as well as to do the monitoring of the technical condition of valuable real estate objects [4].

According to RSPHCM, the following regions and cities of Russia are affected by transport vibration: St. Petersburg, Orenburg, Udmurtia, Ingushetia, the Altai Republic and Altai Region, Mordovia, Krasnodar and Khabarovsk Regions, Arkhangelsk, Bryansk, Kostroma, Samara, Voronezh, Orel and Kaliningrad region, Kaluga, Smolensk, Tomsk. The examples are the House of the merchant

Ogloblin in Izhevsk, the House of Ovchinnikov and the House of Grakhov in Votkinsk (Udmurtia), the unsatisfactory condition of more than 40 objects of monumental art in Orenburg (Orenburg region) [12].

2.2. Ways to reduce the effects of transport vibrations

Among the existing methods of reducing the negative consequences which transport vibrations influence on the physical condition of historical and cultural monuments, two directions can be distinguished (figure 4) [13].

There are many examples of the application of these methods. In Vyborg, a ring road was built in the historical center to reduce the flow of cars. In Astrakhan the tram was almost completely eliminated to reduce the vibration of rail vehicles. The use of various types of insulators is actively practiced in England (Croydon) [14].

3. Results and Discussion

Transport fluctuations are divided into several groups according to the source of vibrations, the nature of the effects and the frequency. Methods to reduce the negative effects which transport vibrations have on the physical condition of historical and cultural monuments can be divided into two directions: reducing the existing transport vibrational impact and preventing such impacts.

4. Conclusion

The problem of the influence of transport vibrations on buildings and structures is currently very acute. The primary task in the process of solving this problem is the preservation of valuable historical objects. Reducing the negative impact of an aggressive environment, such as transport vibrations on cultural heritage object, in the global sense lies in the development of the concept of "smart" cities and using of advanced technologies.

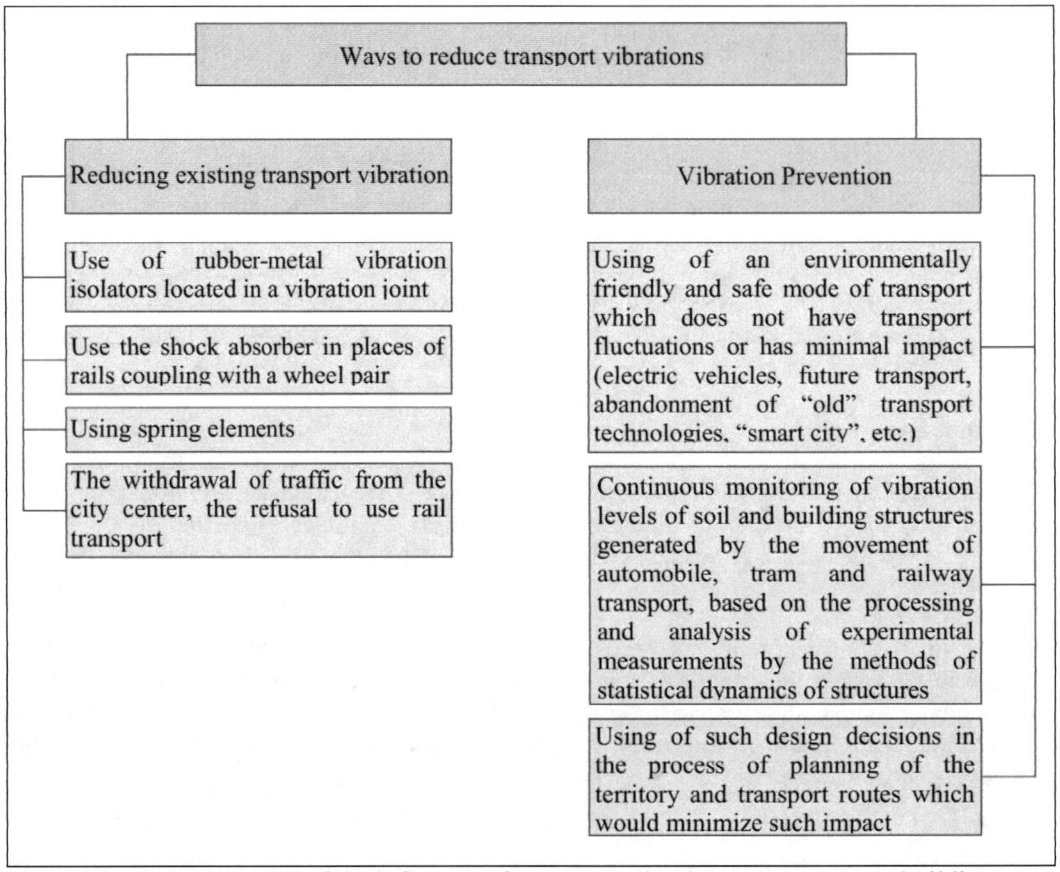

Figure 4. Decrease of the influence of transport vibrations on monument buildings.

References

[1] *All-Russian Society for the Preservation of Historical and Cultural Monuments. The state of historical and cultural heritage*, available at: http://www.voopik.ru/our-heritage/status-cultural-heritage/

[2] *Ecology and cultural heritage*, available at: http://ohrana-bgd.ru/okrsrd/okrsrd3_18.html

[3] Alimov S G 2006 Assessment of the effect of transport vibration on the construction of buildings-monuments of architecture: the example of Vladivostok, PhD thesis, Far Eastern State Technical University

[4] Skachkova M E, Lepihina O Y and Ignatova V V 2018 Journal of Physics: Conference Series *Information support of monitoring of technical condition of buildings in construction risk area 1015* 42056–42056

[5] Statistical collection *Transport in Russia* 2018 (Federal State Statistics Service) pp 9-79

[6] Bykowa E and Gerasimova I 2019 Land *Land plot selection rationale for the location of linear facilities* **8(4)** 1-17

[7] Roselli I, Fioriti V, Bellagamba I, Mongelli M, Tatì A, Barbera M, Cianetti M M and G de Canio WIT Transactions on Ecology and the Environment *Urban transport vibrations and cultural heritage sites in Rome: the cases of the Temple of Minerva Medica and of the Catacomb of Priscilla* ed C A Brebbia, Wessex Institute, J J Sendra (Boston: WIT Press) pp 335-345

[8] Gao X, Zhang L, Liu G and Ba Z 2017 The International Archives of the Photogrammetry, Remote Sensing and Spatial Information Sciences *Monitoring of vibrations for the conservation of cultural heritage : the case of the summer palace in Beijing* pp 253-257

IOP Conf. Series: Materials Science and Engineering **817** (2020) 012009 doi:10.1088/1757-899X/817/1/012009

[9] Machine I R, Koryazhmina Yu A and Sychkina E N 2015 Construction and architecture. Experience and modern technologies *Effect of transport vibration on buildings and structures* **4** 1-4
[10] GOST RF 52892-2007 *Vibration and shock. Vibration of buildings. Measurement of vibration and assessment of its impact on the structure* 2007 (Standartinform, Moscow)
[11] Hełdak M and Bykowa E 2017 Journal of Ecological Engineering *Construction of public roads at the meeting point of different legislation systems* **18(6)** 86-94
[12] *Typology of the destruction of cultural monuments*, available at: https://design.wikireading.ru/12939
[13] *Transport and the Historic Environment*, available at: https://historicengland.org.uk/advice/planning/infrastructure/planning-and-transport/
[14] Ovsyannikova T Yu, Skuridina Yu B and Kotova O V 2009 Bulletin of Tomsk State University. Economics *Risk assessment of loss of immovable objects of cultural heritage of a company* **3(7)** 61-70

IOP Conf. Series: Materials Science and Engineering **817** (2020) 012010 doi:10.1088/1757-899X/817/1/012010

Evaluation of small-sized skidders performance

G V Grigorev*, I N Dmitrieva, E V Kotenev
Institute of Technological Machines and Forest Transportation, *St. Petersburg State Forest Technical University*, 5 Institutskiy Lane, St. Petersburg 194021, Russian Federation

*Corresponding e-mail: vtl-lta@mail.ru

Abstract. The aim of the work is to further develop theoretical description of interaction between small-sized forestry vehicles and soils, to study productivity and indicators of interaction of wheeled and tracked vehicles with weaker forest soils. According to the calculation results it was established that the LMT-3 wheeled skidder when working on weakly-bearing soils is able to provide productivity when skidding trunks within 43 m^3 per shift, with the trunks to be skidded for the tops. Productivity of small-sized tracked vehicle LMT-6 is slightly lower and makes up to 39 m^3 per shift when trunks are skidded for the tops. For both tractor models, the productivity of skidding is reduced by approximately 5 m^3 per shift when the trunks are to be skidded for the bottoms.

1. Introduction

When working on low-bearing soils, small-sized skidding tractors with wheeled and tracked mover are used for skidding. The weight of such machines is approximately 4-6 tons, the volume of skidded timber packs does not exceed 2 m^3. Geometric parameters of the mover and the ratio of sides of the contact patch with the ground in such machines differ from the heavier equipment, which used to be considered in research on cross-country ability and rutting of forest machines [1-5]. Among the modern studies related to determination of tractive performance of small-sized forestry machines, note the work [6]. In [6] there have been developed mathematical models predicting the rut depth and tractive performance of a small-sized wheeled vehicle working in conditions of a swampy ground, the interaction of the propulsor with turf cover lying on a deformable base of unlimited thickness has been considered.

The purpose of our study is the further development of the theoretical description of the interaction of small-sized forestry vehicles with soil, the study of productivity and tractive performance of wheeled and tracked vehicles with weaker forest soils.

2. Results and Discussion

Let us consider tractive performance of two small-sized tractors - universal tracked skidder LMT-6 and universal wheeled skidder LMT-3.

Universal small-size skidder LMT-6 is designed for skidding trees, trunks and assortments. LMT-6 tractor is operated on low-bearing capacity soils during forest harvesting, development of clearings during construction of roads, communication lines and power lines, oil and gas pipelines, cleaning of territories from bushes, during forest park operations [7]. Technical characteristics of the tractor are presented in Table 1. Universal small-sized wheeled skidder LMT- is designed for skidding trees and trunks during forest harvesting, development of clearings during the construction of roads, communication lines and power lines, oil and gas pipelines, cleaning up areas from forest vegetation, in conditions of man-made pollution of forests and in areas of natural disasters [7].

Table 1. Technical characteristics of the universal small-sized tracked skidder LMT-6 and wheeled skidder LMT-3 [7].

Parameter	Value	
	LMT-6	LMT-3
Engine power, kW (HP)	18.7 (25)	18,7 (25)
Speed of movement, km/h	up to 9.0	up to 25.0
The pulling force of the winch, kN	16	20
Operating weight, kg	2 610	3 400
Overall dimensions, mm		
- length	3 100	5 450
- width	1 925	2 215
- height	2 350	2 350

To calculate the tractive performance of the tractors, first it is necessary to determine the depth of the rut h, formed when the tractor passes through the forest soil, for this purpose we use following equation, obtained on the basis of [5, 8-10]:

$$h = \frac{JpaBp_s}{(p_s - p)(E(E - Jp))^{0.5}} \operatorname{arctg}\left(\frac{E(H - h)}{aB(E(E - Jp))^{0.5}} \right), \tag{1}$$

where J is a parameter that takes into account the ratio between the length and width of the tractor mover's contact with the ground, a is a parameter that takes into account the thickness of the soft soil layer, H is the thickness of the soft soil layer, B is the width of the tractor mover, p is the pressure of the tractor mover on the ground (average), E is the general deformation module of the ground, p_s is the bearing capacity of the ground.

Parameters J, a we will find by formulas [5, 8-10]:

$$J = \frac{0.03 + LB^{-1}}{0.6 + 0.43LB^{-1}}, \tag{2}$$

$$a = 0.64 \frac{B + H}{H}, \tag{3}$$

where, L is the length of the contact patch.

For wheeled tractor [4]:

$$L = \frac{d}{2}, \tag{4}$$

where, d is the diameter of the wheel.

For a tracked tractor the length of the contact spot is approximately equal to the length of the track's supporting surface.

The bearing capacity of the soil is calculated using the formulas [1, 10]:

$$\begin{cases} p_s = p_{s0}\alpha_Z \\ p_{s0} = 0.5K_1N_1\gamma B + N_2\gamma h + K_3N_3C \\ K_1 = \dfrac{L}{L+0.4B}; K_3 = \dfrac{L+B}{L+0.5B} \\ N_1 = \dfrac{1-S^4}{S^5}; N_2 = \dfrac{1}{S^2}; N_3 = \dfrac{2(1+S^2)}{S^3}; S = \mathrm{tg}\left(\dfrac{\pi}{4}-\dfrac{\varphi}{2}\right) \\ \alpha_Z = 1+\dfrac{H^*h}{2H\cdot(H-h-0.25H^*)}; H^* = 0.707\exp\left[\left(\dfrac{\pi}{4}+\dfrac{3\varphi}{4}\right)\tan\dfrac{3\varphi}{4}\right]B\cos\dfrac{3\varphi}{4}\tan\varphi \end{cases}, \quad (5)$$

where p_{s0} is the bearing capacity of the ground at the unlimited thickness of the soft layer, α_z is the coefficient of the soft layer of the ground, K_1, K_3 are the coefficients of the ratio of the length and width of the contact spot of the mover with the ground, N_1, N_2, N_3 are the coefficients of the internal friction angle of the ground, C is the specific adhesion of the ground, γ is the specific weight of the ground, S, H^* are auxiliary designations.

Force of soil resistance to tractor movement is calculated according to the formula [9, 10]:

$$F_r = B\int_0^h p\,dh. \quad (6)$$

The mover's adhesion force to the ground is based on the formula [9]:

$$F_T = \frac{BG^2L(p\,\mathrm{tg}\,\varphi + C)}{(G-C)^2} - S\cdot\frac{BCGL^2}{2t_g(G-C)} - \frac{1}{S}\cdot\frac{(p\tan\varphi + C)^2 BG^2 t_g}{(G-C)^3} \cdot \ln\left(1+S\frac{L(G-C)}{t_g(p\tan\varphi + C)}\right) \quad (7)$$

where S is the slipping ratio, t_g is the grouser spacing.

Let us calculate the coefficient of resistance to tractor movement by the formula [9, 10]:

$$\varphi_r = \frac{F_r}{BL}. \quad (8)$$

Find the tractor's coupling factor with the ground using the formula [9, 10]:

$$\mu = \frac{F_T}{BL}, \quad (9)$$

and in determining F_T, the slipping ratio S is a variable.

At calculation the following characteristics of weakly bearing soil [11, 12] are accepted: the module of general deformation E = 0.4 MPa, specific adhesion of soil C = 0.005 MPa, angle of internal friction of soil $\varphi = 12^{\circ}$, specific weight of soil γ = 0,0075 MN/m^3, thickness of soft layer H = 0.73 m, the module of shift G = 0.22 MPa. The parameters of the mover are as follows: wheel width or track width of small tractor B = 0.3 m, track support length L = 3 m, wheel diameter d = 1.2 m. Maximum ground pressure of the mover was limited to $p_{\max}$ = 0.045 MPa.

The procedure for calculation is as follows. The program Maple 2017 has been used to solve the problem. When the initial data are received, the pressure range is set into intervals with the specified step (0.0045 MPa). At each value of pressure p the system of equations (1) - (4) is numerically solved (command *fsolve*). As a result we get the table of correspondence of pressure p and track depth h. According to the table, using the trapezium method, we find an approximate value of the integral by formula (6), then we calculate the resistance coefficient φ_r by formula (8). Then, for each value of p, we determine the maximum function (7) by means of the *Maximize* command of the package of built-in commands Optimization and the corresponding slipping factor value S. We find the coefficient of adhesion by the formula (9).

The results of the calculation of the traction and coupling properties of tractors are shown in figures 1 and 2 and in table 2.

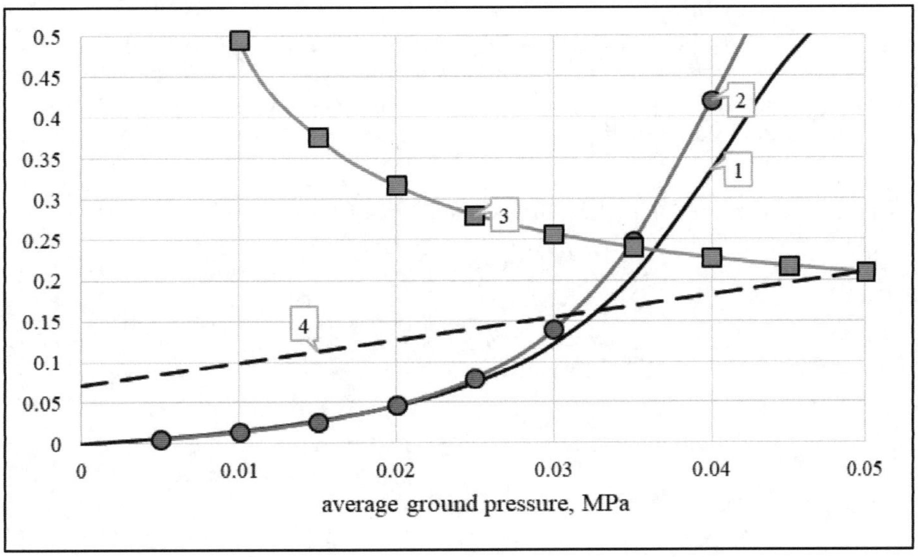

Figure 1. Rutting and tractive performance of small-sized wheeled skidder.

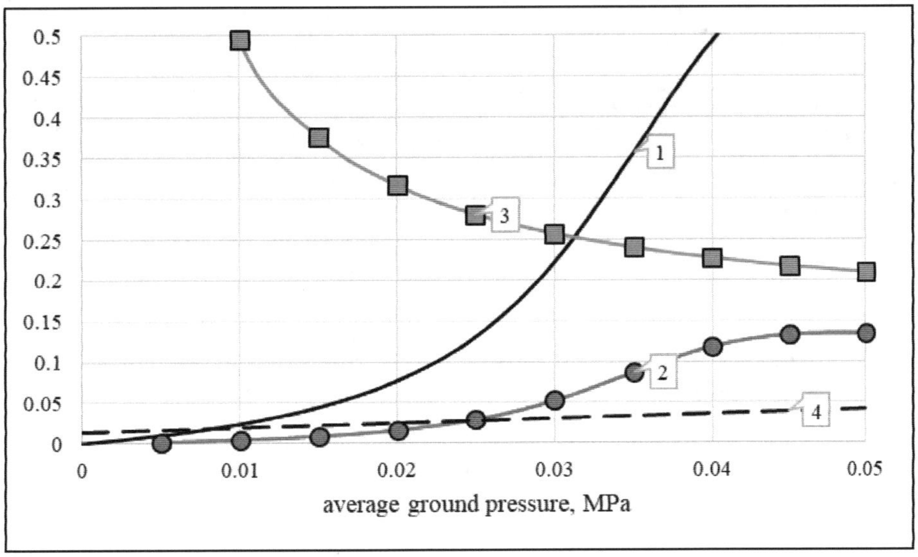

Figure 2. Rutting and tractive performance of small-sized tracked skidder: 1 – rut depth; 2 – rolling resistance coefficient; 3 – net thurst coefficient; 4 – slip ratio.

Basing on tables 1,2 at the accepted initial data we shall establish, that average pressure of a small-size tracked tractor LMT-6 on a ground makes p = 0.015 MPa, small-size wheeled tractor LMT-3 makes p = 0.02 MPa. The numerical results of calculation of tractive performance for two considered models of tractors are summarized in table 2.

Table 2. Traction-coupling properties of small-size tractors LMT-3 and LMT-6 at work on weaker forest soil.

Tractor model	Rut depth h, m	Rolling resistance coeff. φ_r	Net thurst coeff. μ	Slip ratio S
LMT-3	0.046	0.047	0.316	0.127
LMT-6	0.045	0.009	0.376	0.023

The results of calculating the productivity of LMT-3 and LMT-6 skidders are presented in table 3 (the method is described in detail in [13], the initial data concerning tractors are presented in tables 1-2).

Table 3. Results of calculating the productivity of LMT-3 and LMT-6 skidders.

Parameter	Dimension	LMT-3	LMT-6	LMT-3	LMT-6
Tractor speed at II gear	m/sec	1.48	1.48	1.48	1.48
Tractor speed at III gear	m/sec	1.87	1.87	1.87	1.87
Permissible load on the cone	kN	9	9	7	7
Towing force of the winch	kN	20	21	16	16
Skidding method	-	by the tops	by the bottoms	by the tops	by the bottoms
Weight distribution coefficient of the skimmed pack between the tractor and the ground	-	0.41	0.73	0.41	0.73
Average skidding distance	m	250	250	250	250
Number of hours per shift	-	8	8	8	8
Average picking rope feed distance	m	20	20	20	20
Average trunk volume	m^3	0.15	0.15	0.15	0.15
Total weight, limited by the tractive force.	kN	25.70	48.84	28.89	60.26
Total weight, limited by the grip	kN	24.21	48.84	28.89	60.26
Total weight, limited by the load capacity	kN	21.72	12.40	16.89	9.64
Total weight, limited by the pulling force of the winch.	kN	20.92	21.96	16.73	16.73
Calculated pack weight	kN	20.92	12.40	16.73	9.64
Weighted average wood density	kg/m^3	859.00	859.00	859.00	859.00
Volume of wood in skidded bundle	m^3	1.70	1.01	1.36	0.79
Tractor average speed	m/sec	1.68	1.68	1.68	1.68
Driving time in the workplace and at idling speed e.g.	sec	149.25	149.25	149.25	149.25
Bundle forming time	sec	590.97	438.24	515.98	388.85
Bundle bounce time	sec	128.04	90.55	109.63	78.43
Bundle skidding cycle time	sec	868.27	678.04	774.86	616.54
Hourly skidding performance	m^3/hour	6.01	4.56	5.38	3.90
Skidding performance per shift	m^3/shift	43.25	32.82	38.77	28.08

3. Conclusion

According to the calculation results it has been established that the LMT-3 wheeled skidder is able to provide the productivity with the stems skidding within 43 m^3 per shift while the stems should be skidded for the tops. Productivity of a crawler small-sized tractor is slightly lower and makes 39 m^3 per shift when whips are skidded behind the tops. For both tractor models, the capacity per lump of skidding is reduced by approximately 5 m^3 per shift.

References

[1] Khitrov E G and Andronov A V 2019 Bearing floatation of forest machines (theoretical calculation) *IOP Conf. Series: Materials Science and Engineering* **695** 012020 DOI: 10.1088/1757-899X/695/1/012020

[2] Khitrov E G and Andronov A V 2019 Comparison of the forest machine mover average and nominal pressure on the soil *IOP Conf. Series: Materials Science and Engineering* **695** 012021 DOI: 10.1088/1757-899X/695/1/012021

[3] Khitrov E, Andronov A, Iliushenko D and Kotenev E 2019 Comparing approaches of calculating soil pressure of forestry machines *Int. Multidisciplinary Scientific GeoConference Surveying Geology and Mining Ecology Management, SGEM 19* (vol 19) issue **3.2** pp 649–655 DOI: 10.5593/sgem2019/3.2/S14.084

[4] Khitrov E, Andronov A, Bogatova E and Kotenev E 2019 Development of recommendations on environmental certification of forestry machinery drives *Int. Multidisciplinary Scientific GeoConference Surveying Geology and Mining Ecology Management, SGEM 19* (vol 19) issue **3.2** pp 689–696 DOI: 10.5593/sgem2019/3.2/S14.089

[5] Khitrov E G and Andronov A V 2019 Mathematical model of interaction between forest machine mover and consolidating soil *J. Phys.: Conf. Ser.* **1177** 012030 DOI: 10.1088/1742-6596/1177/1/012030

[6] Dmitrieva M N 2018 *Modeling of interaction of the wheel drive of the small forest machines with the weakly-bearing ground* [in Russian - Modelirovanie vzaimodejstvija kolesnogo dvizhitelja malogabaritnikh maschin so slabonesuschschimi pochvogruntami], PhD thesis: 05.21.01 (Arkhangelsk)

[7] LZM world [Electronic resource]. Access mode: www.mir-lzm.ru/skidder/lmt_3.html. Access date: October 21, 2019

[8] Kochnev A and Khitrov E 2018 Theoretical models for rut depth evaluation after a forestry machine's wheel passover *Int. Multidisciplinary Scientific GeoConference Surveying Geology and Mining Ecology Management, SGEM 18* pp 1005–12 DOI: 10.5593/sgem2018/3.2/S14.128

[9] Ivanov V, Stepanishcheva M, Khitrov E and Iliushenko D 2018 Theoretical model for evaluation of tractive performance of forestry machine's wheel *Int. Multidisciplinary Scientific GeoConference Surveying Geology and Mining Ecology Management, SGEM 18* pp 997–1004 DOI: 10.5593/sgem2018/3.2/S14.127

[10] Larin V V 2007 *Methods for predicting the support ability of multi-axle wheeled vehicles in the terrain [Metody prognozirovaniya i povysheniya opornoj prohodimosti mnogoosnyh kolesnyh mashin na mestnosti*], diss. doct. tech. sciences: 05.05.03 (Moscow) (in Russ.)

[11] Khitrov E G, Andronov A V, Martynov B G, Spiridonov S V 2019 Interrelations of various soil types mechanical properties *IOP Conf. Series: Journal of Physics: Conf. Series* **1177** 012032 DOI: 10.1088/1742-6596/1177/1/012032

[12] Khitrov E, Kochnev A, Ivanov V and Stepanishcheva M 2018 Linking the deformation moduli and cone indices of forest and peatland soils *Int. Multidisciplinary Scientific GeoConference Surveying Geology and Mining Ecology Management, SGEM 18* pp 297–304 DOI: 10.5593/sgem2018/3.2/S13.039

[13] Kochegarov V G, Bit Yu A, Menshikov V N 1990 *Technology and machines for logging works*

IOP Conf. Series: Materials Science and Engineering **817** (2020) 012010 doi:10.1088/1757-899X/817/1/012010

[in Russian - Tekhnologia i maschini lesozagotovok], textbook for Higher Educational Institutions (Moscow)

Permissible ground pressure of wheeled forestry vehicles depending on the ground conditions

G V Grigorev[*], I N Dmitrieva, E V Kotenev

Institute of Technological Machines and Forest Transportation, *St. Petersburg State Forest Technical University*, 5 Institutskiy Lane, St. Petersburg 194021, Russian Federation

*Corresponding e-mail: vtl-lta@mail.ru

Abstract. The aim of the work is to investigate interrelations between wheeled vehicle pressure on the ground and its cross-country ability. To comply with the requirement of ecological compatibility of the mover, which is that the rut depth after the passage of the vehicle should be less than 0.1 m, the pressure of the wheeled mover when working on moderate forest soils should not exceed 0.07 MPa (a forwarder 8x8 with a loaded weight up to 31 tons), on weak forest soils the pressure should not exceed 0.025 MPa (a forwarder 8x8 with a loaded weight up to 11 tons). On moderate soil to ensure cross-country ability, the pressure should be limited to 0.095 MPa (a forwarder 8x8 with a loaded weight up to 42 tons) (slipping ratio reaches 0.2, the depth of the rut is 0.18 m, the operating speed of the machine 1.5 m/s). To ensure the wheeled vehicle cross-country ability, the average pressure when working on weak forest soils should be limited to 0.04 MPa (a forwarder 8x8 with its loaded weight of up to 18 tons) (slipping coefficient reaches 0.136, rut depth of 0.253 m, the operating speed of the machine 2.8 m/s).

1. Introduction

Average ground pressure p is one of the most important characteristic of the influence of the wheeled machine engine on the bearing surface. Ground pressure is determined by the depth of the track h formed during the passage of the machine, which, in turn, depends on the ground resistance to the movement of the machine F_r. The value of the propulsor's pressure on the ground determines the normal stress σ in it, this determines the tangential stress τ on the contact spot of the propulsor with the bearing surface, therefore, the adhesion force of the propulsor with the surface F_T and slip ratio S. Taken together, the motion resistance and slipping ratio determine the operating speed of the machine v. The difference between the net thurst coefficient μ and the rolling resistance coefficient φ_r, the drawbar pull coefficient φ_p, is a characteristic of the machine's cross-country ability P, and it also determines the height of a single obstacle z that the wheeled propeller can overcome.

The aim of the study is to investigate the relationship between the wheel vehicle pressure on the ground and the performance of the vehicle's cross-country ability.

2. Results and Discussion

As input parameters of soil we will accept: general deformation module E, specific cohesion of soil C, internal friction angle φ, specific weight γ, shear module G, thickness of deformable soil layer H. Soil will be divided into three categories by physical and mechanical properties, the values of which are presented in table 1 [1-3].

IOP Conf. Series: Materials Science and Engineering **817** (2020) 012011 doi:10.1088/1757-899X/817/1/012011

To calculate the tractive performance of the wheeled mover, first it is necessary to determine the depth of the track *h*, formed when the tractor passes through the forest soil, for this purpose we use the transcendental equation [4, 5]:

$$h = \frac{JpaBp_s}{(p_s - p)(E(E - Jp))^{0.5}} \operatorname{arctg}\left(\frac{E(H - h)}{aB(E(E - Jp))^{0.5}} \right), \tag{1}$$

where *J* is a parameter that takes into account the ratio between the length and width of the tractor mover's contact with the ground, a is a parameter that takes into account the thickness of the soft soil layer, *H* is the thickness of the soft soil layer, *B* is the width of the tractor mover, *p* is the pressure of the tractor mover on the ground (average), *E* is the general deformation module of the ground, p_s is the bearing capacity of the ground.

Table 1. Physical and mechanical properties of forest soil by categories [1-3].

Parameter	Category I (bearing soil)	Category II (moderate)	Category III (weak soil)
E, MPa	3	1	0.4
C, MPa	0.025	0.011	0.0053
φ, °	16.7	13.7	11.6
γ, MN/m^3	0.0096	0.0084	0.0075
H, m	0.28	0.47	0.73
G, MPa	0.28	0.24	0.22

Parameters *J*, *a* we will find by formulas [4, 5]:

$$J = \frac{0.03 + LB^{-1}}{0.6 + 0.43LB^{-1}}, \tag{2}$$

$$a = 0.64\frac{B + H}{H}, \tag{3}$$

where, *L* is the length of the soil contact stain.

For wheeled tractor [6, 7]:

$$L = \frac{d}{2}, \tag{4}$$

where, *d* is the diameter of the wheel.

Ground bearing capacity is calculated by the formulas [8]:

$$\begin{cases} p_s = p_{s0}\alpha_Z \\ p_{s0} = 0{,}5K_1N_1\gamma B + N_2\gamma h + K_3N_3C \\ K_1 = \dfrac{L}{L + 0{,}4B}; K_3 = \dfrac{L + B}{L + 0{,}5B} \\ N_1 = \dfrac{1 - S^4}{S^5}; N_2 = \dfrac{1}{S^2}; N_3 = \dfrac{2(1 + S^2)}{S^3}; S = \operatorname{tg}\left(\dfrac{\pi}{4} - \dfrac{\varphi}{2}\right) \\ \alpha_Z = 1 + \dfrac{H^* h}{2H \cdot (H - h - 0.25H^*)}; H^* = 0.707\exp\left[\left(\dfrac{\pi}{4} + \dfrac{3\varphi}{4}\right)\tan\dfrac{3\varphi}{4}\right] B\cos\dfrac{3\varphi}{4} tg\varphi \end{cases}, \tag{5}$$

where p_{s0} is the load-bearing capacity of the ground at the unlimited thickness of the soft layer, α_z is the coefficient of account of the thickness of the soft layer of the ground, K_1, K_3 are the coefficients of account of the ratio of length and width of the spot of contact of the mover with the ground, N_1, N_2, N_3

are the coefficients of account of the angle of internal friction of the ground, C is the specific cohesion of the ground, γ is the specific weight of the ground, S, H^* are auxiliary designations.

Force of soil resistance to tractor movement is according to the formula [9]:

$$F_r = B\int_0^h p\,dh\,. \tag{6}$$

The force of cohesion of the mover with the ground is according to the formula [9, 10]:

$$F_T = \frac{BG^2L(p\,\mathrm{tg}\,\varphi + C)}{(G-C)^2} - S\cdot\frac{BCGL^2}{2t_g(G-C)} - \frac{1}{S}\cdot\frac{(p\tan\varphi + C)^2 BG^2 t_g}{(G-C)^3}\cdot \ln\left(1 + S\frac{L(G-C)}{t_g(p\tan\varphi + C)}\right), \tag{7}$$

where, S is the slip ratio, t_g is the grousers spacing.

Let us calculate the coefficient of resistance to tractor movement by the formula [9, 10]:

$$\varphi_r = \frac{F_r}{BL} + \varphi_{r0}\,, \tag{8}$$

where, φ_{r0} is a correction for power loss caused by tyre hysteresis, soil stickiness and other factors not directly related to rutting.

We will find the coefficient of tractor's traction with the ground using the formula [9, 10]:

$$\mu = \frac{F_T}{BL}\,, \tag{9}$$

and in determining F_T, the slipping factor S is a variable.

Let us define the coefficient of traction by the formula [9, 10]:

$$\varphi_p = \varphi_r - \mu\,. \tag{10}$$

A negative traction factor indicates a loss of the vehicle's passability.

We will find the height of the threshold single obstacle that the mover can overcome at $\varphi_p > 0$ by formula [8]:

$$z = \frac{d}{2}\cdot\left(1 - \frac{1}{\left(1+\varphi_p^2\right)^{0.5}}\right). \tag{11}$$

The operating speed of the wheeled mover is determined by the formula [9, 10]:

$$v = \frac{N\eta}{\varphi_r G_w n}\,, \tag{12}$$

where, N - engine power of the machine, η - transmission efficiency, G_w - related load per wheel, n - number of wheels of the machine.

The complex criterion of cross-country ability is estimated on the equation [9, 10]:

$$P = \frac{\varphi_p \varphi_r^0 v}{\varphi_p^0 \varphi_r v^0}\,, \tag{13}$$

where the "0" index means an indicator defined in reference conditions (in reference conditions $P = 1$, the closer the P value is to zero, the lower is the machine's passability assessment).

IOP Conf. Series: Materials Science and Engineering **817** (2020) 012011 doi:10.1088/1757-899X/817/1/012011

The pressure range is set into intervals with a given step Δp. At each pressure p value the system of equations (1) - (4) is numerically solved. As a result, we obtain the table of correspondence of pressure p and track depth h. From the table, using the trapezium method, we find an approximate value of the integral by formula (6), then we calculate the resistance coefficient φ_r by formula (8). Then, for each value of p, we numerically determine the maximum function (7) and the corresponding value of the slipping factor S. We find the coefficient of adhesion by the formula (9). We calculate the coefficient of traction by formula (10), then we determine the height of the threshold obstacle by equation (11). The working speed of the machine is based on expression (12). Using the formula (13) we get a complex assessment of the vehicle's cross-country ability.

For calculation in our study we will take the following initial data: wheel width $B = 0.7$ m, wheel diameter $d = 1.6$ m, pitch $t_g = 0.14$ m, vehicle ground pressure $p = 0.01 \dots 0.1$ MPa, pressure step $\Delta p = 0.01$ MPa, energy loss factor not related to rutting $\varphi_{r0} = 0.03$, engine power $N = 0.15$ MW, transmission efficiency $\eta = 0.97$ MW. For reference conditions: machine speed $v^0 = 6$ m/s, driving resistance coefficient $\varphi_r^0 = 0.03$, thrust coefficient $\varphi_p^0 = 0.97$. The results of calculation are presented in figures 1-8.

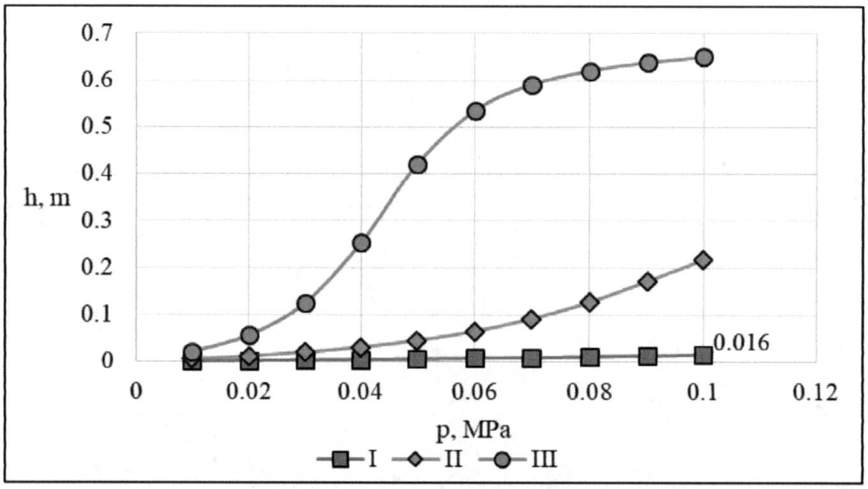

Figure 1. Rut depth as a function of vehicle ground pressure and soil category.

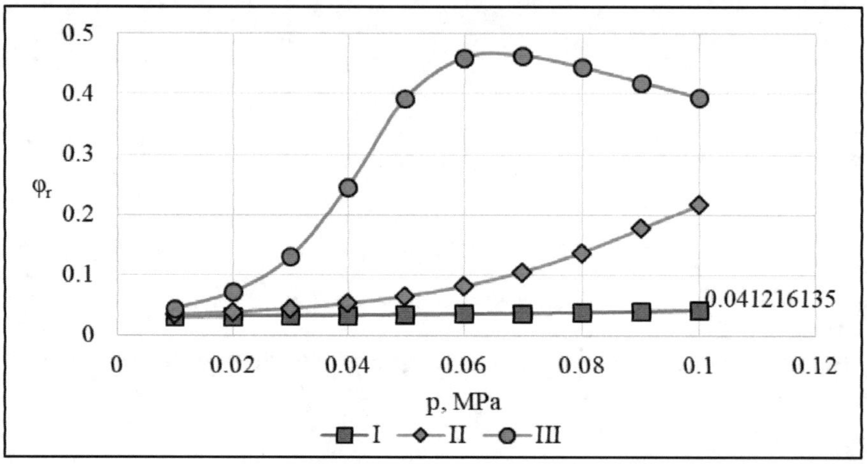

Figure 2. Rolling resistance coefficient as a function of vehicle ground pressure and soil category.

IOP Conf. Series: Materials Science and Engineering **817** (2020) 012011 doi:10.1088/1757-899X/817/1/012011

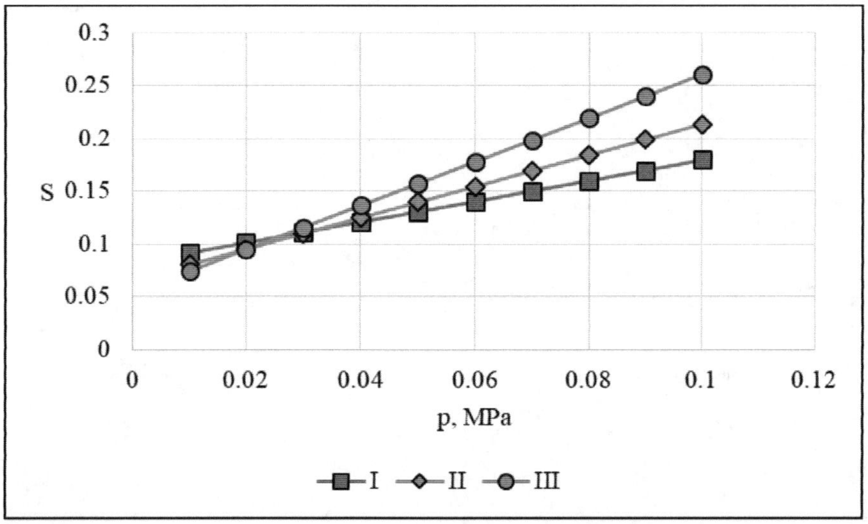

Figure 3. Slip ratio at which maximum traction is achieved as a function of vehicle ground pressure and soil category.

The slip ratio, which provides the maximum adhesion of the propulsor to the ground, almost linearly depends on the average pressure of the propulsor on the ground, its value varies from 0.1 at a pressure of 0.02 MPa (machine weighing 9 tons) to 0.18-0.21 at a pressure of 0.1 MPa (machine weighing 45 tons).

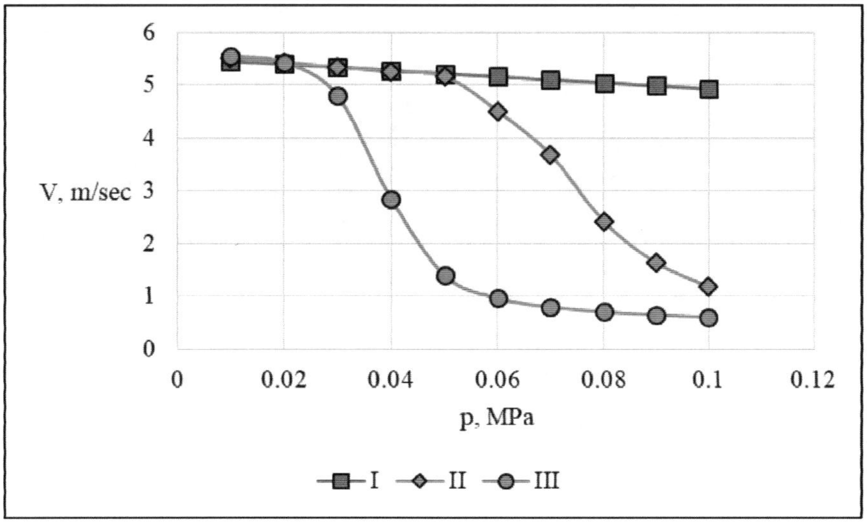

Figure 4. Maximum speed of the machine as a function of vehicle ground pressure and soil category.

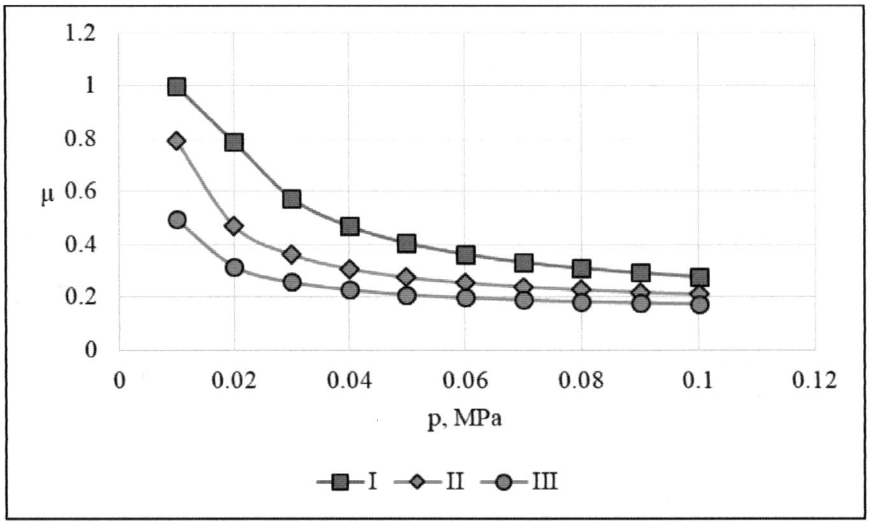

Figure 5. Net thurst coefficient as a function of vehicle ground pressure and soil category.

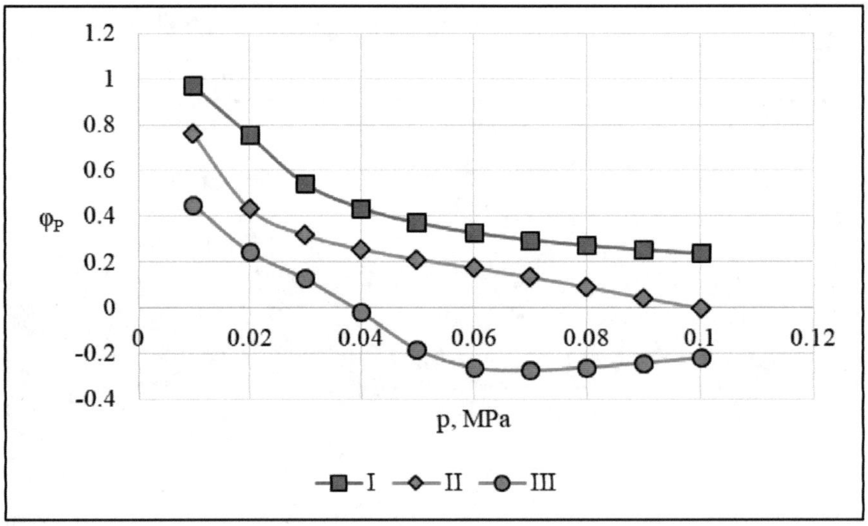

Figure 6. Drawbar pull coefficient as a function of vehicle ground pressure and soil category.

For a quantitative assessment of the vehicle cross-country ability the graphs of the complex parameter of cross-country ability, allowing to compare the stock of the vehicle traction factor in specific conditions in comparison with the reference conditions - figure 7, 8.

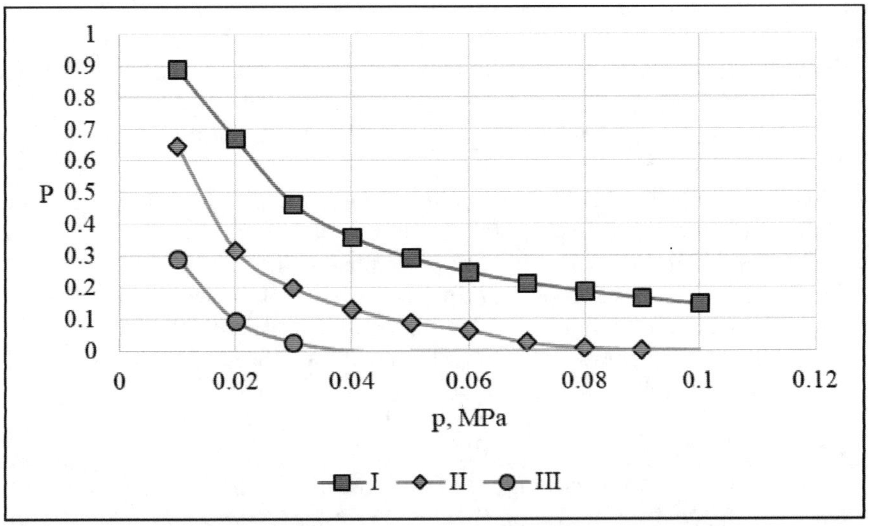

Figure 7. Complex parameter of cross-country ability as a function of vehicle ground pressure and soil category.

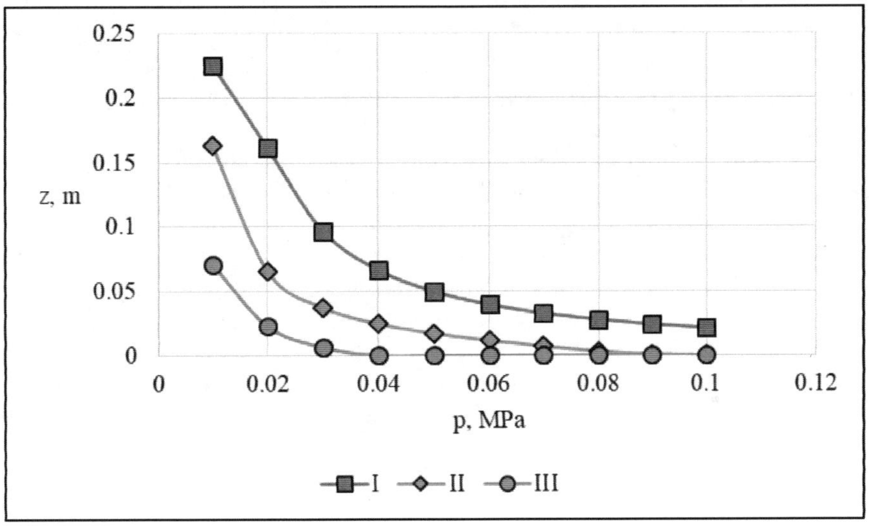

Figure 8. Threshold height of a single obstacle overcome by the mover as a function of vehicle ground pressure and soil category.

The height of a single threshold obstacle that the wheeled propeller is capable of overcoming drops sharply as the average ground pressure of the propeller increases. For this reason, it is advisable to equip wheeled vehicles with removable cross-country vehicles such as mono-trackers on cluttered logging sites.

3. Conclusion

Analysis of the calculation results allows us to draw the following conclusions and recommendations for the selection of wheeled mover (on the example of wheeled forwarder 8x8, with another configuration of the mover should be recalculated in proportion to the change in the ratio of the number of wheels):

IOP Conf. Series: Materials Science and Engineering **817** (2020) 012011 doi:10.1088/1757-899X/817/1/012011

– To comply with the requirement of environmental friendliness of the vehicle, which is that the rut depth after the passage of the machine should be less than 0.1 m [11, 12], the wheeled vehicle pressure when working on soils of medium strength (II category) should not exceed 0.07 MPa (weight of the machine with a load of 31 tons), on a weak-bearing soil (category III) - no higher than 0.025 MPa (weight of the machine with a load of 11 tons). On firm soils of I category pressure limitation on depth of a track is not actual.

– On soils of II category, to ensure cross-country ability, the pressure should be limited to 0.095 MPa (weight of the machine with a load up to 42 tons) (slipping coefficient will be 0.2, track depth 0.18 m, operating speed of the machine 1.5 m/s). To ensure the wheeled propulsor support cross-country ability, the average pressure when working on soils of III category should be limited to 0.04 MPa (the weight of the machine with a load of up to 18 tons) (slipping coefficient will be 0.136, the depth of the track 0.253 m, the operating speed of the machine 2.8 m/s). On soils of the first category the pressure on cross-country ability is not limited.

References

[1] Khitrov E G, Andronov A V, Martynov B G and Spiridonov S V 2019 Interrelations of various soil types mechanical properties *IOP Conf. Series: Journal of Physics: Conf. Series* **1177** 012032 DOI: 10.1088/1742-6596/1177/1/012032

[2] Khitrov E, Kochnev A, Ivanov V and Stepanishcheva M 2018 Linking the deformation moduli and cone indices of forest and peatland soils *Int. Multidisciplinary Scientific GeoConference Surveying Geology and Mining Ecology Management, SGEM 18* pp 297–304 DOI: 10.5593/sgem2018/3.2/S13.039

[3] Khitrov E G and Andronov A V 2019 Bearing floatation of forest machines (theoretical calculation) *IOP Conf. Series: Materials Science and Engineering* **695** 012020 DOI: 10.1088/1757-899X/695/1/012020

[4] Khitrov E G and Andronov A V 2019 Mathematical model of interaction between forest machine mover and consolidating soil *J. Phys.: Conf. Ser.* **1177** 012030 DOI: 10.1088/1742-6596/1177/1/012030

[5] Kochnev A and Khitrov E 2018 Theoretical models for rut depth evaluation after a forestry machine's wheel passover *Int. Multidisciplinary Scientific GeoConference Surveying Geology and Mining Ecology Management, SGEM 18* pp 1005–12 DOI: 10.5593/sgem2018/3.2/S14.128

[6] Khitrov E G and Andronov A V 2019 Comparison of the forest machine mover average and nominal pressure on the soil *IOP Conf. Series: Materials Science and Engineering* **695** 012021 DOI: 10.1088/1757-899X/695/1/012021

[7] Khitrov E, Andronov A, Iliushenko D and Kotenev E 2019 Comparing approaches of calculating soil pressure of forestry machines *Int. Multidisciplinary Scientific GeoConference Surveying Geology and Mining Ecology Management, SGEM 19* (vol 19) issue **3.2** pp 649–655 DOI: 10.5593/sgem2019/3.2/S14.084

[8] Larin V V 2007 *Methods for predicting the support ability of multi-axle wheeled vehicles in the terrain* [in Russian – Metody prognozirovaniya i povysheniya opornoj prohodimosti mnogoosnyh kolesnyh mashin na mestnosti], Doctoral thesis. 05.05.03, Moscow

[9] Ivanov V, Stepanishcheva M, Khitrov E and Iliushenko D 2018 Theoretical model for evaluation of tractive performance of forestry machine's wheel *Int. Multidisciplinary Scientific GeoConference Surveying Geology and Mining Ecology Management, SGEM 18* pp 997–1004 DOI: 10.5593/sgem2018/3.2/S14.127

[10] Lukhminskiy V A 2018 *Perfection of the models and methods for forecast of the tracked forest machine cross-country ability* [in Russian – Soverschenstvovanije modelej I metodov prognozirovanija prokhodimosti gusenichnikh lesnikh maschin], PhD thesis: 05.21.01, Arkhangelsk

[11] Khitrov E, Andronov A, Bogatova E and Kotenev E 2019 Development of recommendations on

environmental certification of forestry machinery drives *Int. Multidisciplinary Scientific GeoConference Surveying Geology and Mining Ecology Management, SGEM 19* (vol 19) issue **3.2** pp 689–696 DOI: 10.5593/sgem2019/3.2/S14.089

[12] Peskov V B 2018 *Improving models for assessing rutting and compaction of soil under the influence of movers of wheeled forest machines* [in Russian – Sovershenstvovanie modelej dlya ocenki koleeobrazovaniya i uplotneniya pochvogruntov pod vozdejstviem dvizhitelej kolesnyh lesnyh mashin], PhD thesis, Arkhangelsk

IOP Conf. Series: Materials Science and Engineering **817** (2020) 012012 doi:10.1088/1757-899X/817/1/012012

Application of Kalman-Bucy filter for vessel traffic control systems in the northern sea route

E Istomin, A Sidorenko, S Stepanov, Y Petrov*, I Martyn

Department of applied computer science, *Russian State Hydrometeorological University*, 79 Voronezhskaya Street, St. Petersburg 192007, Russian Federation

*Corresponding email: yaroslav.petrov025@gmail.com

Abstract. The article considers the method of ice density concentration prediction based on the implementation of the Kalman-Bucy filter using geospatial data. The developed algorithm of data filtering allows us to assess the effective use of this approach in predicting the density of ice in the Arctic. The key aspects of the model application for the construction of optimal and safe routes of vessels in the waters of the Northern sea route are described.

1. Introduction

Currently, the development of the Northern sea route (NSR) is a priority for the development of the Russian transport system in the Arctic [1]. When developing a control system for such a long territorial object, it is necessary to mind about management of the natural risks in the conditions of climate change [2-4], including the influence of black carbon [5, 6]. It is also necessary to take into account the issues of environmental safety related to large sea objects [7, 8]. When solving these problems, it is natural to use advanced information technologies (IT) including the issues of information collection [9]. The development of such IT-technologies is the main goal of this study.

The Russian Federation transport systems are characterized by an enormous scale and developed infrastructure network. Current trends of information technology (IT) improvement together with scientific approaches determine one of the highest priority ways to develop the RF transport infrastructure and manage the optimal safe processes for organizing ship navigation in the Northern Sea Route waters.

To make managerial decisions on conducting business in the Arctic, the timely and relevant information is required that takes into account the regional climatic conditions, expressed in certain parameters. The operational ice information is needed to select the optimal sailing routes for ships, covering climate change and the ice area in the Arctic. For each of the indicators, our own mathematical calculation models were applied. The probability of the event occurrence can be estimated on the basis of known methods for assessing the probability processes characteristics.

With complete certainty of the initial data, the predicting problems of system changes are usually solved on the basis of classical statistics methods (ordinary least-squares method, maximum likelihood, etc.). Optimal filters are also part of these methods; among them, the most universal is the Kalman-Bucy filter. It is a recursive algorithm for weighted smoothing and time series forecasting. The Kalman-Bucy filter allows you to update the assessment of the current state and predict their future values as new data arrives. Let's consider the implementing principle of the optimal Kalman-Bucy filter.

IOP Conf. Series: Materials Science and Engineering **817** (2020) 012012 doi:10.1088/1757-899X/817/1/012012

2. Methods and Materials

It is assumed that there are such time intervals at which the parameter under study (wind force) is a stationary random process. It may be considered as an accidental event that wind force does not exceed a given value of the Beaufort wind force scale. For the known random process, you can determine the probability of wind speed couldn't' not exceed 11 Beaufort number (wind speed 28.5 - 32.6 m/s) using the following (1):

$$P_i(t_3) = P\left[S(t_3)/S(t_0)\right] = S_0 \le \delta_i, \qquad (1)$$

where, $P_i(t_3)$ - probability of wind speed couldn't exceed the i-th Beaufort scale number; δ_i - boundary of the Beaufort scale number at the *i*-slice; S_0 - initial value of wind force; $S(t_3)$ - random function of changing wind force at a given time t_3; $S(t_0)$ - random function of changing wind force at the beginning of time t_0.

Using the estimates [10, 11] (2) was obtained:

$$P_i(t_3) = \Phi\left[\frac{\delta_i - m_s - r_s(0,t_3)\times(s_0 - m_s)}{\sigma_s\sqrt{1 - r_s^2(0,t_3)}}\right] - \Phi\left[\frac{-m_s - r_s(0,t_3)\times(s_0 - m_s)}{\sigma_s\sqrt{1 - r_s^2(0,t_3)}}\right], \qquad (2)$$

where, m_s is the mathematical expectation of a random process; σ_s - standard deviation; $r_s(0,t_3)$ - autocorrelation function.

For the study, at least 2 such points x and y are taken, so that the mutual correlation between them is identically equal to zero $r_{x,y} \equiv 0$. If more points are taken, then the mutual double correlation is identically equal to zero $r_{x,y} = r_{x,z} = r_{z,y} ... \equiv 0$. According to the "minimization method χ^2", the values of unknown parameters were chosen so that Pearson's distance would be the smallest. Once the parameters and the minimum distance have been determined, the question is whether it is small enough. Overall sequence of actions [12]:

1. Selecting the model (H_0 hypothesis).
2. Selecting number position and determining the vector of observed frequencies O_i.
3. Assessing unknown model parameters and building confidence intervals for them (Pearson's minimum distance).
4. Calculating the expected frequencies E_i.
5. Comparison of Pearson's found distance with the largest critical value χ^2_{max}, which is still considered plausible, and compatible with H_0. The value χ^2_{max} determines from the known statistics, solving the equation $P(\chi^2_n > \chi^2_{max}) = 1 - \alpha$, where α - the "level of significance" or "criterion size" or "the magnitude of the error of the first kind" (typical value of 0.05).

If $\chi^2 > \chi^2_{max}$, the H_0 hypothesis is rejected, otherwise it accepted. In $\alpha \cdot 100\%$ of cases (i.e. rarely enough), this method of checking hypothesis H_0 will result in a "first kind error": the H_0 hypothesis will be rejected incorrectly.

3. Results and Discussion

A universal Kalman-Bucy filter is used to predict probabilistic values of ice density. The input data for the filter are a priori estimates of the mathematical expectations of the coefficients a_{ij} and their covariance matrix. The mathematical expectation of the coefficients (formula 3) is defined at this case as follows: $a0 \to -a2 \to a1 \to -a1 \to a2 \to -a0$ (since the function returns values in this order):

$$\left.\begin{aligned} &n = 1; a0 = \left[p1(n) + p2(n) + p3(n) \right] / 3; \\ &n = 2; a1 = \left[p1(n) + p2(n) + p3(n) \right] / 3; \\ &n = 3; a2 = \left[p1(n) + p2(n) + p3(n) \right] / 3; \\ &Mx = \left[a0 \quad a1 \quad a2 \right]; \end{aligned}\right\} \tag{3}$$

The coefficients $p1, p2, p3$ are formed using the least-squares method. The covariance coefficient matrix (4):

$$\left.\begin{aligned} &AfK = [p1; p2; p3]; \\ &Kt = \mathrm{cov}(AfK); \\ &Kt1 = \mathrm{cov}(AfK); \end{aligned}\right\} \tag{4}$$

To construct the covariance matrix, it is necessary three- times to approximate the time dependence of the parameter in order to obtain estimates of the coefficients a_{ij} using the methods of exponential smoothing or least squares [13].

The result of applying the Kalman-Bucy filter is shown in figure 1-4.

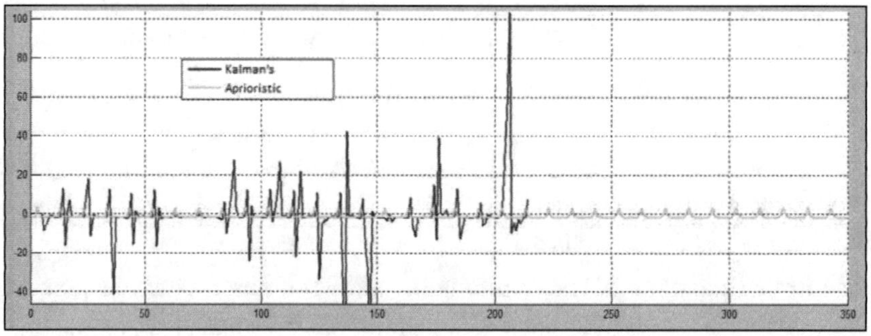

Figure 1. The indication of temperature on water surface coefficient.

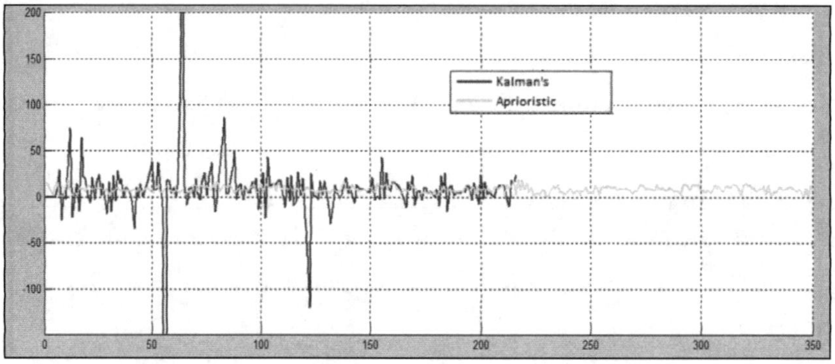

Figure 2. The indication of wind speed coefficient.

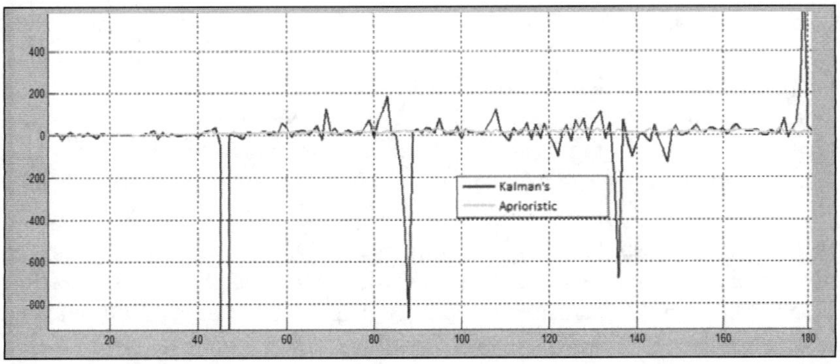

Figure 3. The indications of water vapor in the atmosphere concentration factor.

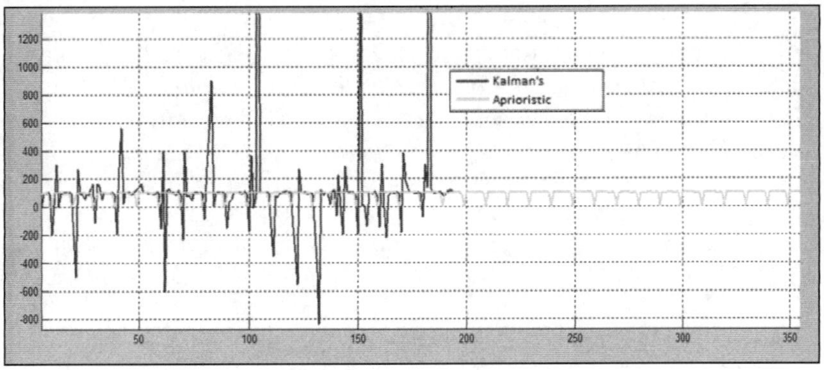

Figure 4. The indication of ice concentration.

4. Conclusions

Graphs 1-4 show that the forecast data using Kalman-Busy filter is close to a priori values, except for outliers (zero values and data errors). Based on the study, it can be concluded that in the presence of sufficient and reliable data, the proposed method is suitable for further deeper consideration.

The implementation of the optimal Kalman-Bucy filter [14] to the vessel's motion control system will display the predicted value of spatially distributed hydro-meteorological data on the intended route of the vessel, using this algorithm to assess the probabilistic factors affecting the vessel and its position on the course. It will help to form an optimal-safe route.

In the future, the Kalman-Bucy filter will take into account the multi-parametric model of probability event assessment. Introducing this filter into the software module will automate the system control of the dynamic positioning and adjustment in accordance with the parameters of the particular vessel movement. Also, this control system should take into account manual commands from the remote control and data of on-board sensors. All of these factors are analyzed to ensure that the ship's trajectory is constantly updated.

Already, the results can help to lay an optimally safe route, which will allow making a rational and optimal decision on ship operations. The model under study can significantly reduce the risks of incidents and provide safer and more efficient navigation in the Northern Sea Route.

References

[1] Rumyantseva E, Gogoberidze G, Abramov V M, Rodin N and Vladimirova G 2017 Priorities and challenges of the state policy of the Russian Federation in Arctic. *Proceedings of the 17th International Multidisciplinary Scientific GeoConference Surveying Geology and Mining Ecology Management, SGEM, (Bulgaria)* vol. 17(52) p 921

[2] Burlov V G, Istomin E P., Abramov V M, Fokicheva A A and Sokolov A G 2018 Risk

management method in parametric geosystems. *Proceedings of the 18th International Multidisciplinary Scientific GeoConference Surveying Geology and Mining Ecology Management, SGEM, (Bulgaria)* vol. 18 (2.2) p 377

[3] Vekshina T V, Abramov V M, Bolshakov V A, Veretennikov V N and Korinets E 2019 Geoinformation technologies for assessing Arctic and Subarctic riverbeds throughput while climate change. *Proceedings of the 19th International Multidisciplinary Scientific GeoConference Surveying Geology and Mining Ecology Management, SGEM, (Bulgaria)* vol.19 (2.1) p 903 DOI: 10.5593/sgem2019/2.1/S08.117

[4] Popov N N, Alexandrova L V, Abramov V M, Gogoberidze G G and Karlin L N 2015 Water exchange between the Pacific and the Bering sea with impact on climate change in the Arctic and Subarctic. *Proceedings of the 15th International Multidisciplinary Scientific GeoConference Surveying Geology and Mining Ecology Management, SGEM, (Bulgaria)* vol. 2(3) p 701

[5] Popov N N, Abramov V M, Gogoberidze G G, Karlin L N and Lednova J A. 2014 Clean technologies development strategy for the national black carbon controlling system in the Russian arctic. *Proceedings of the 14th International Multidisciplinary Scientific GeoConference Surveying Geology and Mining Ecology Management, SGEM, (Bulgaria)* vol. 2(4), c. 313-320

[6] Malakhova J A, Abramov V M, Gogoberidze G G, Karlin L N, Lednova J A and Berboushi S V 2014 Variability of particulate matter in saint-petersburg megacity air within climatic time scale. *Proceedings of the 14th International Multidisciplinary Scientific GeoConference Surveying Geology and Mining Ecology Management, SGEM, (Bulgaria)* vol. 2(4), c. 599-606

[7] Ershova A, Shilin M, Zhigulsky V, Chusov A, Abramov V, Bagrova T. and Popov N. 2018 Environmental Safety of the Nord Stream 2 Marine Gas Pipeline (Russian Section) *2018 IEEE/OES Baltic International Symposium* DOI: 10.1109/BALTIC.2018.8634858

[8] Shilin M B, Ershova A A, Zhigulsky V A and Abramov V M 2018 Geo-ecological support of optimization for the route of the Nord Stream-2 marine gas pipeline. *Proceedings of the 18th International Multidisciplinary Scientific GeoConference Surveying Geology and Mining Ecology Management, SGEM, (Bulgaria),* vol. 18 (5.1) p 423

[9] Isaev A, Gogoberidze G, Abramov V M, Lednova J, Karlin L and Khaimina O 2014 Main results of summer oceanographic surveys in the eastern Gulf of Finland in the framework of the Topcons project. *Proceedings of the 14th International Multidisciplinary Scientific GeoConference Surveying Geology and Mining Ecology Management, SGEM, (Bulgaria)* vol. 3 issue 5 p 253

[10] Istomin E, Petrov Y, Stepanov S, Kolbina O and Sidorenko A About technology of risk management in forestry. *IOP Conference Series: Earth and Environmental Science*. 316. 012011. 10.1088/1755-1315/316/1/012011

[11] Istomin E, Petrov Y, Stepanov S, Kolbina O and Sidorenko A Model of optimum integration of diverse geodata for the benefit of management of forestry. *IOP Conference Series: Earth and Environmental Science.* 316. 012013. 10.1088/1755-1315/316/1/012013

[12] Kalman R E 1960 A New Approach to Linear Filtering and Prediction Problems, Transactions of the ASME – *Journal of Basic Engineering,* No. 82 (Series D), pp 35-45

[13] Sorenson *H W* 1970 *Least-Squares estimation: from Gauss to Kalman* // IEEE Spectrum **7** p 63-68

[14] Welch G and Bishop G 1995 "An Introduction to the Kalman Filter," *University of North Carolina, Department of Computer Science*, TR 95-041

IOP Conf. Series: Materials Science and Engineering **817** (2020) 012013 doi:10.1088/1757-899X/817/1/012013

Study of intra-day dynamics of currents in the area of the navigable strait of Baltiysk to adjust the movement of water transport

E Istomin, I Martyn, Y Petrov*, **S Stepanov, A Sidorenko**

Department of applied computer science, *Russian State Hydrometeorological University*, 79 Voronezhskaya Street, St. Petersburg 192007, Russian Federation

*Corresponding email: yaroslav.petrov025@gmail.com

Abstract. The aim of this work is to determine the characteristics and dynamics of flows navigable waters of the Vistula lagoon, strait of Baltiysk and coastal areas of the Baltic sea. The research was conducted by employees of the Russian State University together with employees of Shirshov Institute of Oceanology of the Russian Academy of Sciences (IO RAS) at release of drifters on the water area of the sea strait with the subsequent processing of the received GPS data with visualization in the MapSource program. The study in the Baltic sea canal was the regularities of traffic flow, identified certain intra-day dynamics of a course, significantly dependent on wind conditions over the water area. In the course of the work performed, an almost complete intra-day analysis of currents in the Baltic shipping strait area was carried out.

1. Introduction

Under the influence of various causes in the surface, deep and bottom layers of the seas and oceans there are different in the nature of the movement of water masses-sea currents. A moving mass of water is hundreds of times denser than air and can generate forces of enormous magnitude. Currents are mainly characterized by direction and speed, and has an effect on the moving vessel [1]. A vessel caught in the current will have a drift from the line of the intended course, with a decrease or increase in speed. An important condition of marine navigation is the knowledge of the speed and direction of the current by the skipper and his account when maneuvering. Despite the difficulty of navigating and maneuvering in the presence of a current, such as when the current is directed across the pier, it can be used to improve the handling of the vessel at low speed and to give the vessel a downforce.

The Baltic sea is of great importance for Maritime transport in Russia. Some of the most important ports of the Baltic sea in Russia are located in the Kaliningrad region, which are of great economic and transport importance for the Russian Federation.

One of such port cities of the Kaliningrad region is the city of Baltiysk, in the area of which the Strait of Baltiysk was studied.

The study of water dynamics, structure, intensity and repeatability of currents in the area of this navigable strait is an actual study, as currently there is a need:

1. in determining the most dangerous situations threatening navigation and development of recommendations for their prevention;
2. in management of the natural risks in the conditions of climate change [2];

3. in clarifying the mechanism of formation of the diffusion basin between the ends of the entrance moles with recommendations for maintaining this basin in a stable state, since its steep slopes pose a threat to the construction of moles together with navigation beacons;
4. in clarifying the value of water exchange between the Vislin Bay and the Baltic sea;
5. in the development of information technologies for the collection and storage of such data [3].

The article deals with the approach of studying the main characteristics of currents in the shipping strait and its possible impact on the maneuverability of the vessel.

2. Methods and materials

Shelf and coastal areas of oceans and seas have complex and diverse natural conditions, which are characterized by intense spatial and temporal variability. Currents in such areas are one of the most complex and dynamic phenomena. In the area of the Baltic canal is one of the main ports of the Kaliningrad region, which in turn makes it relevant to study the nature of currents in the area of the shipping strait [4].

Recently, the circulation studies based on satellite altimetry have become widespread. This method is promising for this area of research, but at the moment satellite altimetry data have rough spatial and temporal resolution and are poorly suited for the study of coastal currents. In addition to the above, there is a problem in comparing the received satellite data with subsatellite information, which is usually absent.

One of the methods of studying surface currents is the method of floats [5]. It is implemented on the Lagrange approach in the framework of classical hydromechanics and observation of the movement of each individual particle. The location of the particle is fixed at regular intervals, which allows you to restore the trajectory and speed of the particle and make schemes of surface water circulation. In practice, a drifting buoy-drifter is used as an observed particle. Global navigation satellite systems are used to position drifters.

In the study, the drifter consists of a float, an underwater sail and a cargo. For cost-effectiveness of the design, data on the location of the drifter were taken from GPS when the drifter was directly detected. The underwater sail represents metal blades, the float from foam plastic for detection of a drifter on water area is fixed, a place for fastening of cargo for a drifter deepening on the necessary horizon of research. The work used deep and surface drifters, due to the different buoyancy, which is achieved by different types of dredgers. On the surface is a buoy to determine the location of the drifter. Visualization of the trajectory of the drifters was performed using MapSource software based on GPS data [6].

There is also another approach in the form of the Euler method. The method is based on the study of the velocity field and wind direction at each point of space occupied by a moving liquid. With the help of Ekman's theory (1) of wave (drift) flows, it is possible to calculate the velocity and direction of the flow from the data of the wind field:

$$\left.\begin{aligned} u &= U_0 e^{-az}\cos(45^o - az),\\ v &= U_0 e^{-az}\sin(45^o - az). \end{aligned}\right\} \quad (1)$$

As a result, the coordinates of the movement of drifters were obtained. Visualization of the trajectory of the drifters was performed using MapSource software based on GPS data.

Studies of the dynamics of currents were carried out for two days (12 and 13 July). When working on July 12, 3 deep and 4 surface drifters were released into the water area, one of them (4M) with a track. July 13 released 3 deep and 3 surface drifters.

At the beginning of the research in the water area, drifters are released to the pre-planned points and remain in free drift for the planned time [7]. Location is determined visually when inspecting the water area and determining the location of foam buoys fixed to drifters with GPS coordinates fixation. One of the most actual problem is a drifters monitoring, because drifters can walking on a long distance (about 300m) from ship boards. Therefore, the output points is a complex task. To eliminate this problem, it is advisable to equip each buoy with a receiver with a possible transmission of coordinates.

The MapSource mapping program allows you to visualize the resulting tracks of the drifter system according to the tracks of each drifter stored by GPS.

3. Results and Discussion

Initially, the boundaries of the area of work were defined-the Baltic canal and the adjacent sections of the Vislin's gulf and the Baltic sea (figure 1). A characteristic feature of this water area is the change of directions of currents depending on the time of day, noted during the observation of buoys, which can be used for navigation in this area [8].

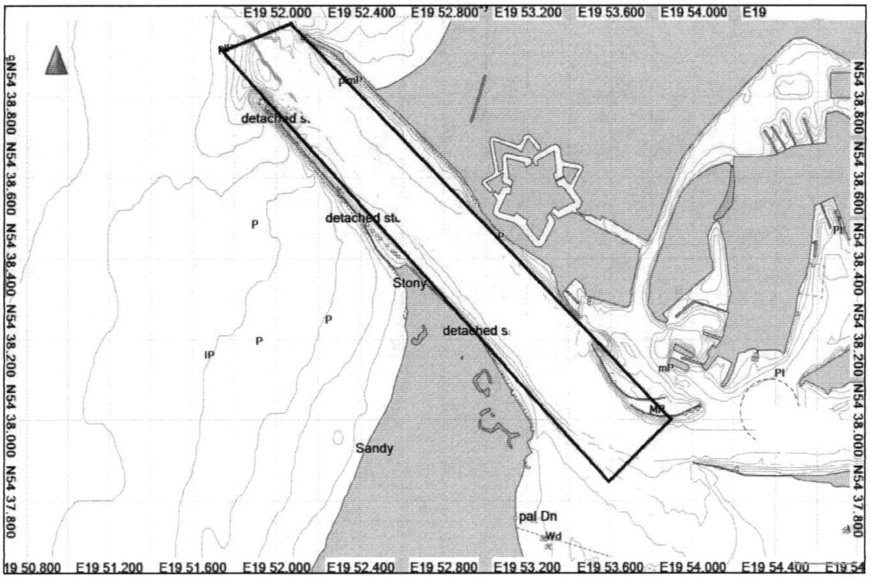

Figure 1. Area of work.

July 12 work was carried out from 18-20 to 20-15 was launched 4 surface and 3 depth drifter. The results are shown in figure 2 and table 1.

July 13 work was carried out from 14-10 to 16-20 was launched 3 surface and 3 depth drifter. The results are shown in figure 3 and table 2.

To analyze the results of the flow characteristics, measurements of meteorological characteristics in the area of work were carried out along the way. A summary table of meteorological characteristics for 12 and 13 July is presented in table 3.

Comparison of the obtained data on the characteristics of the currents with the data of the weather report is presented in figure 4 and figure 5.

IOP Conf. Series: Materials Science and Engineering **817** (2020) 012013 doi:10.1088/1757-899X/817/1/012013

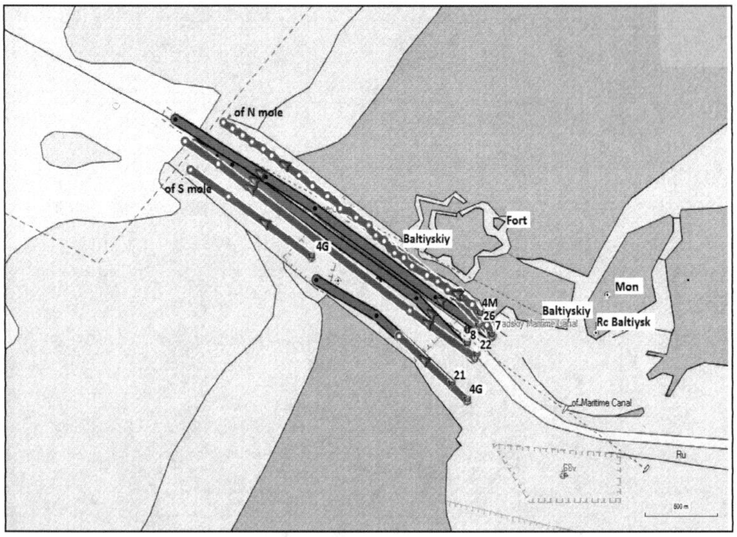

Figure 2. Trajectories of drifters in the strait of Baltiysk 12.07.

Table 1. Characteristics of the strait of Baltiysk current 12.07.

№ drifter's	average speed cm / sec	direction °
	surface	
4M	7.4	315
4G	15.56	311
8	17.32	312
7	24.03	317
	deep	
21	9.07	317
22	10	310
26	16.11	309

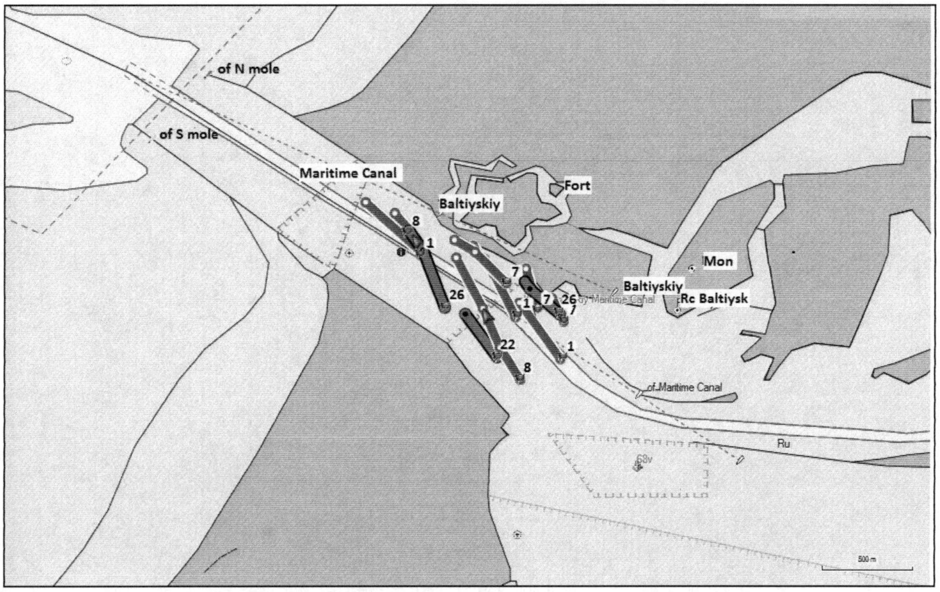

Figure 3. Trajectories of drifters in the strait of Baltiysk 13.07.

Table 2. Characteristics of the strait of Baltiysk current 13.07.

№ drifter's	average speed cm / sec	direction °
	surface	
1	7.36	333
7	3.94	347
8	17.58	339
	deep	
22	8.47	325
26	15.28	330

Table 3. Meteorological characteristics in the area of the strait of Baltiysk 13.07on 12.07 and 13.07.

Wind			
date	time	direction °	speed cm / sec
12.07	18:00	90	87
13.07	15:00	202	83
13.07	18:00	225	77

IOP Conf. Series: Materials Science and Engineering **817** (2020) 012013 doi:10.1088/1757-899X/817/1/012013

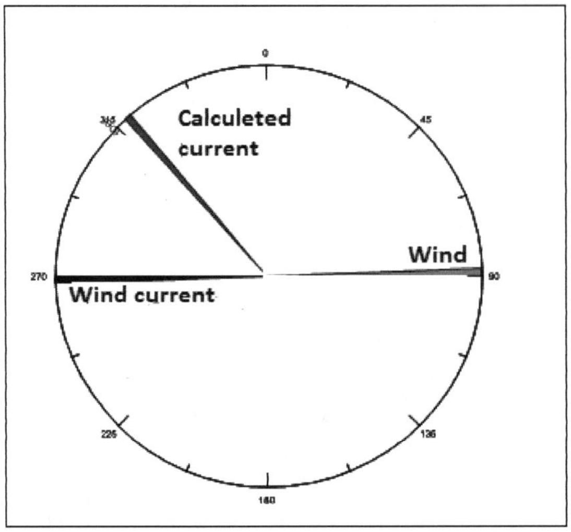

Figure 4. Wind direction and currents according to drifters and wind (12.07).

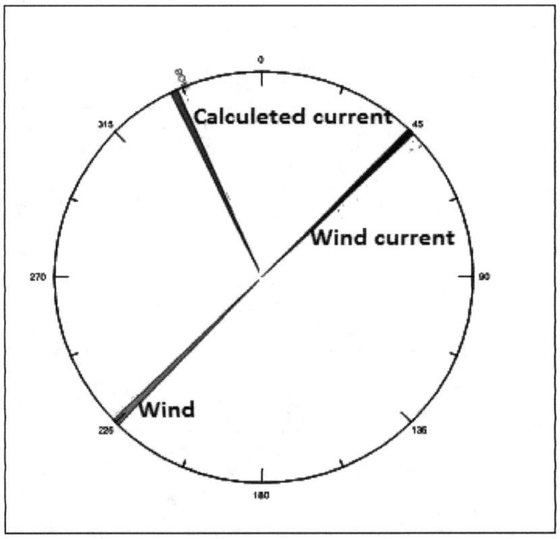

Figure 5. Wind direction and currents according to drifters and wind (13.07).

Buoys located in the study area Baltic canal were observed the patterns of traffic flow in the strait that it was decided to explore with drifters. Since intraday changes were observed-in the morning the current is directed to the strait, in the evening from the strait, it was decided to conduct studies at different times of the day: morning, afternoon and evening. It was not possible to obtain morning data due to bad weather conditions, so the analysis of the results was carried out on the received day and evening observations. From figure 2 (evening measurements) it can be seen that the current is directed from the strait, since the direction of the arrows showing the movement of water in the sea strait in figure 2 has a direction from the Vislin Bay to the Baltic sea. The speed of surface and deep drifters does not differ significantly, is about 15 cm/s. in figure 3 (daytime), the direction of the flow towards the strait is noted, as well as the turn of the current to the North, in consequence of which the current spreads across the strait. For this reason drifters at measurements on July 13 constantly washed ashore. The speeds of surface drifters slightly exceed the speeds of deep drifters. The surface velocities are approximately 7 cm \ s, the flow velocities in the deep layers vary within 10 cm\s. from the studies conducted, it follows that in this period the flow velocity in the evening is slightly higher than the value of the velocities obtained in the daytime. It is noted that the flow in the evening is more uniform, compared with the flow recorded in the daytime [9-10].

When comparing these characteristics of the currents with the data of the weather reports we can assume that the speed and direction of currents has a direct effect wind influence [11], as the data obtained from a weather station shows that the wind speed on 12 July was higher than July 13, that could increase the speed of the current on July 12, compared to the flow rate on July 13. In spite of the fact that currents have a big dependence on wind influence, they not always have absolutely direct dependence. In figure 4, it can be seen that the measured direction of the current on this day does not completely coincide with the direction of the classical wind flow. In figure 5, the calculated and obtained flow directions have a greater difference [12].

The obtained results of currents in the strait can be used for recommendations to skippers when passing the strait of Baltiysk. If the ship is late to put the rudder to the right, the stern will begin to attract to the shore, and the bow will deviate to the opposite side of the strait. A stream of water hitting the starboard side will deflect the bow even further to the left. In such a situation, it becomes difficult to straighten the ship with the help of the rudder. It will be necessary to stop the ship to reduce the impact of interaction between the shore and the ship. Otherwise, the vessel may crash into the opposite Bank and then turn against the current (figure 6).

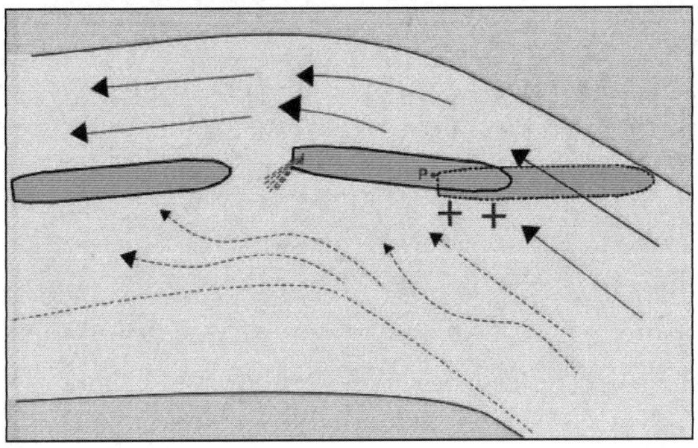

Figure 6. Influence of the counter current on the controllability of the vessel.

In the case of a passing current, the speed relative to the water is less than the speed of the same vessel going against the current. If the ship belatedly put the rudder to the right, it will turn across the strait. The stern will most likely begin to attract to the right Bank. The flow of water will hit the port side and prevent the turn to the left. However, this phenomenon affects the vessel much less than the counter current (figure 7).

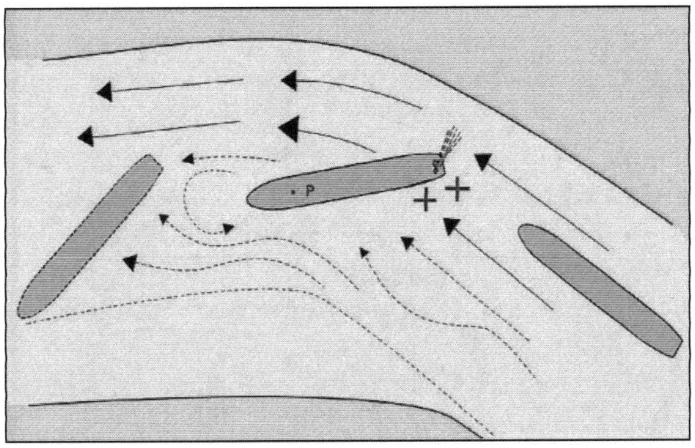

Figure 7. The influence of the passing current on the management of the vessel.

IOP Conf. Series: Materials Science and Engineering **817** (2020) 012013 doi:10.1088/1757-899X/817/1/012013

4. Conclusion

The result of this work is clearly seen that in the evening for directed out of the strait, and in the morning for sent to the strait, during the day there is a change of direction of flow. This rotation of the flow in the strait can be observed in figure 3, where the flow is directed across the strait. Having carried out almost complete day-to-day research, it can be assumed that the current in the strait of Baltiysk does have day-to-day variability, which was observed by buoys located in this strait.

The velocities of surface and deep currents differ slightly and vary within 10-15 cm / s.

The probable cause of intraday change of direction of the current in the wind regime, formed over this area, however, it is worth noting that the wind impact has no absolute influence on the formation of currents strait [12-13].

The obtained data on the nature of currents in the strait contribute to better awareness of skippers when passing this strait. Can be useful in making decisions maneuvering the vessel depending on the intraday dynamics of currents in the strait.

Acknowledgments

We are grateful for the opportunity to work together with the employees of Shirshov Institute of Oceanology of the Russian Academy of Sciences (IO RAS).

References

[1] Dubravin V F and Nagornova H H 2007 On estimates of water exchange through the Baltic Strait [in Russian – Ob ocenkax vodoobmena cherez Baltijskij proliv] *Innovations in science and education*-2007: SB. Tr. Vol. Conf. KSTU Ch 1

[2] Istomin E, Petrov Y, Stepanov S, Kolbina O and Sidorenko A About technology of risk management in forestry. IOP Conference Series: Earth and Environmental Science. 316. 012011. 10.1088/1755-1315/316/1/012011.

[3] Istomin E, Petrov Y, Stepanov S, Kolbina O and Sidorenko A. Model of optimum integration of diverse geodata for the benefit of management of forestry. IOP Conference Series: Earth and Environmental Science. 316. 012013. 10.1088/1755-1315/316/1/012013.

[4] The Hydrometeorological regime of the Vistula lagoon 1971 Eds N N Lazarenko and A Mayevsky (Leningrad: Gidrometeoizdat) p 6

[5] Tolstosheev A P 2010 A metyod of estimation of the results of reconstruction of the trajectories of drifting buoys (Physical Oceanography vol 19) **6** pp 358-365

[6] Korotaev G K, Oguz T, Nikiforov A and Koblinsky C 2003 Seasonal, interannual, and mesoscale variability of the Black Sea upper layer circulation derived from altimeterdata (J Geophys Res Vol 108) **4** pp 22-31

[7] Gill A 1986 Dynamics of atmosphere and ocean. [in Russian – Dinamika atmosfery` i okeana] *Mir* vol 1 pp 130-190

[8] Chubarenko B, Babakov A 2016 Sediment transport near the Vistula spit (Baltic sea) *Atlantic Branch of P P Shirshov Institute of Oceanology of Russina Academy of Sciences* Kaliningrad

[9] Ullgren J E, Emidio A, Gammelsrod T and Hoguane A M Observations of strong ocean current event offshore Pemba, Northern Mozambique *Operational Oceanography* Vol 9 pp 55-66

[10] Graeme C Hays 2017 Ocean current and marine life *Current Biology* vol 27) pp 470-473

[11] Joseph B Klemp, Richard Rotunno and William C Skamarock On the dynamics of gravity current in a strait (J Fluid Mechanics vol 269) pp 169-198

[12] Omstedt A, Elken J, Lehmann A, Leppäranta M, Meier H E M, Myrberg K and Rutgersson A Progress in physical oceanography of the Baltic Sea during the 2003–2014 period *Progress in Oceanography* vol 128 pp 139-171

[13] Ichikawa K, Imawaki S and Ishii H Comparison of surface velocities determined from altimeter and drifting buoy data *Oceanography* vol 51 pp 729-740

IOP Conf. Series: Materials Science and Engineering **817** (2020) 012014 doi:10.1088/1757-899X/817/1/012014

Problems of technological processes selection in logging

V Kalyashov[1], A Smirnova[2], N Tjurin[3], L Gromskaya[3], T Antonova[3]

[1]Descriptive Geometry and Engineering Graphics Department, *Saint Petersburg State University of Architecture and Civil Engineering*, 4 2-nd Krasnoarmeyskaya Street, Saint Petersburg 190005, Russian Federation
[2]Department of forest policy, Economics and management**,** *St. Petersburg State Forest Technical University*, 5 Institutskiy Lane, St. Petersburg 194021, Russian Federation
[3]Department of Industrial Transport, *St. Petersburg State Forest Technical University,* 5 Institutskiy Lane, St. Petersburg 194021, Russian Federation

*Corresponding email: vit832@yandex.ru

Abstract. The article deals with current issues of the logging enterprises effective functioning with assortment technology for wood harvesting in the Leningrad region of the North-West Federal District of Russia. A survey of the production activity of a number of enterprises that use logging technology based on harvesters and forwarders was carried out. The results of technical and economic assessment of this technology usage are presented, the reasons of the decrease in the efficiency of the technological process of the logging are identified, the alternative assortment technologies are considered, and recommendations on selection of technological processes and systems of logging machines are made.

1. Introduction

The relevance of the study is that a number of works are devoted to the introduction of logging sorting technology based on multi-operation machines of foreign production without a solid justification of the effectiveness of their application, which often leads to erroneous technological decisions in choosing systems of logging machines [1, 2]. In this paper, an assessment of the effectiveness of the assortment technology of timber harvesting is considered and the problems of its implementation are identified by the example of logging in the Leningrad region in the near future.

The purpose of this study is to identify the reasons of the decrease in logging efficiency, both dependent and independent of the applicable technological process of logging, in particular assortment technology, as well as the development of recommendations for the selection of technological processes and systems of logging machines.

2. Methods and Materials

Research methods include simulation of the machine systems functioning, comparative analysis and modeling of logging efficiency indicators in assortment technology, a systematic approach, and abstract logical methods for substantiating recommendations on the choice of technological options.

3. Results and Discussion

The results of the research are to substantiate the reasons of the decrease in the efficiency of logging with the use of assortment technology for a five-year period (2014-2018). The calculation of performance indicators was carried out for logging production with a volume of 100 thousand m3 completed material, in forests, the species composition of which is as follows: 40 % - spruce, 30 % - pine, 20 % - birch, 10 % - aspen, an average volume of a whip of 0.36 m^3, the average distance of wood removal 50 km, and 75 % of commercial wood. These forest growing conditions are the most typical for logging in the Leningrad region [3].

During comparative assessment of the technological processes of assorted wood logging, the following technological variant was adopted as the base one: for logging operations machine technology based on John Deere 1270E harvesters and John Deere 1510 forwarders is used, for wood transportation and the lower warehouse the KAMAZ-timber truck 65111 with the Palfinger Velmash VM10L74 hydraulic manipulator and a three-axle trailer, the forklift truck LT-72B type are used. In the calculations it was assumed that only 85% of the total volume is logged by machine complexes according to the Harvester-forwarder system, and 15% of the volume (mainly large trees) is logged by fellers using chain saws with logging by tractors with sliding choker equipment such as TLT-100A and then bucking of whips to assortments at loading areas of the cutting area followed.

The comparative results of calculating the cost, profit and profitability of the production and sale of round timber (final products of forest harvesting) in rubles per $1m^3$ for 2018 and 2014 [4], as well as the indices of changes in these indicators are given in table 1. The calculations were performed in accordance with industry recommendations [5].

Table 1. Cost price, profit and profitability of the production and sale of round timber.

Cost item	**Expenses, (rub./m³) 2014**	**Expenses, (rub./m³) 2018**	**Indexes of change indicators (2018 /2014)**
1. The rent for the wood	115	181.3	1.576
2. Forestry costs	120	132	1.1
3. Remuneration of production workers	124.1	157.6	1.27
4. Contributions to social funds (30%)	37.2	47.2	1.27
5. Costs of preparation and development of production	73.6	101.2	1.37
6. The cost of maintenance and operation of equipment	210	302.4	1.44
7. Logging transport services	132.2	191.8	1.44
8. Maintenance costs of logging roads	73.8	101.1	1.37
9. Shop expenses	50.9	64.6	1.27
10. General running costs	64.3	81.7	1.27
11. Other operating expenses	32.5	41.3	1.27
12.Production cost	1033.6	1402.2	1.356
13. Loading assortments into Russian Railways wagons	52.6	75.7	1.44
14. Advertising and other business expenses	88.8	112.8	1.27
16. Cost of commodity production, (full)	1175	1590.7	1.353
17. The average selling price of 1 m 3	1403.5	1684.2	1.2
18. Profit from sales of products	228.5	93.5	0.41
19. Product profitability, %	19.45	5.88	0.30

During the above-mentioned five-year period the cost of timber harvesting has increased in all cost items. Moreover, the largest increase in costs (57.6%) occurred under the item “Payment for standing timber”. The reason for such a significant increase of rent was the decision of the government of the Russian Federation on the approval of the raising factor to the minimum rates of rent [6], which is 2.17 in 2018, which is significantly higher than in previous years.

The second item, which significantly affected the cost of harvesting – "Foresty cost", it is generally not characteristic of logging production. According to the current Forest Code [7], the activities related to reforestation, forest management, removal of cutting areas, safety and protection of forests on forest areas leased are assigned to the tenant. Forestry costs are usually included in the cost of round timber and financed from the profits from the sale of timber.

The amount of costs for the above articles is practically independent of the type of used technology, but in general it leads to lower profits and a decrease in sources of investment in new equipment, the other cost items more or less depend on the adopted technology and the system of logging machines.

One of such cost items is “The costs of maintaining and operating logging equipment”. So, using the harvester-forwarder system of assorted wood logging in the technological process, the cost of operating harvesters and forwarders per 1 m^3 of harvested wood was 210 rubles in 2014, and it increased by 44% and became 302.4 rubles in 2018. Thus, the share of costs of this item in the cost of harvesting round timber using the technology "harvester + forwarder" is most significant and amounts to about 20%.

This is due to the fact that in the Russian forest market the harvesters and forwarders produced by John Deere, Ponsse and Komatsu manufacturers (so called “Big Three” forestry equipment manufacturers) are the most popular, which in addition to some differences in design and performance have a common negative side for Russian loggers: the high cost of equipment and services, which eventually leads to an increase depreciation charges and costs during operation process.

So the costs for the removal of wood from the upper warehouse to the lower ones are also significant, which in 2018 it was about 293 rubles per 1m^3 and increased over five years by 42%. At the same time the logging transport services for timber removal was 191.8 rubles per 1 m^3 (transport component), the cost of maintaining forest roads was 101.1 rubles per 1 m^3 (road component). The reason for such high costs is the lack of year-round roads, which is covered by the construction of winter roads at the expense of cost. The problem of construction of logging roads is very acute, it is comprehensively discussed, but solved too slowly, so we cannot expect a rapid reduction in costs.

As a result, the prime cost of round timber has grown by 35.3% over five years. The annual cost increase was about 7%, which exceeds the rate of inflation. As of 2014, the level of profitability of products amounted to 20.45% with the average current sales price of 1 m^3 of harvested wood in 1403.5 rubles. The prices for selling round timber for 2018 have not yet been published in the open press, so the changes of the final result are possible: an annual increase of sale prices of timber products in accordance with the rate of inflation growth would lead to a reduction in profitability by 2018, up 6.8%, and the increase in the price level, the cost would maintain the profitability at the same level.

However, the forest market is developing according to its own laws, so if you do not consider investing in new foreign equipment, it is very risky to predict a sharp increase in prices for forest products and expect the payback of such equipment in the future. It turns out from the foregoing that the purchase of expensive imported equipment will be able to afford only large timber enterprises, medium and small businesses should focus on the use of domestic machinery and equipment.

As an alternative to the basic process described above using the “harvester + forwarder” machine systems (hereinafter - option 1), on the basis of a survey of the production conditions of some enterprises, it was selected several options for the most common technological processes of assortment harvesting and considered timber and given their comparative technological and economic assessment for the conditions of the Leningrad region.

A brief description of the variants of technological processes and machine systems is presented in table 2.

IOP Conf. Series: Materials Science and Engineering **817** (2020) 012014 doi:10.1088/1757-899X/817/1/012014

Table 2. Variants of technological processes and machine systems.

Type of main cutting operations, sequence of their execution (technological process) and characteristics of machine systems according to options				
1	2	3	4	5
Harvester[a]+ Forwarder[b]	CS[c]+Forwarder[b]	CS[c]+CS[c]+ChS[h]+ CS[c]+HM[e]	FB[d]+DFGS[g] + CBBM[f]+HM[e] (foreign machines)	FB[d]+DFGS[g] + CBBM[f]+HM[e] (domestic machines)
Felling of trees, cutting branches, bucking whips on assortments at the cutting sites – Harvester[a] John Deere 1270E; Logging and piling assortment – forwarder[b] John Deere 1510; Loading assortments – timber truck KAMAZ-65111 with HM[e] Palfinger Velmash VM10L74	Felling of trees, pruning, bucking whips on assortments in the apiary – CS[c] Husqvarna 372XP; Logging, piling and loading assortment – forwarder[b] John Deere 1510	Felling of trees - CS[c] Husqvarna 372XP, cutting branches – CS[c] Husqvarna 372XP; skidding whips – ChS[h] MSN-10, bucking of whips for assortments on loading platforms of a cutting site – CS[c] Husqvarna 372XP; loading assortments onto timber trucks – timber truck KamAZ 43118 with HM[e] OMTL 97-04	Roll felling trees – FB[d] John Deere 853M; skidding of packs of trees – DFGS[g] John Deere 748L; Trimming branches and bucking of whips for assortments – CBBM[f] John Deere 2154D with a Attachment HTH622B; sorting, stacking and loading of assortments onto timber trucks – timber truck KamAZ 43118 with HM[e] OMTL 97-04	Roll felling trees – FB[d] LP-19 with a mill and a Quadco drive (Canada); skidding of packs of trees – DFGS[g] LT-187; Trimming branches and bucking of whips for assortments – CBBM[f] LP-19B with a Waratah harvester Attachment; sorting, stacking and loading of assortments onto timber trucks – timber truck KamAZ 43118 with HM[e] OMTL 97-04

[a] Harvester – feller-delimbing and slitting machine (Wheeled Harvester).
[b] Forwarder – tractor for skidding round timber (skidding of assortments).
[c] CS – gasoline powered Chain Saw.
[d] FB – tracked Feller Buncher.
[e] HM – hydraulic manipulator mounted on the chassis of a logging vehicle.
[f] CBBM – machine for cutting branches and bucking trees (tracked harvester-processor).
[g] DFGS – skidding tractor with pincer grip package of trees (Skidder with Dual-Function Grapple).
[h] ChS – skidding tractor with sliding cable equipment (Cable skidder).

The calculation of technical and economic indicators for the options is done by modeling. In that case, if the designed machines and equipment are of foreign production, then all the technical and economic indicators of the operation of machines and equipment are taken according to actual data without the involvement of foreign data. The comparability of the indicators of the compared options is ensured by the volume of production, the analogy of the technological operations performed, the main operating conditions, methods for determining the cost and natural indicators, i.e. compared options are given to the same features.

Technical and economic indicators of the compared variants of technological processes are brought to the technological cost of logging operations without taking into account rent and costs of forestry works, the value of which practically does not depend on the technological process, and are summarized in table 3.

Table 3. Technical and economic indicators of the compared variants of technological processes

Name of indicator	Options for technological processes and systems of logging machines (according to table 2)				
	1	2	3	4	5
1. Leading machine (mechanism) in the system	Harvester *John Deere 1270E*	Forwarder *John Deere 1510*	Cable skidder MSN-10	Feller Buncher *John Deere 853M*	Feller Buncher *LP-19*
2. Labor intensity of cutting works, man-hours per 1 thousand m^3	283.6	589.5	701.3	307.3	351
including the main works	177.5	481.5	565.9	187.9	236.7
on preparatory and auxiliary works	106.1	108.0	135.4	119.4	114.3
3. Labor productivity, m^3 per 1 person-hour	3.53	1.7	1.43	3.25	2.85
including on the main works	5.63	2.08	1.77	5.32	4.22
4. Remuneration with deductions to the Social Insurance Fund, rub. on $1m^3$	113.5	229.8	273.4	119.8	136.8
5. Equipment maintenance and operation costs, RUB per 1 m^3	378.6	277.6	173.6	270	265.3
6. Technological cost of cutting operations, rub per 1 m^3	562.5	507.5	447.0	389.8	402.1
7. Capital intensity, rub per 1 m^3	1124	403.8	208.1	520.8	283.6

The comparative analysis of various technological processes allowed to estimate efficiency of their application on logging works and to identify advantages and disadvantages of each of them.

In contrast to the basic 1st option, the technological process in the second option involves the use of Chain saws on felling trees, delimbing and bucking whips. This technology takes place in medium-sized enterprises, as a result the labor intensity of cutting operations increases in more than 2 times and, accordingly, labor productivity decreases, which leads to an increase in wage intensity. So wages with contributions to social insurance funds increased here to 228.8 rubles per $1m^3$ of harvested wood compared to the first option (113.5 rubles). At the same time, there is a decrease in the cost of maintenance and operation of equipment, which ultimately leads to a decrease in the technological cost of the workpiece by 55 rubles per 1 m^3. The second option shows a significant advantage in the need for capital investments. So the capital intensity of the process is 403.8 rubles per $1m^3$, which is 2.8 times lower compared to the first option.

The third variant of the technological process, organized with the use of mainly domestic equipment, is characterized by a significant labor intensity and labor intensity of the process. The values of these indicators are the highest and are respectively 0.701 man-hours and 273.4 rubles per $1m^3$ of harvested wood. According to these indicators, this version of the technology loses to the first two. However, the advantage of this option is a significant reduction in the cost of maintenance and operation of equipment and, accordingly, the technological cost, which is 447 rubles per $1m^3$, which is lower by 115.5 rubles. compared with the first option and 60.5 rubles – compared with the second. In addition, under this option, the need for capital investments is reduced by 5.4 times compared to the first option and by 2 times compared to the second. These indicators are economically very significant for small and medium-sized business.

The fourth and fifth options provide for the use of Feller bunchers in the assortment process. The difference between these options is that the fourth option involves the use of systems of machinery and equipment of foreign production, and the fifth involves the machinery of domestic production.

The pattern of change in performance indicators is the same here, they are slightly inferior to option 1 in terms of labor productivity and, accordingly, in labor input and salary intensity, however, the technological cost price for them is much lower in comparison with other options. In addition, these options require significantly less investment in equipment compared to the first option. So the fourth option is higher priority in comparison with the first and second at cost, it is lower by 172.7 rubles, respectively. and 117.4 rubles per 1 m^3, and especially in terms of capital intensity, whose performance is 2.16 times lower compared to the first option. A comparison of two options with domestic equipment (the third and fifth) shows the advantage of the fifth in all performance indicators, except for capital intensity, which is slightly higher.

4. Conclusion

The choice of the technological option for logging operations should be made taking into account the capabilities of the forest user according to three criteria: the complexity of the work, the technological cost, and the capital intensity of the process.

Studies have shown that the use of machinery and equipment of foreign production in logging operations in the current state of the forest market is not effective for all conditions, and it is practically inaccessible for small and medium-sized businesses. Therefore, at present, the necessity of forestry engineering development in Russia is particular relevance, and first of all, it is very important to organize mass production of those types and items of domestic equipment that have proved their effectiveness in operation.

References

[1] Gigel T 2017 During the transition to market relations logging suffered the most *Lesprominform* [in Russian - Lesprominform] **7(129)** pp 28-29

[2] Irtuganov T 2017 Measures for the development of the timber industry *Lesprominform* [in Russian - Lesprominform] **8(130)** pp 20-23

[3] Gilts N R, Fedorov V V, Vasyukov V A and Demin K K 1986 *Selective logging* (Moscow: Forest industry) p 192

[4] Shcherbin A M, Chernobaeva E A and Smirnova A I 2015 Efficiency of technological processes in logging *Proceedings of the faculty of Economics and management* [in Russian – Sbornik Trudov Fakulteta Ekonomiki] (St Petersbrg: SPbFTU) pp 157-163

[5] *Methodical recommendations (instruction) on planning, accounting and calculation of the cost of production of the timber industry complex* 2006 (Moscow: Moscow state forest University) p 256

[6] Collection of legislation of the Russian Federation 2017 *Decree of the Government of the Russian Federation of 11.11.2017 #1363 «About the coefficients of the rates of payment per unit volume of forest resources and the rates of payment per unit area of a forest plot owned by the federal»* vol 47

[7] *Forest Code of the Russian Federation: Federal Law of 04.12.2006 No. 200-FZ (ed. of 27.12.2018)* available at: https:// http://www.consultant.ru/document/cons_doc_LAW_64299/

IOP Conf. Series: Materials Science and Engineering **817** (2020) 012015 doi:10.1088/1757-899X/817/1/012015

Deviation of resulting load of forestry vehicle from the normal to the bearing surface

E G Khitrov*, D A Iliushenko, A V Kalistratov

Institute of Technological Machines and Forest Transportation, *St. Petersburg State Forest Technical University*, 5 Institutskiy Lane, St. Petersburg 194021, Russian Federation

*Corresponding email: yegorkhitrov@gmail.com

Abstract. The purpose of the article is to develop and analyze relationships for calculating the ratio of the normal pressure of forestry vehicle on the bearing surface and the shear stress caused by the vehicle's slipping. The study bases on provisions of off-the-road locomotion theory and soil mechanics. The study uses methods of mathematical analysis and applied mathematics for calculations and numeric data processing. The calculations were performed using Maple 2017 software. Calculations show that the ratio of shear stress caused by slipping of the vehicle to the normal pressure on strong forest soil is an average of 17%, reaching 22%; on bearing soil, the ratio averages 22%, reaching 25%; on a weaker forest soil, the ratio is on average 24%, reaching 29% when cutting of the soil layers occurs. It is established that the deviation of the resulting vehicle load from the normal to the bearing surface, at which the cut of the soil layers begins, is 12 - 16 °. The deviation of the resulting load caused by the vehicle slip results in a decrease in the bearing capacity of forest soil by 8.5-10.5%.

1. Introduction

Under influence of a forestry machine mover, several types of deformation arise in the soil: 1) compression of the soil under the influence of the mover normal pressure [1-3], 2) a shift in the soil layers caused by shear stress when the mover is indented into the soil [1], and 3) a shift in the soil caused by the mover slipping [1]. Almost all mathematical models of the interaction of forestry vehicles with the surface, proposed earlier, take into account compression and the shift when the mover is indented into the soil [4, 5]. It was also found that tangential stress along the contact patch increases resulting load on the bearing surface and reduces bearing capacity of the soil, while increasing the thickness of its deformable layer in the direction of the resulting vehicle's impact [4, 5]. As a result, there is an increase in the total deformation of the soil. Unfortunately, to date, issues of calculating rutting taking into account the ratio of normal and tangential loads have not been fully developed; relations are known that allow qualitative taking into account the influence of an increasing resulting load, but its value is calculated using mostly graphoanalytically [5, 6], which complicates implementation of modelling interaction of vehicles with forest soils.

The aim of the study is to develop and analyze dependencies to calculate the ratio of the normal pressure of the mover of a forest machine and the shear stress caused by its slipping. The study uses methods of mathematical analysis and applied mathematics for calculations and numeric data processing. The calculations were performed using Maple 2017 software.

IOP Conf. Series: Materials Science and Engineering **817** (2020) 012015 doi:10.1088/1757-899X/817/1/012015

2. Results and Discussion

The shear stress distributed over the contact patch is determined by the formula [6]:

$$\tau_x = \left(\frac{1}{\tau_m} + \frac{t_g}{Gj} \right)^{-1}, \tag{1}$$

where, t_g is the grousers spacing, j is the soil shear strain, τ_m is the maximum soil shear resistance, G is the soil shear modulus.

The maximum soil shear resistance will be found by the formula [6]:

$$\tau_m = q \operatorname{tg}\varphi + C\xi, \tag{2}$$

where q is the average normal pressure along the contact patch, C, φ are inner cohesion and the angle of internal friction of the soil, respectively, ξ is a parameter that takes into account the decrease in shear resistance when cutting the soil occurs [6]:

$$\xi = \left(1 - \frac{j}{t_g}\right) \text{Heaviside}\left(\frac{j_0 - j}{t_g}\right), \tag{3}$$

where, Heaviside (...) is the Heaviside function, j_0 is the shear deformation at which the soil cut occurs [6]:

$$j_0 = \frac{\tau_{m0} t_g}{G}\left(\left(\frac{G}{C}\right)^{0.5} - 1\right), \tag{4}$$

where, τ_{m0} is the shear resistance without taking into account a possible cut of the soil [6,7]:

$$\tau_{m0} = q \tan\varphi + C. \tag{5}$$

The shear strain is usually determined taking into account horizontal coordinate x, counted from the beginning of the contact patch along its length [4]:

$$j = Sx, \tag{6}$$

where, S is the slip ratio.

Average value of the shear stress distributed along the contact patch, we find as the integral [4, 5, 7]:

$$\tau = \frac{1}{l}\int_0^l \tau_x dx. \tag{7}$$

The resulting load on the soil p and the angle of its deviation from the normal to the bearing surface β will be found by following formulas [1, 4, 5]:

$$p = \sqrt{q^2 + \tau^2}, \tag{8}$$

$$\beta = \arctan\frac{\tau}{q}. \tag{9}$$

Figure 1 shows calculation results for the average tangential stress according to formula (7) for three categories of forest soil with the mover parameters: q = 0.05E (approximately corresponds to half of the bearing capacity of forest soil p_s), d = 1.333 m, $l = d / 2$ (indicative values for a wheeled forestry mover [8]). For strong forest soil (I) E = 3 MPa, for a bearing forest soil (II) E = 1 MPa, for a weak forest soil (III) E = 0.4 MPa [9-12]. Hereinafter, we will evaluate the physic-mechanical properties of the soil using the formulas [9-12]:

$$C = 10.774E^{0.7737}, \tag{10}$$

$$\varphi = 13.669E^{0.1818}, \tag{11}$$

$$G = 0.244E^{0.117}. \tag{12}$$

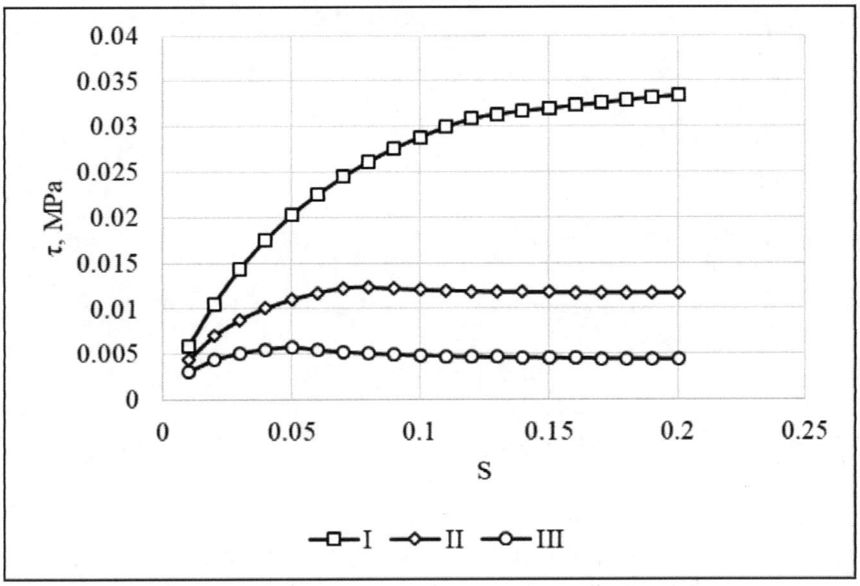

Figure 1. The average tangential stress along the contact patch.

Figure 2 shows the results of calculating the angle of deviation of the resulting load from the normal to the bearing surface.

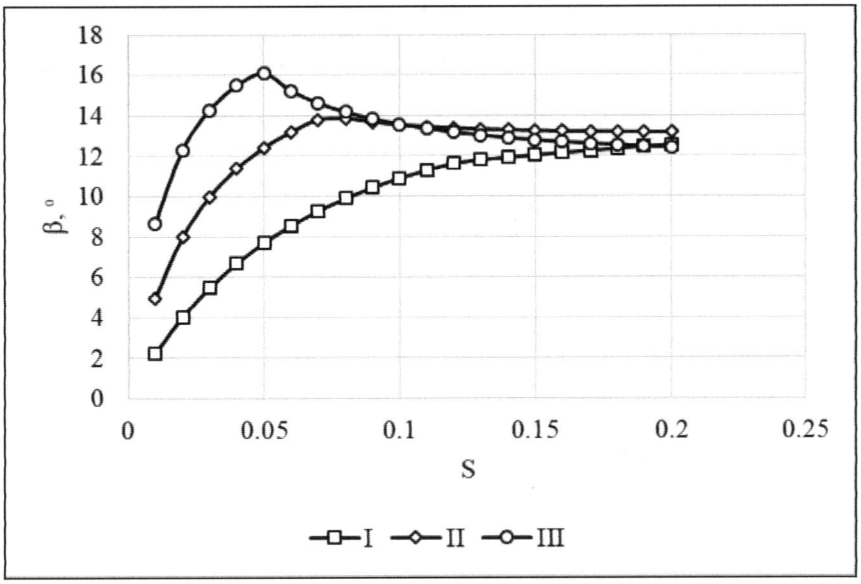

Figure 2. The angle of deviation of the resulting load from the normal to the bearing surface.

Figure 3 illustrates the ratio of the average tangential stress and the normal pressure along the contact patch.

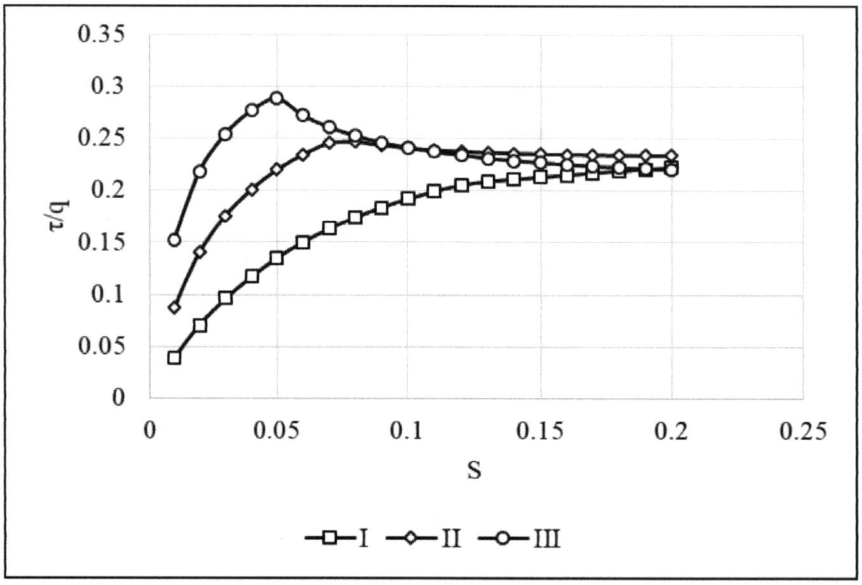

Figure 3. The ratio of the average tangential stress and the normal pressure along the contact patch.

Calculations show that the percentage of shear stress on the contact patch caused by slipping of the mover to normal pressure on strong forest soil is an average of 17%, reaching 22%; on a bearing forest soil the ratio averages 22%, reaching 25%; on a weak forest soil, the ratio of shear stress and normal pressure is on average 24%, reaching 29% at the beginning of the soil layers cutting.

We express the slip ratio corresponding to the beginning of a decrease in shear resistance due to the occurrence of the soil cut, according to formulas (3) - (6), then:

$$S = \frac{1}{l} \cdot \frac{t_g (q \tan\varphi + C)}{G} \left(\left(\frac{G}{C} \right)^{0.5} - 1 \right). \tag{13}$$

Note that the slip ratio according to formula (13) on a weak forest soil and on a bearing forest soil corresponds to the loss of vehicle passability, since the shear resistance begins decreasing. According to calculations, on strong forest soils after the beginning of its cut, shear resistance does not change significantly.

For S, limited by formula (13), integral (7) has the form:

$$\tau = \frac{G}{2t_g (G-C)^2} \left(SlC^2 - CGlS + 2Gqt_g \operatorname{tg}\varphi + 2CGt_g \right) - \frac{G^2 t_g}{Sl (G-C)^3} \left(C^2 - 2Cq \operatorname{tg}\varphi - q^2 + \frac{q^2}{\cos^2\varphi} \right)$$

$$* \ln\left(1 + Sl \frac{G-C}{t_g q \tan\varphi + Ct_g} \right). \tag{14}$$

Substituting expression (13) into formula (14), we obtain the shear stress corresponding to the beginning of the cut of the soil layers due to slipping of the mover:

$$\tau = (q \tan\varphi + C)\zeta, \tag{15}$$

where, ζ is dimensionless parameter determined by the shear properties of the soil:

$$\zeta = \frac{-G^3 \ln\left(\frac{C}{G} + \left(\frac{G}{C}\right)^{0.5} - \left(\frac{C}{G}\right)^{0.5}\right) - (G-C)\left(\left(C^2 - CG - G^2\right)\left(\frac{G}{C}\right)^{0.5} - \frac{C^2}{2} + \frac{3G^2}{2}\right)}{(G-C)^3\left(\left(\frac{G}{C}\right)^{0.5} - 1\right)}. \quad (16)$$

Basing on formula (9), taking into account (15), (16), we obtain expressions for the angle of deviation of the resulting load from the normal to the bearing surface at the beginning of the soil cut:

$$\beta = \operatorname{arctg}\left(\left(\operatorname{tg}\varphi + \frac{C}{q}\right)\zeta\right). \quad (17)$$

We note that formulas (15) - (17) are invariant with respect to geometrical parameters of the contact patch, but it should be mentioned that the value of the slip ration corresponding to the beginning of the cut of the soil depends on the length of the contact patch and the grousers spacing t_g, formula (13).

The results of calculations by formula (16) are presented in figure 4.

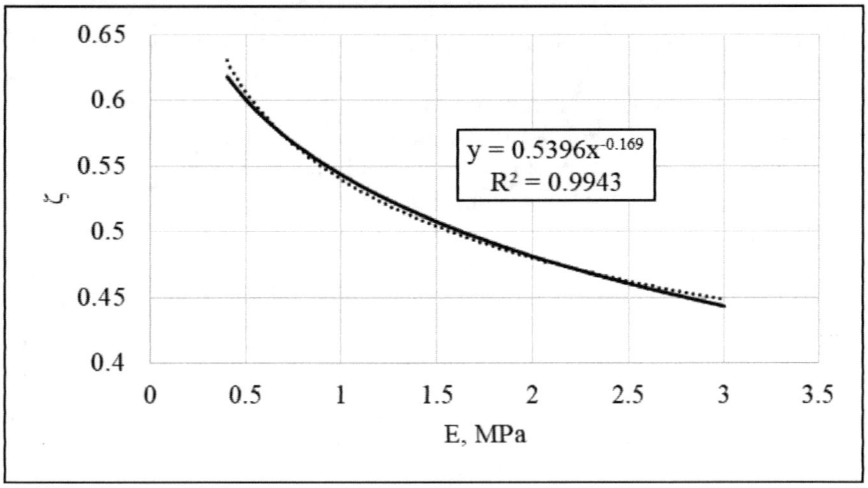

Figure 4. The parameter ζ in the equation for calculating the average tangential stress corresponding to the beginning of the cut of soil layers.

Taking into account expressions (10) - (12), for a forest soil, the dimensionless parameter as a function of the angle of deviation of the resulting load, determined by the shear properties of the soil, is almost exactly determined by the value of the total strain modulus E:

$$\zeta = 0.54 E^{-0,17}. \quad (18)$$

The estimation of the deviation angle at the beginning of the cut of soil layers is illustrated in figure 5 (the calculation was performed at q = 0.05E).

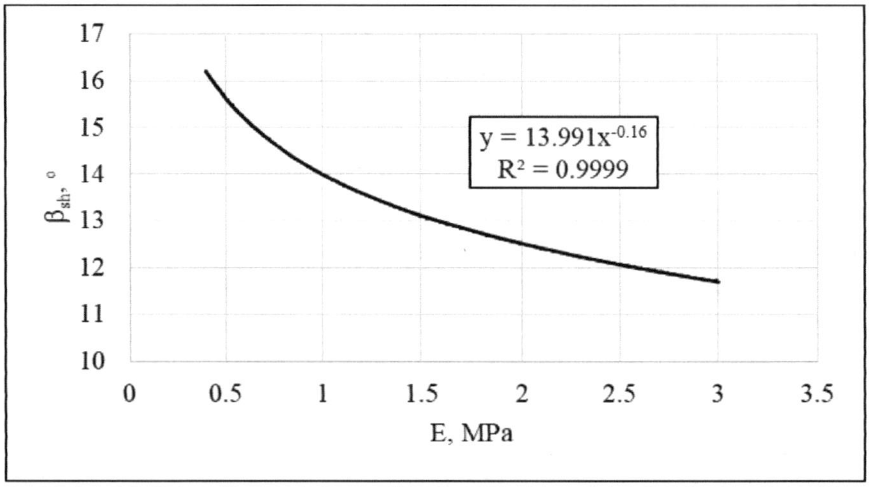

Figure 5. The angle of deviation of the resulting load from the normal to the bearing surface at the beginning of the cut of the soil.

As the graph shows, that the deviation of the resulting load from the normal to the bearing surface, at which the cut of the soil layers begins, is 12 - 16 °. This range can be considered as a limit, since with a greater slip ratio, the shear resistance of a weak and bearing forest soil decreases, while the resistance of a strong forest soil changes slightly (figure 1).

Let us evaluate the effect of the deviation of the resulting load from the normal on the bearing capacity of the soil. The bearing capacity is calculated by the formulas [5]:

$$\begin{cases} p_s = p_{s0}\alpha_Z \\ p_{s0} = 0{,}5K_1K_{\beta 1}N_1\gamma b + N_2\gamma h + K_3K_{\beta 3}N_3C \\ K_1 = \dfrac{l}{l+0{,}4b};K_3 = \dfrac{l+b}{l+0{,}5b} \\ K_{\beta 1} = \dfrac{L}{L+0{,}4B};K_{\beta 3} = \dfrac{L+B}{L+0{,}5B} \\ N_1 = \dfrac{1-S^4}{S^5};N_2 = \dfrac{1}{S^2};N_3 = \dfrac{2\left(1+S^2\right)}{S^3};S = \mathrm{tg}\left(\dfrac{\pi}{4}-\dfrac{\varphi}{2}\right) \\ \alpha_Z = 1 + \dfrac{H_\varphi h}{2\dfrac{H}{\cos\beta}\cdot\left(\dfrac{H}{\cos\beta}-h-0{,}25H_\varphi\right)};H_\varphi = 0.707\exp\left[\left(\dfrac{\pi}{4}+\dfrac{3\varphi}{4}\right)\tan\dfrac{3\varphi}{4}\right]b\cos\dfrac{3\varphi}{4}tg\varphi \end{cases}, \quad (19)$$

where, p_{s0} is the bearing capacity of the soil with unlimited thickness of the deformable layer, α_Z is the coefficient taking into account the thickness of the deformable layer of the soil, $K_{\beta 1}$, $K_{\beta 3}$ are coefficients that take into account the angle of deviation of the resulting load from the normal to the bearing surface, K_1, K_3 are coefficients that take into account the contact patch geometry , N_1, N_2, N_3 are coefficients that take into account the internal friction of soil particles, S, H_φ are auxiliary notations, H is the thickness of the deformable layer of soil, γ is the specific gravity of the soil [9-12]:

$$H = 0.4714E^{-0.479}, \quad (20)$$

$$\gamma = 0.0084E^{0.1168}. \quad (21)$$

The calculation results according to formulas (19) for β according to expression (17) and for $\beta = 0$ are presented in figure 6 (calculations were performed at $h = 0.5$H).

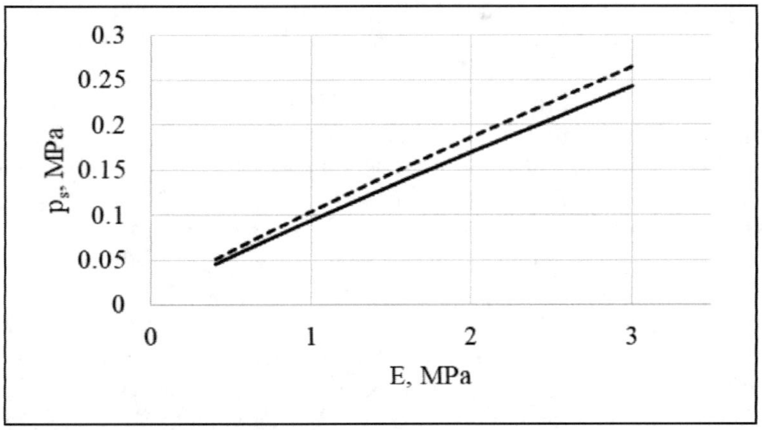

Figure 6. Bearing capacity of forest soil (solid line – no deviation of resulting load from the normal to bearing surface; dash – deviation of the resulting load).

Analysis of the calculation results shows that the deviation of the resulting load from the normal to the bearing surface caused by slipping of the mover leads to a decrease in the bearing capacity of the forest soil by 8.5-10.5%.

3. Conclusion

Analysis of results of calculating indicators of interaction of a wheeled mover with forest soil before a start of cutting the soil layers, subsequent loss of the mover trafficability with cutting the bearing surface with the grousers, allowed the following conclusions:

– Calculations show that the percentage of shear stress on the contact spot caused by slipping of the mover to normal pressure on strong soil is an average of 17%, reaching 22%; on a bearing forest soil, the ratio averages 22%, reaching 25%; on a weak forest soil, the ratio of the shear stress and the normal pressure is on average 24%, reaching 29% at the beginning of the soil layers cutting.

– Deviation of the resulting load from the normal to the bearing surface, at which the soil layers cutting begins, is 12 - 16 °. This range can be considered as a limit, since with a greater slip ratio the shear resistance of a weak and bearing forest soil decreases, the resistance of a strong forest soil changes slightly (figure 1).

– The deviation of the resulting load caused by slipping of the mover leads to a decrease in the bearing capacity of the forest soil by 8.5-10.5%.

References

[1] Bozhbov V E 2015 *Increasing skidding process efficiency by justifying wheeled forwarders' workload* [in Russian – Povyshenie effektivnosti processa trelevki putem obosnovaniya rejsovoj nagruzki forvarderov], PhD thesis: 05.21.01. Arhangel'sk

[2] Porsinsky T, Pentek T, Bosner A and Stankic I 2012 Ecoefficient timber forwarding on lowland soft soils *In Global Perspectives on Sustainable Forest Management.* Ed C A Okia; (Rijeka InTech) pp 69-69

[3] Jarkko L 2013 *Design parameter analysis of the bogie track surface pressure in peatland forest operations*, master of science thesis (Tampere)

[4] Khakhina A M 2018 *Methods for predicting and increasing passability of wheeled forestry machines* [in Russian – Metody prognozirovaniya i povysheniya prohodimosti kolesnyh lesnyh mashin], Doctoral thesis: 05.21.01. Arhangel'sk

[5] Larin V V 2007 *Methods for predicting the support ability of multi-axle wheeled vehicles in the terrain* [in Russian – Metody prognozirovaniya i povysheniya opornoj prohodimosti mnogoosnyh kolesnyh mashin na mestnosti], Doctoral thesis: 05.05.03 Moscow

[6] Agejkin Ya S 1981 *Passability of vehicles* [in Russian – Prohodimost' avtomobilej] Moscow p 232

[7] Khitrov E, Andronov A, Iliushenko D and Kotenev E 2019 Comparing approaches of calculating soil pressure of forestry machines *Int. Multidisciplinary Scientific GeoConference Surveying Geology and Mining Ecology Management, SGEM 19* (vol 19) issue 3.2 pp 649–655 DOI: 10.5593/sgem2019/3.2/S14.084

[8] Saarilahti M 2002 *Modelling of the wheel and tyre. Tyre and soil contact – Survey on tyre contact area and ground pressure models for studying the mobility of forest tractors Soil interaction model*, Appendix Report No.5 pp. 1-43

[9] Grigorev I V 2006 *Reducing the negative impact on the soil of wheeled skidders by substantiating their movement modes and technological equipment* [in Russian – Snizhenie otricatel'nogo vozdejstviya na pochvu kolesnyh trelevochnyh traktorov obosnovaniem rezhimov ih dvizheniya i tekhnologicheskogo oborudovaniya] Saint-Petersburg

[10] Khitrov E G, Andronov A V, Martynov B G and Spiridonov S V 2019 Interrelations of various soil types mechanical properties *IOP Conf. Series: Journal of Physics: Conf. Series* **1177** 012032 DOI: 10.1088/1742-6596/1177/1/012032

[11] Kochnev A and Khitrov E 2018 Theoretical models for rut depth evaluation after a forestry machine's wheel passover *Int. Multidisciplinary Scientific GeoConference Surveying Geology and Mining Ecology Management, SGEM 18* pp 1005–12 DOI: 10.5593/sgem2018/3.2/S14.128

[12] Peskov V B 2018 *Improving models for assessing rutting and compaction of soil under the influence of movers of wheeled forest machines* [in Russian – Sovershenstvovanie modelej dlya ocenki koleeobrazovaniya i uplotneniya pochvogruntov pod vozdejstviem dvizhitelej kolesnyh lesnyh mashin], PhD thesis Arkhangelsk

Bases of transport development organization of leased forest plots taking into account climate

T Kovalenko[1*], N Vokhmyanin[2], A Krivonogova[2]

[1]Department of Industrial Transport, *Saint Petersburg State Forest Technical University*, 5 Institute lane, Saint Petersburg 194021, Russian Federation
[2]Department of applied mechanics and engineering graphics, *Saint Petersburg State Forest Technical University*, 5 Institute lane, Saint Petersburg 194021, Russian Federation

*Corresponding email:kovalenko.stl@spbftu.ru

Abstract. The article raises the issues of zoning of rental forests taking into account climatic factors for the purpose of rational transport development of forests. The basis of the methodology for taking into account the influence of climatic factors in the organization and planning of forest transport processes through the average speed of timber trucks is proposed. The results of studies are given that allow taking into account the climatic parameters of the area where the enterprise is located.

1. Introduction

Insufficient provision of rental forests with transport network continues to be one of the key problems of the modern forests of Russia. The solution of this question is the most important stage in the program of work of the modern timber processing enterprise. According to the existing forest legislation [1] the enterprise must have a forest development plan, the preparation of which is impossible without the development of an optimal transport network of forest roads. Without taking local conditions into account as fully as possible, the transport development strategy seems unlikely.

One of the factors that have the greatest impact on forest roads is the climate. A significant part of the forest resource bases of logging enterprises is located in areas with unfavorable soil and hydrological conditions, which leads to seasonality in the work of logging and forest transportation sites of enterprises. Thus, the identification of the degree of climaticinfluence factors on the processes of logging and forest transport operations is a very important and urgent problem, especially in recent years, when abnormal climate fluctuations cause significant interruptions in the work of enterprises, associated primarily with the late onset of frost or heavy snow. Abnormally hot summer, in turn, leads to an increase in the number of forest fires, prevention and control of which are also the problem of tenants.

2. Methods and Materials

2.1. The main provisions of the optimization of forests transport development taking into account climatic factors

The density of the transport network of a logging enterprise is traditionally considered one of the most important indicators of its production activities. In modern realities, this parameter continues to remain significantly less than the normative values, which is due to both economic and legislative difficulties. On the other hand, the transport component of the enterprise is increasingly beginning to have an unstable climatic situation observed in recent years. Therefore, when solving the problems of optimizing the location of transport routes in rental forests, one should pay attention not only to traditional restrictions, such as soil and hydrological conditions of the area, the availability of local road-building materials, technological features of forest transport, but also to the climate of the region where this company is located.

Transport and technological processes are carried out in the open air, being under the direct influence of all climatic factors and weather phenomena. Climate and weather significantly affect both the state of the driving surfaces on which the forest machines move, the components, assemblies and systems of the machines themselves, as well as the emotional and psychological state of the driver (operator). In addition, weather and climate determine not only seasonal changes in the subject of labor (whole tree), but also the conditions of the logging and transport process, which determines its safety and effectiveness. In order to organize transport and technological processes in such a way as to ensure efficient and safe execution of all operations in different weather and climate conditions, it is necessary to understand and evaluate these effects and dependencies.

The theory of organization of logging and transport processes taking into account the influence of climate is a qualitative and quantitative description of the nature and patterns of changes in the parameters and characteristics of processes occurring under the influence of climatic factors. Using the methodology of the system approach, the structural diagram of the system can be represented in the form of the following subsystems: "environment" - "operator (driver)" - "forest machine" - "driving surface".

The peculiarity of the impact of weather and climatic factors on the functioning of the logging and transport system, shown in table 1, is that one part of the meteorological elements simultaneously affects all or several subsystems, while the other only affects some of them.

Table 1. The impact of weather and climate factors on the cutting and transport system.

Weather and climate factors	Operator (Driver)	Forest machine (mechanism)	The object of labor (whole tree, stem)	Driving surface (road)	Performance indicators
Atmospheric pressure	P_t, t_r, V_i	N_e, q	-	-	C_t
Temperature	P_t, t_r, V_i	N_e, q, U, K_s	t_s, A_p	ω, φ	T_c, C_t, C_r
Humidity	P_t, t_r, V_i	N_e	t_s	ω, φ	C_t, C_r
Visibility	P_t, t_r, V_i	W, U, K_s	K_r	S_v	T_c, C_t, C_r
Rainfall	P_t, t_r, V_i	W, U, K_s	K_r	S_v, ω, φ	T_c, C_t, C_r
Snow	P_t, t_r, V_i	W, U, K_s	K_r	S_v, ω, φ	T_c, C_t, C_r
Wind	P_t, t_r, V_i	U, K_s	K_r	ω	T_c, C_t, C_r
Snowstorm	P_t, t_r, V_i	W, U, K_s	K_r	S_v, ω, φ	T_c, C_t, C_r
Ice	P_t, t_r, V_i	U, K_s	K_r	φ	C_t, C_r
thunderstorm	P_t, t_r, V_i	K_s	K_r	-	T_c, C_t

where, P_t – psychological tension, t_r – reaction time, V_i – information processing speed, N_e– engine power, q – fuel consumption, U – stability, controllability,K_s– security,W – visibility,t_s – storage time,A_p– specific work of cutting and sawing, K_r– roll safety,ω– resistance to movement,φ– adhesion coefficient,S_v – visibility distance,T_c – continuity,C_t – transport component of cost,C_r– road component of cost.

At the end of the 70s at the Department of Land Transport Forest of Forest Engineering Academy Professor B AIlyinwith his team began research work aimed at developing the applied aspects of logging meteorology [2]. The main objective of these studies was the development and justification of logging technologies and transport operations, which make it possible to take into account and effectively use the weather and climatic conditions of the region, as well as to reduce damage from dangerous meteorological phenomena.

As a result of the initial analysis, it was found that the climatic factors have the greatest impact on the forest transport process, which primarily determines the seasonality of logging. In turn, in the forest transport process, the most dependent on weather and climate are secondary roads the performance of which is determined, first of all, by soil moisture and its deformation modulus. Therefore, to improve the efficiency of forest dirt roads, the rational organization of the entire forest transport process, knowledge of the laws governing changes in soil moisture and their water-thermal regime is necessary.

In the forest industry, up to the present time, there are no long-term systematic observations of soil moisture of forest roads. However, agroclimatology has accumulated a large long-term statistical material of observations of soil moisture in the open field. These data, as can be seen from the studies [2], can also be used to determine the estimated soil moisture on summer dirt roads.The possibility of such an approach is based on a close correlation between the soil moisture of the open field and the soil moisture of the secondary roads soil bed at their location in the same soil-hydrological conditions. The difference will be only in the absolute values of soil moisture in the field and in the roadbed, due to the presence of drainagesystem and construction of the canvas. This difference can be taken into account by the transition correlation coefficient from soil moisture in the field to soil moisture of the subgrade.

When creating an optimization system logging processes of development of forests taking into account the influence of climatic factors is most appropriate to start with zoning of the territory of the forest resource base and forest transport routes, taking into account seasonality. It is necessary to take into account not only the requirements for transport, but also to take into account the features of logging [3], to ensure the continuity of the chain "logging - forest transport".

2.2. Basic principles of forest zoning

As the main principles of the enlarged zoning of forest areas in their transport development, the following principles can be specified:

• fragmentation of the forest by insurmountable obstacles in the form of railways not equipped with crossings, pipelines, water protection zones, etc;

• allocation of zones of unconditional winter transport, determined by obstacles, that can be overcome only in the winter period. These include large rivers crossing the forest, where there are no year-round crossings, long swamps and lakes;

• allocationof seasonal removal zones.

As already mentioned, the seasonal zoning of transport accessibility should be linked to the seasonal planning of logging operations [4], thus there is a possibility of close synchronization of harvesting and transportation of wood, which allows for a more rhythmic supply of timber to the enterprise and reduce losses from long-term storage of wood at forest terminals [5].

Depending on the bearing capacity [4], forest soils can be divided into 4 groups, on the basis of which there are four zones of seasonal transport accessibility according to soil conditions:

I. only winter timber removal;

II. transportation in the winter and summer;

III. transportation in the winter and summer-autumn period;

IV. year-roundtransportation.

In the spring, it is most expedient to transport the timber along year-round roads or work from stocks collected at terminals on paved roads and accumulated during winter harvesting, as such wood is less prone to spoilage andquality decrease.

It is preferable to conduct territorial zoning according to seasonal transport accessibility for the entire rental area in order to draw up a development plan for the entire rental period.

Thus, when setting the mentioned task of transport optimizationit is necessary to rely on the following key points:

• The planning period is 10 years.
• The iterative step in planning is 1 month.
• The forest is considered as fragmented.
• The nodes of the transport network are forest terminals, organized at highways and branches.
• Year-round and seasonal branches, highways and public roads are considered as components of the transport network.
• The transport network is "pulsating".
• The relevance of the road is determined by the cost of transportation, consisting of road, transport and climate components.

2.3. Accounting speed parameters in the system to optimize the transport of forest development

The speed of the timber truck [6, 7] can act as a central indicator of the state of the road and be used as a criterion in systems for optimizing forest transport processes in the development of leased forest.

For a complex assessment of influence of various climatic factors, the average annual security factor of the estimated speed should be used [8]. Taking as a basis the data of climate directories, it is necessary to obtain the probability of occurrence of meteorological phenomena. The calculated dependence for determining the average annual security factor of the estimated speed in this case is as follows (1):

$$\overline{K}_{as} = \int_x K_{cs}(x) \cdot A_T \cdot P(x) dx, \quad (1)$$

where, $K_{cs}(x)$ –the coefficient of security of the calculated speed of the meteorological factor$x$;A_T – time action meteorological factoroperator; $P(x)$ –probability of action of the meteorological factor x.

The greatest influence of climatic factors is affected by forest roads with transitional type of coating, as well as secondary roads [1, 9]. During the period of increased moisture, such roads, especially with a dirt surface, are prone to rutting.

Thus, the speed indicator is advisable to use to assess the degree of patency of forest roads in real time. Having established the relationship between the high-speed regime of the forest road and climatic factors, we get the opportunity to quickly respond to the negative effects of climate, timely restricting movement. On the other hand, it becomes possible to predict the possible state of the transport network, having on hand an accumulated array of speed data (for both light and heavy road trains, taking into account the road structures used for transportation) and the climatic parameters of the region, by establishing possible critical periods when there is a danger of the destruction of the road by timber trucks.

The basis of the proposed methodology was the study of the data of the waybills of the timber road trains operated by OAO Timber Industry Complex Kipelovo (The Vologodskaya region). Records of the local weather station (atmospheric pressure, air humidity, rainfall, wind speed, average daily air temperature) and the main climatic parameters for the VologodskayaRegion (albedo, average soil temperature) were used as a source of meteorological data [10, 11].

The studied road trains were divided into two groups: "Group 1" (based on MAZ typevehicles) and "Group2 " (based on Ural type vehicles). The roads were divided into the following categories: 1 - year-round main roads; 2 - branches of summer action; 3 - branches of winter action.

Using correlation analysis in the STATGRAPHICS system, the factors that have the strongest influence on the average speed of movement were selected. These include: time factor, road category, surface albedo, average soil temperature and average air temperature.

Based on the results of the analysis, the following conclusions can be drawn about the influence of these climatic factors on the average speed:

1. It is advisable to consider separately the model for timber trucks of Group 1 and Group 2;

2. A steady increase in the average speed for timber trains of Group 1 is observed in the warm period of the year (from May to September), with a significant decline in the autumn period and a slight increase in the winter months (figure 1);

3. For timber trucks of Group 2, a relatively stable average speed is observed during the period from November to February, with a significant decline in April and a slight increase in the summer months (figure 2);

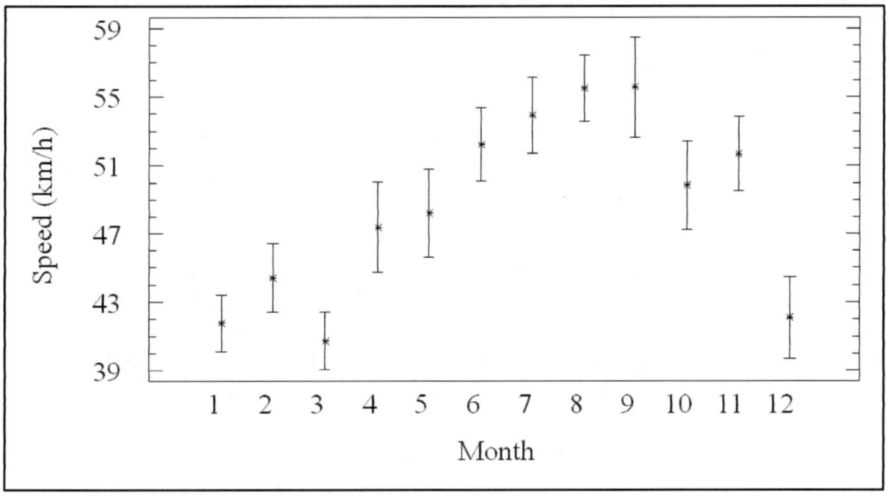

Figure 1. Change in the average speed of the group 1 timber trucks during the year.

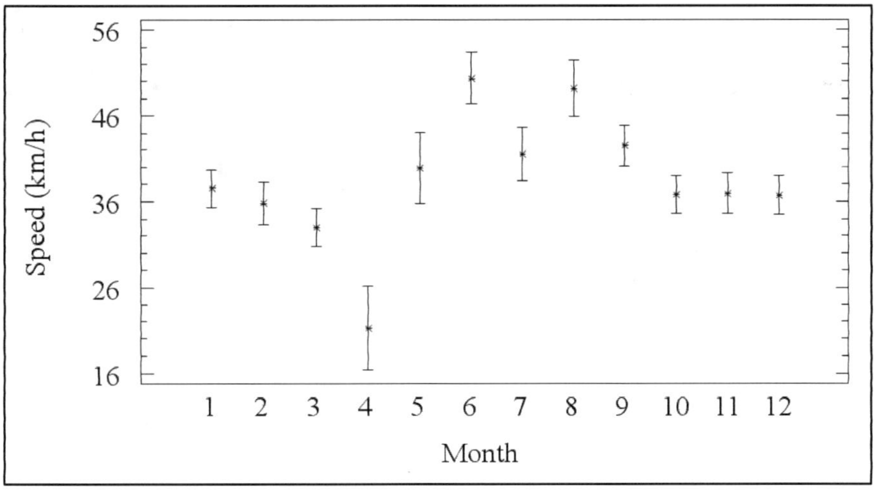

Figure2. Change in the average speed of group 2 timber trucks during the year.

4. Timber trucks of the 1st group tend to decrease the average speed depending on the simplification of the road structure, while for road trains of the 2nd group the speeds of movement along highways and "winter roads" differ slightly, and a noticeable decrease in the average speed is clearly visible for the summer branches;

5. With an increase in the average soil temperature, the average speed of the Group 1 timber trucks increases more intensively than that of the group 2 timber trucks, which, in turn, clearly show a decline in the average speed at about zero temperatures.

6. With the increase in the average air temperature, the average speed of the Group 1 timber trucks changes little, however, when moving to a positive temperature range, a jump in the average speed is clearly noticeable, which is obviously due to a changein the vehicle's adhesion to the road surface;

7. With an increase in the average air temperature, in the sub-zero temperature range, the average speed of Group 2 timber trucks gradually decreases. When crossing 0° C, a gradual increase in the average speed of the trains.

At the next stage of the research, equations were obtained that describe the influence of climatic factors.

- for Group 1 timber trucks (km/h):

$$V_1 = 0.162814 \cdot a + \frac{35.8485}{k_r} + 0.0443855 \cdot t_s^2 - 0.025966 \cdot t_a^2 \quad (2)$$

- for Group 2 timber trucks (km/h):

$$V_2 = 0.192589 \cdot a + \frac{26.5996}{k_r} + 0.0409256 \cdot t_s^2 - 0.0412448 \cdot t_a^2 \quad (3)$$

where, a – surface albedo, %;k_r – coefficient taking into account the category of (1.0 - highways; 2.0 - summer branches; 3.0 - winter branches); t_s–average soil temperature, °C; t_a – average air temperature, °C.

3. Conclusions

The theory of organizing logging and transport processes taking into account the influence of climate should be based on a qualitative and quantitative description of the nature and patterns of changes in the parameters and characteristics of logging and transport processes occurring under the influence of climatic factors.

The greatest impact of meteorological factors on the functioning of the logging and transport system is carried out through the interaction of the machine with the driving surface. The most dependent on the weather and climate are secondary roads, the performance of which is determined primarily by the moisture content of the soil and its deformation modulus.

To improve the efficiency of forest dirt roads, rational organization of the entire forest transport process requires knowledge of the laws of soil moisture changes, their water-thermal regime. Statistical processing of observations of soil moisture in different climatic zones and weather conditions will allow to establish the regularity of changes in moisture reserves for the year.

Seasonal zoning of transport accessibility of rental forests should be linked with seasonal planning of logging operations, thus there is a possibility of close synchronization of harvesting and transportation of wood.

The speed of the timber truck can be used as a criterion in systems for optimizing forest transport processes in the development of leased forests and act as a central indicator of the state of the forest road.

It is expedient to consider the differentiated models of the account of influence of climatic factors on average speed of the movement of the timber carrier, depending on its group and category of the forest road.

When creating optimization systems for the transport development of forests, it is necessary to take into account the influence of climatic factors through the average speed of trains, comparing it with the normative established for this section of the road, which makes it possible to judge the state of the road structure and allow timely removal of the forest road from operation.

References

[1] Federal Law Forest Code of the Russian Federation 200-FZ from 04.12.2006

[2] Tyurin N 1994 Design of logging transport processes taking into account the impact of climate. / *Land transport of forest* [in Russian –Proektirovanie lesosechno-transportnyh processov s uchetom vliyaniya klimata](Saint Petersburg: SPbFTA) pp 53-57

[3] Kosarev V 1991 Forest meteorology. [in Russian – Lesnaya meteorologiya] (Moscow: Ekologiya) p 176

[4] Zasuhin D 2004 Recommendations on protection of forest soils from damage during logging operations in the Republic of Komi [in Russian – Rekomendacii po zashchite lesnyh pochv ot povrezhdeniya pri provedenii lesozagotovitel'nyh rabot v RespublikeKomi] (Syktyvkar)

[5] Shchepin B 1976 The effect of the storage time of stems on the quality of round timber *Forest exploitation and timber smelting* [in Russian – Vliyanie prodolzhitel'nosti hraneniya hlystov na kachestvo kruglyh lesomaterialov] (Moscow: Lesoekspluataciya I lesosplav) pp 11-12

[6] Vasil'ev A 1976 Road conditions and traffic safety in difficult weather conditions [in Russian – Sostoyanie dorogi bezopasnost' dvizheniya v slozhnyh pogodnyh usloviyah] (Moscow: Transport) p 224

[7] ODN 218.0.006-2002Rules of diagnostics and assessment of the state of roads *State road service (Rosavtodor) of the Ministry of transport of Russia* [in Russian – Pravila diagnostiki i ocenki sostoyaniya avtomobil'nyh dorog: ODN 218.0.006-2002] (Moscow: Informavtodor) p 138

[8] Vasil'ev A 2004 Reference encyclopedia of the road Builder. Repair and maintenance of roads [inRussian – Spravochnaya enciklopediya dorozhnika. Remont i soderzhanie dorog] (Moscow: Transport) vol 2p 507

[9] Kovalenko T 2009 The account of climatic factors at the organization of forest transport processes in the logging enterprise *Land transport of forest: Materials of scientific and technical conference* [in Russian – Uchet klimaticheskih faktorov pri organizacii lesotransportnyh processov v lesozagotovitel'nom predpriyatii] (Saint Petersburg: Saint Petersburg State Forest Technical Academy) pp 105-111

[10] Reference to the climate of the USSR, Issue 1, The Arkhangelsk and Vologda regions. Karelian and Komi Autonomous Soviet Socialist Republic, Part 1, Solar radiation, radiation balance and sunshine. [in Russian – Spravochnik po klimatu USSR. Arhangel'skayai Vologodskaya oblasti. Karel'skaya I Komi ASSR. Solnechnaya radiaciya, radiacionnyj balans I solnechnoe siyanie] (Leningrad: Gidrometizdat) p 97

[11] Reference to the climate of the USSR, Issue 1, The Arkhangelsk and Vologda regions, Karelian and Komi Autonomous Soviet Socialist Republic. Part 2 Air and soil temperature[in Russian – Spravochnik po klimatu SSSR. Arhangel'skaya i Vologodskaya oblasti. Karel'skaya I Komi ASSR. Temperatura vozduha I pochvy] (Leningrad: Gidrometizdat) p 361

Assessment of infrastructure of forest land

V Kovyazin*, A Romanchikov, A Kitcenko

Department of Engineering Geodesy, *Saint-Petersburg Mining University*, 2, 21 line of Vasilyevsky Island., St. Petersburg 199106, Russian Federation

*Corresponding email: vfkedr@mail.ru

Abstract. The article discusses the main directions of forest policy in the field of creating forest fund infrastructure, provides foreign experience in accounting for infrastructure and comparative indicators. An analysis of the transport infrastructure of the forest fund was carried out, as well as a methodology for assessing the infrastructure. Based on the data obtained, conclusions are drawn and ways to solve the existing problems in the forest sector are proposed.

1. Introduction

To date, the methodology for cadastral valuation of forest land is imperfect and has several disadvantages. The legal concept of the territory of the forest background does not coincide with the land allotment. Assessment of forest land is associated with the transfer of land to lease, to establish the exact rental rate, in some cases, to assess the pricing policy of the rental rights of the land of the forest fund, which directly include forests. The cost of forest land depends on many factors, one of which is the degree of development of forest infrastructure [1].

One of the significant drawbacks is that the state cadastral valuation of forest lands does not take into account the infrastructure of forest territories. Forest infrastructure involves the construction of forest roads, timber depots and similar facilities. In general, in the constituent entities of the Russian Federation, the degree of development of forest infrastructure, which includes: forest roads, fire reservoirs, bridges, hydraulic structures for draining the territory, etc., is very low today, despite the fact that it is necessary to ensure rational use, protection, protection, reproduction of lands of the forest fund.

Therefore, the issues of the correct accounting of forest infrastructure objects in the cadastral valuation of forest lands and, in general, the improvement of the cadastral valuation of forest lands are especially relevant. To take into account the infrastructure of the forest fund in the cadastral valuation of forest lands, it is necessary to first assess the infrastructure itself [2].

2. Methods and Materials

The formation and development of forest infrastructure is the basis of the socio-economic development of the regions. However, today the problems that exist in forestry are related to the forest fund infrastructure. Recent decades have seen a decline in production in the forestry industry. For example, in developed countries, much attention is paid to the development of infrastructure on forest lands (figure 1).

The high availability of forest roads in countries that are large loggers is due to the fact that the availability of forest resources helps to make thinning operations on time, which in turn increases the productivity of forests. T S Lobovikov conducted studies on the restoration of forest areas after intensive logging and fires, which showed that in Europe the forest area increased by 54% (spruce forests replaced by small- leaved forests), and the rental Soviet model of the economic mechanism in forestry adopted in Russia did not even provide compensation cut down coniferous forests [3]. According to A P Petrov, the current situation arose due to the neglect of the experience of other foreign countries in forest management [4].

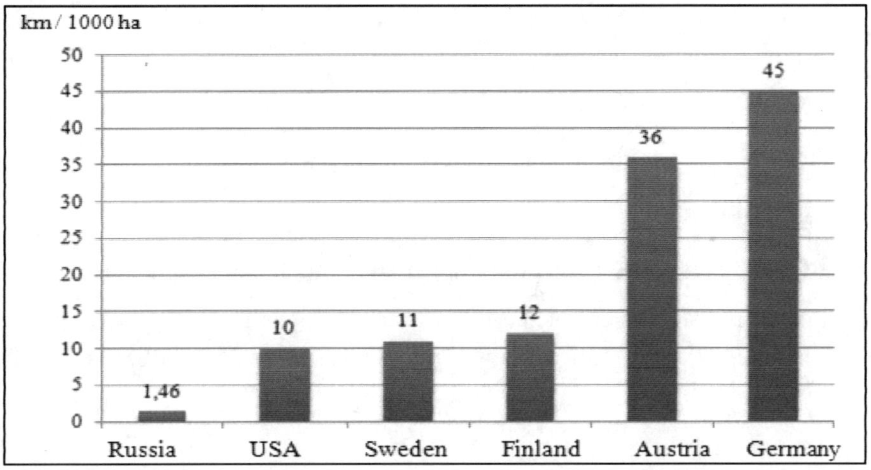

Figure 1. Forest roads in some countries of the world.

The main problems are insufficient financing (more than 60% of industrial funds are worn out) of the road transport network. A large area of forests of the Russian Federation is located in remote parts of the region, which increases the costs of production enterprises and the concentration of timber processing enterprises in specific areas. We can conclude that the insufficient development of the forest fund infrastructure affects the forestry economy.

One of the features of the forest sector of the economy is that it is dominated by large loggers (tenants), and small and medium-sized businesses are being forced out of this sector due to the fact that it is difficult for such enterprises to get additional investments for creating forest roads, forest conservation and protection. Including the creation of forest and timber processing infrastructure.

A poorly developed road network is the cause of incomplete forest development, depletion of forest resources in some regions and the inviolability of forest resources in other parts of the country. In many ways, the reason for the low developed road network is the natural features and landscape of the country's regions. However, in order to increase the accessibility of forests, the Strategy for the Development of the Forest Complex of the Russian Federation for the period until 2020 was adopted, which was aimed at ensuring the achievement of indicators to increase the density of the forest road network and the quality of their creation. On average, the need for the creation of new forest roads in the regions varies from 2.2 thousand km to 9.3 thousand km. When evaluating the transport network, it is necessary to consider not only the length and density of the network of motor roads, but also their quality. Due to the climatic characteristics of many regions, forest roads in some constituent entities of the Russian Federation become unusable after a season [5].

The profitability of the construction of the forest road is estimated by the optimal density of the road network, which is achieved by minimizing the cost of construction and operation of the road, as well as reducing the distance of transportation. Data on the required annual growth of forest roads are presented in figure 2.

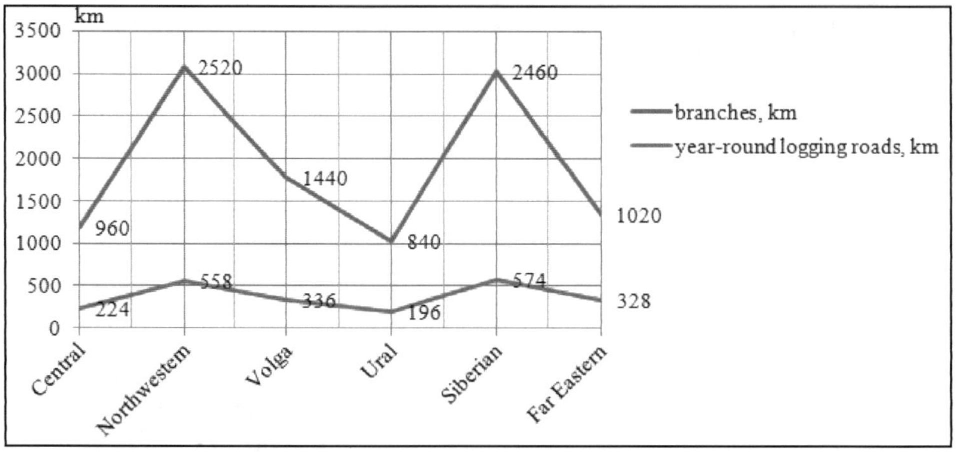

Figure 2. Forecasted need for an increase in the length of forest roads.

However, for an objective assessment of the need to create forest roads, it is necessary to study and analyze in detail other economic indicators, such as volume of cuttings, reforestation, in order to fulfill which the road network is mainly being created [5].

In addition, an important component is the economic income received from the forest sector. This aspect is significant as it provides priority funding for the development of forestry. The costs of financing the forest sector significantly exceed revenues from the use of forest land, forming a huge hole in the country's budget. Based on the analysis, it can be concluded that the current forest management system is inefficient and represents a loss-making industry for the Russian economic sector at the present stage [6]. To change the situation, a reorganization in the system of approaches to the assessment of forest resources, the distribution of responsibilities of tenants of forest plots and the state, as well as in the development of methods of control and responsibility for the distribution of material resources and the quality of the created infrastructure of the forest fund is required. Particularly affected by the lack of transportation routes and poor quality roads are the multi-forest areas where the main target raw materials are concentrated today.

A separate issue is given to the classification of infrastructure according to various criteria. The most common is the classification of infrastructure based on action. According to this classification, the infrastructure can be: industrial, social, transport, engineering, information, military, market, innovative, tourism, etc. [7]. Forest infrastructure is most often understood as forest roads, timber depots, and other structures and structures necessary for the effective functioning of forestry as a branch of production. According to Article 14 of the Forest Code of the Russian Federation [8], the forest processing infrastructure includes facilities for processing harvested wood and other extracted forest resources, bioenergy facilities. In addition to forest infrastructure and forest processing facilities, it is allowed to create objects in the forest fund that are not related to the creation of forest infrastructure.

To improve the methodology for cadastral valuation of forest lands, in this article, the authors propose to take into account the infrastructure of the forest fund, and also provides a methodology for assessing infrastructure [9].

The cadastral valuation model includes many infrastructure factors (the presence of forest infrastructure and land infrastructure to increase forest productivity, recreational use and the mineral resource complex and the purpose of forest use), which affect the cadastral value of forest land. After selecting the factors affecting the infrastructure of the lands of the forest fund, it is necessary to take these factors into account in the methodology for assessing forest lands, taking into account the degree of development of their infrastructure. The next stage of the assessment involves the establishment of a coefficient of infrastructure development. To establish the coefficient of infrastructure development, it is indispensable to link the previously defined indicators.

O V Kharionovskaya to assess the level of infrastructure development by regions, in the course of the study, she suggested using the following indicators: length of roads per 1 thousand hectares, forest fund area per forestry worker; number of non-governmental forestry enterprises. In the work of O V Kharionovskaya also proposed a ranking system for assessing infrastructure indicators. In our work, as a result of the correlation and regression analysis, the following indicators affect the cadastral value of forest lands to the greatest extent: the length of roads per 1 thousand hectares, the number of timber enterprises, and the annual costs of creating the forest fund infrastructure [9].

To calculate the coefficient of infrastructure development, the following methodology is proposed. Firstly, the territorial unit of the cadastral assessment should not be the subject of the Russian Federation or the forest tax region, but at least the region of the RF subject, the forest tax region, in the best case, if the initial data are sufficient, the territory of the forestry.

For each indicator, a rank is determined (r_r, r_c, r_{fe}, respectively), calculated as a fraction of its maximum value, which characterizes the position of the region in the distribution of the indicator's values in the general series (formula 1). In this case, the application of the normalized coefficient in the best way reflects the relationship between the values:

$$r_i = \frac{x_i - x_{\min}}{(x_{\max} - x_{\min})} \quad (1)$$

where, r_i – rank of infrastructure indicator (r_r, r_c, r_{fe}); x_i – the value of the infrastructure indicator for the district; $x_{\max}$ – the maximum value of the indicator in a row.

Based on the values of the ranks for each of the three indicators for each region of the region, the coefficient of infrastructure development is determined using the arithmetic mean (2):

$$k_i = \frac{r_r + r_c + r_{fe}}{i}, \quad (2)$$

where, k_i – the coefficient of development of the district infrastructure; r_r – rank along the length of the roads; r_c – rank of the costs of creating the forest fund infrastructure; r_{fe} – rank by the number of timber enterprises in the region; i - the number of investigated parameters.

Further, the classification of district forestries by the level of infrastructure development is carried out. Forestry groups with high, medium and low levels of forest fund infrastructure are identified. The division into intervals is carried out according to (3):

$$i = \frac{(x_{\max} - x_{\min})}{n} \quad (3)$$

where, $x_{\max}$ – the largest value of the grouping characteristic in the aggregate; $x_{\min}$ – the smallest value, n - the number of groups.

Next, the classification of district forestries by the level of infrastructure development will be carried out. Forestry groups with high, medium and low level of forest fund infrastructure have been identified. The division of the obtained values into intervals is carried out according to (3).

Based on the values of the ranks according to the indicators of the forest area, the coefficient of infrastructure development is determined as the arithmetic mean value [9].

3. Results and Discussion

The Leningrad Region was chosen as the subject for assessing the infrastructure of the forest fund. The industry of the region is represented by a wide range of mining, processing and manufacturing enterprises [10].

The products of chemical processing of wood and oil are a constant component of Russian exports, as well as scarce timber: softwood sawlogs, raw materials for the pulp and paper industry and other

products. In the Leningrad Region, there are wooden house-building enterprises, pulp and paper mills, woodworking and sawmills, furniture and plywood-slab factories, and biofuel production facilities.
One of the most important conditions for the intensification of forestry and forest management is infrastructure. Forest roads provide access to forest resources and allow year-round supply of wood to enterprises.

The economic effect of the increase in transport routes is achieved at different stages of forestry activities. Leningrad Region is the largest transport and logistics hub of the North-West Federal District. The share of the transport and communications industry in the structure of the gross regional product is 15.9% [10].

The infrastructure on the lands of the forest fund affects the rate of technological growth. Transport infrastructure is of great importance for the forest fund. It includes automobile and railway roads, parking of vehicles, tunnels, bridges, trestles, crane and access roads (railway or automobile) [11]. The length of forest roads in the forestries of the region are shown in table 1.

Table 1. The length of forest roads in forestries of the Leningrad region.

Name of the local forestry	The length of forest roads, km / thousand ha	District Name
Vsevolozhskoye	22.1	Vsevolozhskiy
Severo-Zapadnoye	15.1	Vyborgskiy
Gatchinskoye	6.5	Gatchinskiy
Lodeynopol'skoye	6.3	Lodeynopol'skiy
Podporozhskoye	6.2	Podporozhskiy
Slantsevskoye	6.1	Slantsevskiy
Kirovskoye	6.0	Kirovskiy
Volosovskoye	5.9	Volosovskiy
Kirishskoye	5.8	Kirishskiy
Luzhskoye	5.6	Luzhskiy
Lomonosovskoye	5.5	Lomonosovskiy
Boksitogorskoye	5.5	Bokstigorskiy
Kingiseppskoye	5.4	Kingiseppskiy

The territory of the Leningrad Region, in accordance with the transport accessibility of forest resources, can be divided into 4 regions: north-western (Karelian Isthmus), central, western and eastern. Differences in growing conditions between the Karelian Isthmus and the rest of the Leningrad Region led to noticeable differences in the species composition of these regions: 81% of the forested lands of the Karelian Isthmus are occupied by conifers. But the stock of ripe and overripe stands of this category is only 29%.

The formation of the forest fund structure of the Karelian Isthmus is undoubtedly influenced by the relatively lower productivity of forests and intensive harvesting, due to the developed road network and the proximity of Finland [11].

The Karelian Isthmus is the only region in the Leningrad Region that has a forest fund with a road network 10.4 km / thousand ha, which is approaching the optimal (10-12 km / thousand ha). This indicator varies widely in the Leningrad Region (table 2).

Table 2. Provision with a road network of the Leningrad regions.

Region of the Leningrad region	Administrative region of the Leningrad region	The coefficient of coverage of the road network, km / thousand ha
Northwest (Karelian Isthmus)	Priozersky	10.4
Central	Vsevolozhskiy, Kirovskiy, Tosnenskiy	9
West	Slantsevskiy, Kingiseppeskiy, Luzhskiy, Volosovskiy	7
Oriental	Kirishskiy, Lodeynopol'skiy, Tikhvinskiy, Volkhovskiy, Bokstitogorskiy, Podporozhskiy	2-3

The average value of the coefficient of provision with forest roads for the forest fund of the region is 5.9 km / thousand ha, which is slightly more than 50% of the standard indicator, which is 11.2 km / thousand ha. In the eastern regions of the region, the most common roads are the so-called "winter roads", which cannot be used for thinning, main use, forest conservation and protection during the most intense summer period [12]. For the forest fund, the optimal length of the road network for the region is 9-11 km / thousand. ha, which, with a uniform distribution of roads over the area, provides quarterly transport accessibility of forests. Such a density of the transport network is less than half of the forestries of the Leningrad Region. Currently, the transportation distance to loading points is from 25.0 km to 100.0 km or more. Therefore, the length of forest roads should be increased 2-2.5 times.

Figure 3 shows the volumes of wood harvesting by region and the number of timber enterprises in them.

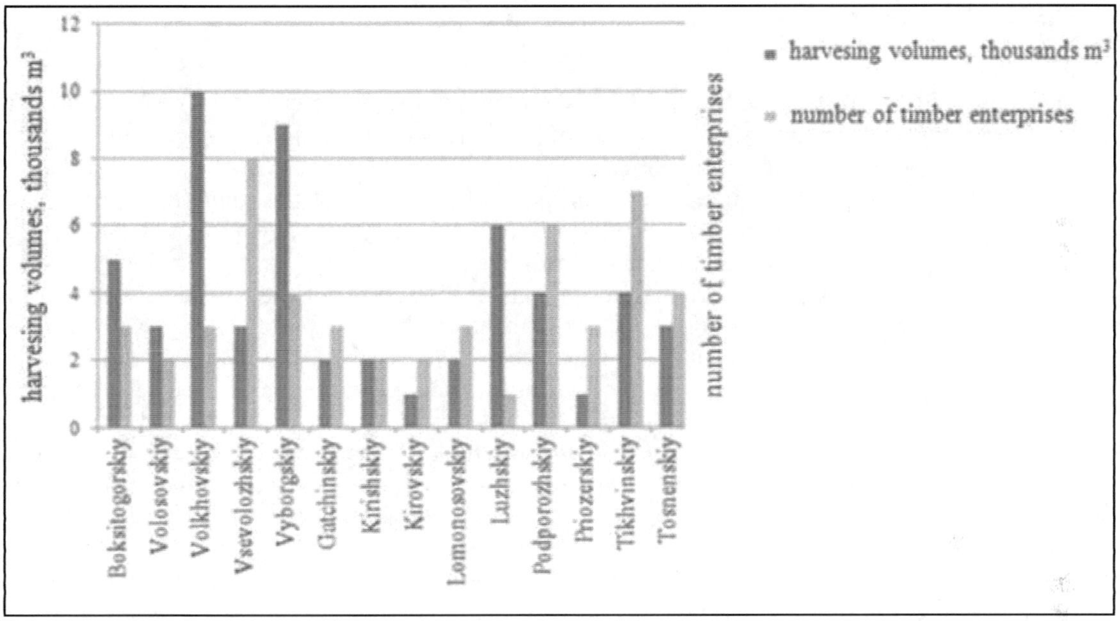

Figure 3. Distribution of non-governmental forestry organizations in the Leningrad Region by the number and volume of logging.

To date, more than 500 timber enterprises have been registered in the Leningrad Oblast, and forests with an area of about 5 million hectares, with an annual volume of harvesting ripe wood of 7524.3 thousands m^3, are leased for timber harvesting [12].

Table 3 presents the ranks for three infrastructure indicators for each district of Leningrad region. Due to the difficulty of collecting primary data, it was decided to do the gradation not in the regions of the Leningrad region, but in the administrative regions.

Table 3. Rank values for the main indicators of district infrastructure development.

The name of the district	Rank			Coefficient of infrastructure development
	length of roads, r_r	the cost of creating forest infrastructure, r_c	timber enterprises, r_{fe}	
Vsevolozhskiy	1.0	0.1	1	0.93
Vyborgskiy	0.7	0.1	0.5	0.60
Priozerskiy	0.4	0.1	0.4	0.32
Tosnenskiy	0.4	0.1	0.4	0.46

Volkhovskiy	0.3	0.1	0.4	0.28
Gatchinskiy	0.3	0.1	0.0	0.28
Lodeynopol'skiy	0.3	0.1	0.8	0.19
Podporozhskiy	0.3	0.1	0.0	0.34
Slantsevskiy	0.3	0.1	0.3	0.05
Kirovskiy	0.3	0.1	0.3	0.30
Volosovskiy	0.3	0.1	0.3	0.19
Kirishskiy	0.3	0.1	0.1	0.12
Luzhskiy	0.3	0.1	0.4	0.35
Lomonosovskiy	0.3	0.1	0.4	0.16
Boksitogorskiy	0.2	0.1	0.0	0.36
Kingiseppskiy	0.2	0.1	0.5	0.07

4. Conclusion

A study of the infrastructure of the forest fund of the Leningrad Region made it possible to determine the main directions of development of forestry. Not in all areas of the Leningrad region, the length of the roads reaches the average value for the region.

Therefore, there is a need to improve the transport infrastructure of the forest fund and finance work for the construction, repair and maintenance of forest roads. Based on the proposed methodology for accounting for infrastructure when assessing forest lands, a comparative and integrated assessment of the development of forestry infrastructure by regions of the subject of the Russian Federation and by forestries will be given. According to the assessment results, groups of districts with different levels of infrastructure development are distinguished. Also, based on the data used for cadastral valuation, it is possible to determine the main problems and possible directions for the development of forestry infrastructure within a specific territory [12].

The data obtained during the cadastral valuation of forest lands according to the methodology proposed in this paper can be used in GIS technologies to visually display the current situation in the forest fund. The need to improve the infrastructure of the forest fund in the constituent entities of the Russian Federation on the basis of the analysis is obvious, first of all, it is necessary to improve the transport system, not only in terms of length, but also in terms of quality. The formation of the forest fund infrastructure and forest processing infrastructure will allow for the stabilization and growth of the share of the forest sector in the Russian economy along with the large and major, today, oil refining and mining sectors.

References

[1] National Duma Committee for Natural Resources, Property and Land Use 2017 *Round-table conference materials: "Exercise of Authority in the Field of Forestry by Government Bodies of Russian Regions: Issues, Aims, and Prospects"* (Irkutsk: National Duma Committee for Natural Resources, Property and Land Use)

[2] Kovyazin V and Romanchikov A 2018 The problem of cadastral valuation of forest land, taking into account forest fund infrastructure *Journal of Mining Institute* 229 98-104

[3] Lobovikov T 1968 Forest as an economic phenomenon *Journal of Issues of forestry economics of the USSR* 126-140

[4] Petrov A P 1993 Economic assessment of forest resources in terms of their lease (licensing) *Journal of Forestry* 4 12-15

[5] Kovyazin V and Romanenko M 2019 Problem of temporary logging roads` inventory as real estate items *IOP Conference Series: Earth and Environmental Science 316* 1-7

[6] Government of Russian Federation 2017 *On Approval of Hunting Infrastructure Objects List (Decree N 1469 11 July 2017)* [Moscow: Government of Russian Federation]

[7] Federation Council 2018 *Condition and Prospects of Forestry Transport and Industrial Infrastructure* (Moscow: Federation Council)

[8] National Duma of Russian Federation 2006 *Forestry Code of Russian Federation* (Moscow: National Duma of Russian Federation)

[9] Kharionovskaya I 2016 Features of the development of forestry infrastructure in the Komi Republic *News of Komi Scientific Center, Ural Branch of the Russian Academy of Sciences* **1** 104-111

[10] Government of Russian Federation 2018 *On Approval of 2030 Forestry Development Strategy (Decree N 1989 20 September 2018 (ed 28 February 2019)* [Moscow: Government of Russian Federation]

[11] Lantsov A 2013 Infrastructure: concept, types and significance *Statistics and Economics* 3 49-54

[12] Demidova P and Gorelikov V 2013 Impact of Common Real Property Land Site Value Unit to Market Value of Non-commercial Horticulture Land in Leningrad Region *Journal of Mining Institute* 204 198-202

IOP Conf. Series: Materials Science and Engineering **817** (2020) 012018 doi:10.1088/1757-899X/817/1/012018

Monitoring of Forest Land Use/Cover Change in Cat Tien National Park, Dong Nai Province, Vietnam Using Remote Sensing and GIS techniques

V F Kovyazin[1*], A Yu Romanchikov[1], Dang Thi Lan Anh[1,2], Dang Viet Hung[2]

[1]Department of Engineering Geodesy, Saint-Petersburg Mining University, 2, 21 line of Vasilyevsky Island., St. Petersburg 199106, Russian Federation
[2]Department of Forest Plant, Vietnam National University of Forestry, Tran Phu street, 54, Trang Bom town, Trang Bom District, Dong Nai Province, Vietnam

*Corresponding email: vfkedr@mail.ru

Abstract. The Cat Tien National Park in the south of Vietnam represents a protected ecological area. It is located in tropical rainforest zone. For monitoring purposes of forest land use/cover change of the park satellite images (Landsat 5, Landsat 7 and Sentinel-2A) taken in 1988, 2003 and 2019 were used. NDVI was used for the estimation of vegetation quality. Forest land use/cover were classified by 5 categories using maximum likelihood classifier algorithm in ENVI 5.3. After classification, differentiation of forest land use/cover was analyzed on every image. Deforestation took place in 1988-2003 periods and broad-leaved forest areas have undergone reduction. Currently, new policies were introduced to reduce deforestation and therefore support free growth of tropical forest areas.

1. Introduction

The object of this investigation commonly known as Nam Cat Tien was founded in 1978 for the purpose of nature landscape reservation. In 1992 national park was reorganized to Cat Tien National Park. Since 2011 it became biosphere reservation as part of the UNESCO program. The area is located between $10^{0}20'50''$ – $11^{0}32'13''$ N and $107^{0}11'13''$ – $107^{0}28'20''$ E and has an area of about 41052.91 ha (figure 1).

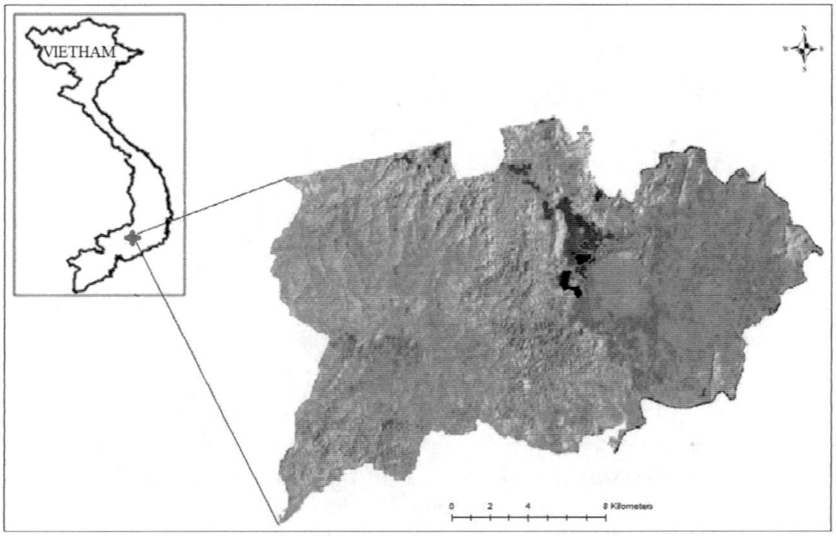

Figure 1. Cat Tien National Park location.

The forest ecosystem of the Cat Tien National Park is virgin tropical rainforest. Satellite image data and forest inventory in 1992 showed that forestlands in the national park are under human impact because of urban population growth in Dong Nai province.

2. Materials and Methods

2.1. Materials

The materials for the research was satellite images and forest inventory maps (table 1). All satellite images were taken in dry season of 1988, 2003 and 2019 with cloud cover less than 10%. Source of images is *https://earthexplorer.usgs.gov/* [1, 2].

Table 1. Details of satellite images used.

Entity ID	Data parameters	Date	Collected by
LT51240521988007BKT00	Landsat-5 Spatial resolution 30 m	07.01.1988	U.S. Geological Survey
LE71240522003072BKT01	Landsat-7 Spatial resolution 30 m	13.03.2003	U.S. Geological Survey
L2A_T48PYT_A019234_201 90227T031435 T48PYT_20190227T030651	Sentinel-2A Spatial resolution 10 m	27.02.2019	European Space Agency
Cat Tien National Park forest inventory maps	Spatial data grouped in different layers Scale 1:20 000	Year 2015	Cat Tien National Park administration

2.2. The research methodology

The research methodology included different stages.

Image preprocessing – correction and improving of satellite images. Radiometric calibration of Landsat-5 and Landsat-7 data was made in ArcGis 10.5. SNAP software with Sen2Cor algorithm was used for atmospheric correction of Sentinel-2A data [3].

Forest lands use/cover classification was provided by supervised maximum likelihood classifier, which proposes normal distribution of vegetation imaging and calculates probability of single pixel affiliation to certain vegetation class [4, 5]. We used ENVI 5.3 for this classification algorithm.

Classification of forest lands use/cover using elements of satellite image interpretation were proposed in this study. Table 2 shows the elements of satellite image interpretation.

Table 2. Elements of satellite image interpretation.

Forest land use/cover category	Forest land use/cover category
Broad-leaved forest	
Mixed forest	
Land without vegetation cover	
Meadows and shrubs	
Wetlands	

Classification accuracy assessment was provided using satellite images. We compared classified images with data from other sources such as Google Earth images and forest inventory maps made in 2015. Confusion matrixes are tables containing comparison of created map with control values [6]. Four accuracy assessment results were collected: user accuracy; producer accuracy; overall accuracy and Kappa index.

User accuracy is total amount of correct pixels in category divided by number of pixels classified in this category. Result is commission error. Producer accuracy is index showing quality of defined vegetation area classification. Kappa index measures the agreement between classification (X) and control values (Y).

We used formula below for Kappa index calculation [7]:

$$\text{Kappa} = \frac{P_0 - P_e}{1 - P_e}, \tag{1}$$

where, P_0 - relative observed agreement among raters; P_e - hypothetical probability of chance agreement.

Kappa value = 1 means complete agreement and kappa = 0 means no agreement between classification results and control data values. We used Confusion Matrix tool in ENVI 5.3 for this accuracy assessment.

Vegetation change detection – The forest land use/cover cover most of the Cat Tien National Park area and therefore NDVI was used for vegetation change detection with formula below [8]:

$$\text{NDVI} = \frac{\text{NIR} - \text{RED}}{\text{NIR} + \text{RED}}, \tag{2}$$

where, NIR – near infrared band value for a cell; RED – redband value for the cell.

NDVI value ranges from – 1 to + 1 [9]. Positive NDVI values are specific for vegetation [10, 11]. They increase with growth of plant biomass [12, 13]. We used next sensor bands for NDVI calculation: B3 (RED), B4 (NIR) [14] for Landsat images and B4 (RED), B8 (NIR) [15] for Sentinel images. We classified vegetation for 1988-2019 period and divided it for 3 categories: low vegetation density ($0<\text{NDVI}<0.2$), middle vegetation density ($0.2<\text{NDVI}<0.5$), high vegetation density ($\text{NDVI}>0.5$) [16, 17].

For the purpose of ***forest lands use/cover change detection*** we have statistics of its area in 1988-2019 period. Based on the three new maps we created in 1988, 2003 and 2019, Arcgis 10.5 was used to portray the dynamics of forest land use/cover change [18, 19] that have taken place in Cat Tien National Park for 31 years. These maps show dynamics of forest lands use/cover for 31 years by categories.

All remote sensing data analysis process is shown on figure 2.

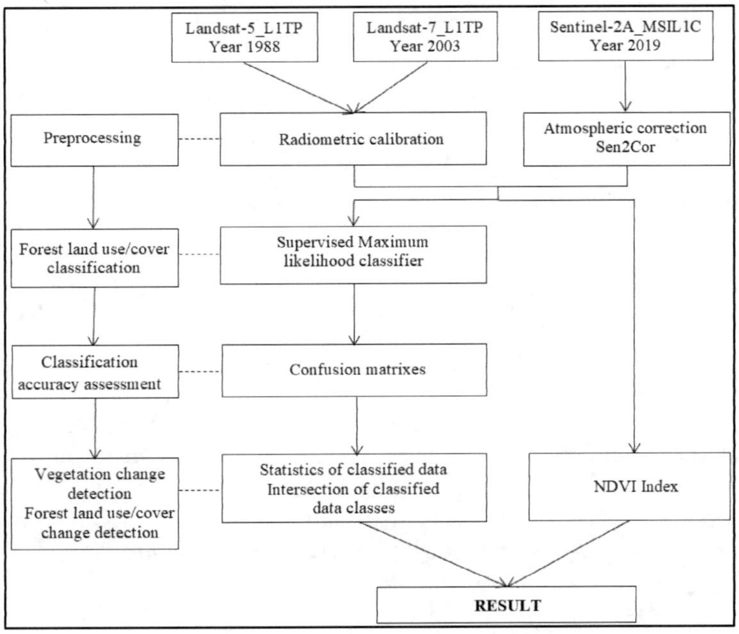

Figure 2. Flow chart of research methodology.

Based on the results of the survey of Landsat-TM images, maps of forest land use/cover of Cat Tien National Park in 1988, 2003 and 2019 were developed (figure 3).

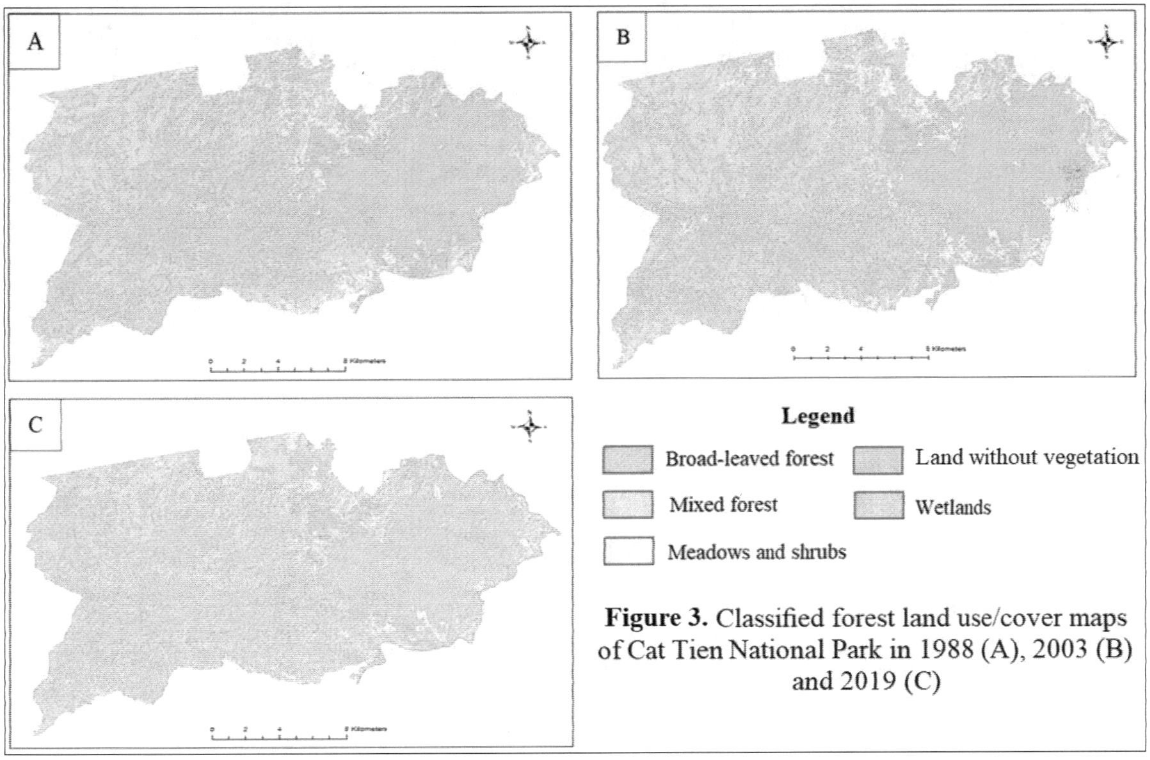

Figure 3. Classified forest land use/cover maps of Cat Tien National Park in 1988 (A), 2003 (B) and 2019 (C)

The accuracy assessment of image classification of different time periods (1988, 2003, 2019) are given in table 3.

Table 3. The results for the accuracy assessment of satellite image classification.

Forest land use/cover categories	Accuracy, %					
	Year 1988		Year 2003		Year 2019	
	User's accuracy	Producer's accuracy	User's accuracy	Producer's accuracy	User's accuracy	Producer's accuracy
Broad-leaved forest	94.00	90.38	96.08	94.23	96.08	98.00
Mixed forest	80.36	88.24	88.46	86.79	90.74	92.45
Meadows and shrubs	79.63	86.00	85.19	90.20	90.38	90.38
Land without vegetation cover	86.00	84.31	88.68	90.38	94.64	91.38
Wetlands	97.67	87.50	97.96	94.12	92.59	92.59
Overall accuracy, %	86.96		91.12		92.88	
Kappa index	0.84		0.89		0.91	

In the table 3 shows that the accuracy of user, producer and overall is quite good, Kappa index indices greater than 0.84 indicate that the classification results have reached a level of significant consistency.

Maps of NDVI index for 1988, 2003 and 2019 show vegetation dynamics over a given period (figure 4). Vegetation inventory on NDVI index maps are shown in figure 5.

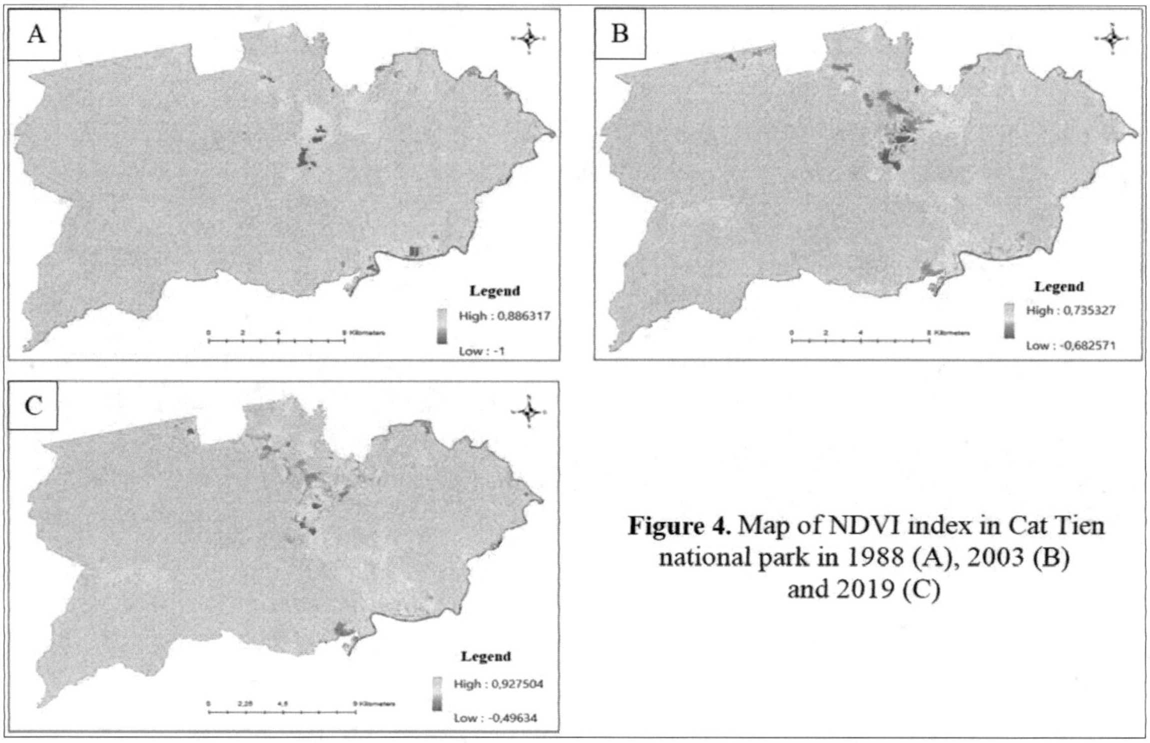

Figure 4. Map of NDVI index in Cat Tien national park in 1988 (A), 2003 (B) and 2019 (C)

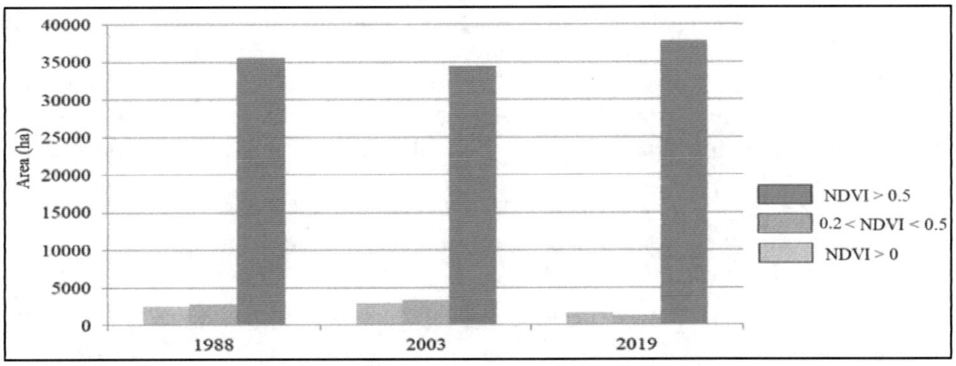

Figure 5. Diagrammatic illustration of vegetation area by category of density.

The analysis showed that over the past 31 years, in Cat Tien has seen a trend towards vegetation restoration in general; between 1988 and 2003, the NDVI index decreased, some forest areas were in a state of degradation. In particular, areas along populated areas were found to have a lower NDVI index, reflecting anthropogenic impacts in these areas. In the period 2003-2019, the NDVI index increased, forest areas are in a state of recovery. On the territory of Cat Tien national park recorded index value NDVI < 0.92, a significant proportion of the area of national park is covered with forest vegetation with high density, indicating that Cat Tien is dominated by tropical moist evergreen broad-leaved forest with a significant supply of wood. Dominant species in forest vegetation are represented by such as: *Afzelia xylocarpa* (Kurz) Craib., *Dalbergia cochinchinensis* Pierre., *Dalbergia oliveri* Gamble ex Prain, *Pterocarpus macrocarpus* Kurz., *Dipterocarpus dyeri* Pierre., *Hopea recopei* Pierre., *Hopea odorata* Roxb., etc. The forest vegetation described in this study is almost similar to that given by other researchers, such as Kuznetsov (2015) [20].

The inventory of forest land use/cover by category in Cat Tien National Park between 1988 and 2019 is shown in table 4.

Table 4. Change in forest land use/cover by category in Cat Tien National Park.

Forest land use/cover categories	Year 1988		Year 2003		Year 2019	
	ha	%	Ha	%	ha	%
Broad-leaved forest	24546.61	59.79	20668.72	50.35	26765.20	65.20
Mixed forest	10770.50	26.24	14136.46	34.43	11161.45	27.19
Meadows and shrubs	2782.49	6.78	3052.62	7.44	1518.10	3.70
Land without vegetation cover	2685.35	6.54	2871.47	6.99	1444.33	3.52
Wetlands	267.95	0.65	323.64	0.79	163.83	0.40
Total	41052.91	100	41052.91	100	41052.91	100

As can be seen from figure 3 and table 4, the dominant type of forest and vegetation cover on protected areas is broad-leaved forests, the share of which is about 59.79%, 50.35%, 65.20%, respectively, for 1988, 2003 and 2019. In the period 1988-2003, the area of broad-leaved forests decreased by 3877.89 ha (9.45%). This is due to illegal logging in Cat Tien. In the period 2003-2019, the area of broad-leaved forests decreased by 6096.49 ha (14.85%). After 13 years towards the end of Vietnam war in 1988, the welfare of the people living in the buffer zone in the Cat Tien was still quite difficult. Due to lack of funds, residents illegally cut down the forests of protected areas to sell wood and grow food crops in the cuttings. Thus, the area of broad-leaved forest was reduced. This created favorable conditions for the growth of mixed forest in the cleared area. Also in the east of Cat Tien, where there is a zone of the main distribution of broad-leaved forest, the forest area has been transformed into land without vegetation cover, meadows and shrubs. In the period 2003-2019, the

IOP Conf. Series: Materials Science and Engineering **817** (2020) 012018 doi:10.1088/1757-899X/817/1/012018

state forestry policy of Vietnam to attract and create conditions for the development of ecotourism in Cat Tien for the inhabitants of the region, especially indigenous communities, was effective.

During this period, the level of well-being of the local population has increased, directly reducing the burden on forest ecosystems, contributing to the development and protection of forest. Along with the good performance of the forest protection work carried out by Cat Tien national park management unit, the situation of deforestation and transformation of forest lands into agricultural fields was quickly prevented, and the number of forest fires decreased. To assess the transfer of forest land use/cover in Cat Tien from one category to another, we have compiled a matrix of their dynamics over the past 31 years (table 5).

Table 5. Matrix of forest land use/cover change in Cat Tien National Park by category from 1988 to 2019.

	Forest land use/cover categories	Area (ha), 2019					Total 1988
		Broad-leaved forest	Mixed forest	Meadows and shrubs	Land without vegetation cover	Wetlands	
Area (ha), 1988	Broad-leaved forest	21315.01	3120.33	77.51	33.22	0.54	24546.61
	Mixed forest	4071.39	6594.18	33.50	68.74	2.69	10770.50
	Meadows and shrubs	706.26	1145.16	898.90	30.23	1.94	2782.49
	Land without vegetation cover	645.84	295.26	495.52	1226.82	21.91	2685.35
	Wetlands	26.70	6.52	12.67	85.32	136.74	267.95
	Total 2019	26765.20	11161.45	1518.10	1444.33	163.83	41052.91

In the table 5 shows that, the area of forest vegetation in Cat Tien national park tended to increase during the study period. The area of meadows, shurbs and land without vegetation cover decreased by 1241.03 ha (3.02%) and 104.13 ha (0.25%), respectively, due to the fact that part of these lands for 31 years was planted with forest crops. The area of wetlands decreased by 1264.39 ha (3.08%). In Cat Tien national park, the dry season lasts six months in a row, resulting in a fairly low water level in the area, which has led to the transformation of part of the wetlands into land without vegetation cover.

3. Conclusions

Long-term studies of forest land use/cover in Cat Tien national park have shown that in the period from 1988 to 2019, they are dominated by tropical moist evergreen broad-leaved forest. The magnitude of forest land use/cover change in broad-leaved, mixed forest, meadows and shrubs, land without vegetation cover and wetlands are respectively 5.40%, 0.95%, 3.02%, 0.25% and 3.08%. For 31 years, the area of forest land use/cover under broad-leaved and mixed forest of Cat Tien national park increased by 2609.54 hectares (6.36%), respectively and area of other categories of forest decreased. Government reforms in Vietnam to improve the management system of protected areas have increased the area of land use/cover with forest vegetation and preserve the development of forest ecosystems in protected areas.

Referenses

[1] Kovyazin V F, Do Quang Huy, Tran Hoang Hieu and Do Hong Hanh 2018 Definition of categories of forest lands in Kim Hy Nature Reserve of Vietnam using GIS technologies [in Russian - Opredelenie kategorij lesnyh ugodij v prirodnom zapovednike Kimhi respubliki Vietnam s primeneniem GIS–Tekhnologij] Forest journal **6/366** pp 39-47 Doi; 10.172.38/issn0536-1036.2018.6/39

[2] Kovyazin V F, Romanchikov A Y and Kitcenko A 2019 Classification of lands infrastructure forest fund// IOP Conference Series: Earth and Environmental Science. Volume 316. 012622

[3] Marina-Ramona Rujoiu-Mare *et al* 2017 Land cover classification in Romanian Carpathians and Subcarpathians using multi-date Sentinel-2 remote sensing imagery European Journal of Remote Sensing, **50:1**, pp 496-508
[4] Kovyazin V F, Usoltsev V, Osmirko A, Tsepordey I and Chasovskikh V 2019 Additive model of Larix sp. forest stand biomass sensitive to temperature and precipitation variables in Eurasia// IOP Conference Series: Earth and Environmental Science. Volume 316. 012074
[5] Ryazanov S S, Sahabiev I A and Galimov D Z 2015 Classification of the land cover of the Kuibyshev reservoir Islands using remote sensing data [in Russian - Klassifikaciya zemnogo pokrova ostrovov kujbyshevskogo vodohranilishcha s ispol'zovaniem dannyh distancionnogo zondirovaniya] Russian journal of applied ecology pp 50-55
[6] Kovyazin V F, Lebedev Yu, Lebedeva T and Romanchikov A Y 2019 Value of Forest Ecosystem Natural Potential in the Areal Regional Richness Structure IOP Conference Series: Earth and Environmental Science Volume 316. 012027
[7] Congalton R G, Oderwald R G and Mead R A 1983 Assessing landsat classification accuracy using descrete multivariate analysis statistical techniques. Photogrammetry Engineering and Remote Sensing Vol 49, No 12 pp 1671-1678
[8] Rouse J W, Haas R H, Schell J A and Deering D W. 1974Monitoring vegetation systems in the Great Plains with ERTS, eds S C Freden, E P Mercanti, and M Becker Third Earth Resources Technology Satellite –1 Syposium. Volume I: Technical Presentations, NASA SP-351, NASA, Washington, D C pp 309-317
[9] Viktorovich B V and Vagitovich A Z 2017 Vegetation indices and their use for mapping mountain landscapes of The Russian Caucasus [in Russian - Vegetacionnye indeksy i ih ispol'zovanie dlya kartografirovaniya gornyh landshaftov Rossijskogo Kavkaza] The electronic scientific journal "Argo". Series: natural and technical Sciences pp 3-21
[10] Kovyazin V F and Dang Thi Lan Anh 2019 Monitoring of forest lands of Kim Hy Reserve in Vietnam with using GIS technologies [in Russian - Monitoring lesnyh zemel' zapovednika «Kim Hy» Vietnama s ispol'zovaniem GIS–Tekhnologij] Astrakhan Bulletin of Ecological Education, Astrakhan, **3** pp 95-102
[11] Manylov I V 2011 Assessment of the accuracy of class recognition in automated processing of aerial photographs [in Russian - Ocenka tochnosti raspoznavaniya klassov pri avtomatizirovannoj obrabotke aerofotosnimkov] Journal of Instrument Engineering. Vol 54 **5** pp 35-39
[12] Labutina I A and Baldina E A 2011 Use of remote sensing data for monitoring of protected areas ecosystems. Methodical manual [in Russian - Ispol'zovanie dannyh distancionnogo zondirovaniya dlya monitoringa ekosistem OOPT. Metodicheskoe posobie] World Wildlife Fund (WWF Russia). UNDP/GEF/ICI project "Biodiversity Conservation in the Russian part of the Altai-Sayan Ecoregion" Moscow p 88
[13] Zholobov D A and Baev A V 2015 Specification of values of normalized vegetative index (NDVI) by transpiration mask overlay method [in Russian - Utochnenie znachenij normalizovannogo vegetativnogo indeksa (NDVI), metodom nalozheniya transpiracionnoj maski] Innovations in science: sat. XLV international. science-pract. Conf. № 5 (42). (Novosibirsk: Sibak) pp 164-185
[14] Haque M I and Basak R 2017 Land cover change detection using gis and remote sensing techniques: a spatio-temporal study on Tanguar Haor, Sunamganj, Bangladesh. The Egyptian Journal of Remote Sensing and Space Sciences **20** pp 251–263
[15] Hankui K Z *et al* 2018 Characterization of Sentinel-2A and Landsat-8 top of atmosphere, surface, and nadir BRDF adjusted reflectance and NDVI differences Remote Sensing of Environment pp. 482-494 doi:10.1016/j.rse.2018.04.031
[16] Dang Viet Hung, Dang Thi Lan Anh and Potokin A F 2019 Monitoring vegetation of Dong Nai Reserve with using GIS technologies (Vietnam) [in Russian - Monitoring rastitel'nosti zapovednika «Dong Nai» s primeneniem GIS – tekhnologij] Materials of the youth

international scientific and practical conference: "Topical issues of the economy" St. Petersburg pp 80-83
[17] Kovyazin V F and Romanchikov A Y 2018 The Problem of Cadastral Appraisal of Forest Lands Taking into Account the Infrastructure of the Forestry Fund. Zapiski Gornogo instituta. Vol 229 pp 98-104 doi:10.25515/PMI.2018.1.98
[18] Kovyazin V F and Nguyen T X 2019 Izvestiya News of the Development of predictive models of transformation of the land of Vietnam [in Russian - Izvestiya Razrabotka prognoznoj modeli transformacii zemel'nyh ugodij Vietnama] Tomsk Polytechnic University. Engineering of geo-resources. Tomsk. Vol 330 **9** pp 221-229
[19] Kovyazin V F and Romanenko M 2019 Problem of temporary logging roads` invento-ry as real estate items IOP Conference Series: Earth and Environmental Science. Vol 316 012024
[20] Kuznecov A N 2015 Structure and dynamics of monsoon tropical forests of Vietnam [in Russian - Struktura i dinamika mussonnyh tropicheskih lesov Vietnama] Thesis for the degree of doctor of biological sciences Moscow p 554

Public easements` establishment for forest fund lands during seasonal roads` designing

V Kovyazin*, M Romanenko, M Khaikine

Saint-Petersburg Mining University, 2, 21 line of Vasilyevsky Island., St. Petersburg 199106, Russian Federation

*Corresponding email: vfkedr@mail.ru

Abstract. The article focuses on question of public easements` establishment for forest areas intended to accommodate seasonal roads as temporary structures. On one hand, seasonal roads are transport routes, which location changes every season. At the same time, they have features of real estate items as they are inseparably linked with the ground surface, and their movement without disproportionate damage to appointment is impossible. In both cases season roads require immediate strategic decisions, concerning their construction and operation. Due to the fact that seasonal roads are distinctive for territories of the northern regions, they are most often located on the forestry fund lands. That's why the question of territory planning while designing of such objects in difficult natural conditions is one of the most difficult in civil engineering, geodesy, land and forest law. Relevance is determined by need of increasing efficiency of the forest fund land use during designing of these real estate items, taking into account their specificity. The aim of research involves advanced technology of designing work`s development for seasonal roads. This technology gives opportunity of first-ever defining possible changes of seasonal roads` boundaries by results of field measurements of physical and mechanical soil properties.

1. Introduction

Today need of solving territorial distribution problems of transport infrastructure objects has been determined by requirements of socio-economic development of any country, as well as the Russian Federation. At the same time, issues related to cadastral works in relation to linear objects remain the most difficult in the sphere of land and property relations. Current situation is justified by considerable length of these objects, and therefore by territorial distribution on a large number of land parcels, which belong to different categories of land, including the forestry fund lands.

Consequently the Government of the Russian Federation proposed to realize the technology of establishing a public easement aimed at simplifying the construction, reconstruction, overhaul and operation of linear objects in land and property complex management (Federal Act No. 341) [1].

Enactment of the above-mentioned law is connected with the fact that process of registration and coordination of linear objects` construction, most often on the forestry fund lands, is a resource-intensive and long-term action. According to art. 21, para 1 of the Forestry Code of the Russian Federation, the construction, reconstruction and operation of objects, which are not related to the forest infrastructure creation on the forestry fund lands, are permitted for electric power lines, communication lines, roads, pipelines and other linear objects, as well as facilities, that are technological part of these objects [2].

IOP Conf. Series: Materials Science and Engineering **817** (2020) 012019 doi:10.1088/1757-899X/817/1/012019

Establishment of the institute of public easement`s law excludes satisfactory settlement under conditions of objects` passage on each particular forested land parcel, as well as allows to ensure construction without using land withdrawal [3]. Moreover, land withdrawal is economically unjustified, because seasonal object doesn`t always prevent use of forested land parcel for its purpose. Currently the concept of public easement has been clarified: in fact, it becomes the right of legal entity to ensure territorial distribution of socially important objects. Besides, modern resource-saving technologies of building and preserving forest roads contribute to mitigating the negative impact of transport on natural ecosystems [4].

Establishment of public easement boundaries contributes linear objects` territorial distribution, provided that interests of all landowners and land users are taken into account by fees and compensatory damages in the result of its establishment. In general terms, this procedure excludes cadastral and registration actions for neighboring forested land parcel.

In spite of the fact, that above mentioned provisions solve all kinds of atypical problems with regard to transport infrastructure objects, the process of construction and operation of seasonal roads on forestry fund lands is not regulated in any statutory instrument. In such case seasonal roads are changeable transport routes of certain direction and are classified as temporary structures.

2. Methods and Materials

The location of seasonal roads is influenced by physical and mechanical soil properties, numerical coefficients of which are determined as result of year-round stationary observations of temperature and humidity modes, base deformation of roadway at specially organized posts and test areas [5]. The field stage of research involves carrying out engineering surveys in the form of geodetic, geological and hydrometeorological works. These studies are included in the list of functions of the Department of Land and Property Relations of The Federal Road Agency and are carried out within its experimental laboratories` competence [6].

As a result of field stage, lists of road axes coordinates in accordance with the operational seasons of 2011-2018 (figure 1) and information about numerical indicators of humidity, stability, freezing depth, bearing capacity of the soil were obtained.

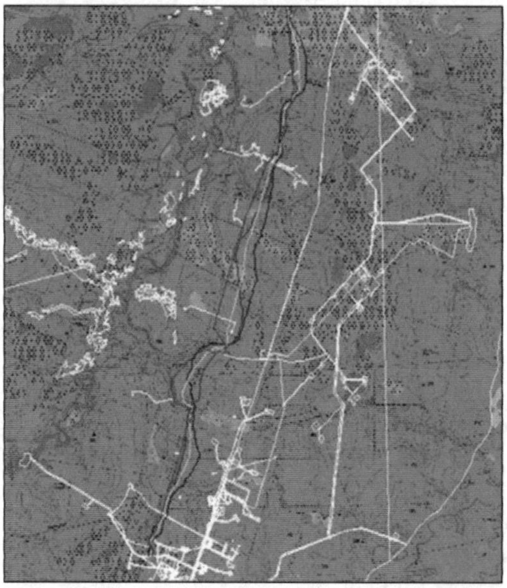

Figure 1. Seasonal roads' lines with reference to the cartographic image of the metric scale 1:200000 and to the cadastral plan of the territory: the red color indicates the line of 2016-2017 years, purple - 2011-2012, blue - 2007-2008, yellow – middle line.

IOP Conf. Series: Materials Science and Engineering **817** (2020) 012019 doi:10.1088/1757-899X/817/1/012019

In-office stage work is carried out with use of spatial analysis method, according to materials of engineering survey and is interpreted taking into account the experience of road designing and construction during past operational seasons. In order to quantify the width of the public easement area, mathematical modeling has been carried out, allowing to predict location of the object. As the final stage with application of analysis of seasonal axes` proximity the construction of buffer zones as public easement`s zones is performed. Coordinates of its boundaries are determined on the basis of soil characteristics obtained as a result of engineering survey.

Thus, the technology is experimental-theoretical and involves geographical prediction method to determine the future position of the seasonal road and geographic information method, with which it`s possible to create data banks on the basis of information received from organizations, which keep track of seasonal roads. The cartographic method provides opportunity to visualize geographical location of the object.

3. Results and discussion

Construction of study objects is carried out on territories, where it`s not feasible to create paved roads. Their roadway is destroyed due to difficult engineering and geological conditions during one operational season. Therefore, seasonal roads have life cycle, which is not specific to other real estate items [7].

There are numerous seasonal roads` superimposements on neighboring land parcels, most often forest fund lands during seasonal roads` designing [8]. As a result of seasonal roads` studying, as a separate type of transport infrastructure, it`s revealed: as for territorial distribution of constant linear objects, for seasonal roads the spatial basis in the form of land parcels is also determined. Taking into account the change of measured points` coordinates of seasonal roads, there is discrepancy between boundaries of land parcels under the road and the object with its changing location.

To optimize land organization for seasonal roads and to plan their rational use we need adjustment of their boundaries. Assuming that the adjustment timing distribution occurs according to Poisson 's law, the probability of seasonal road boundary stability will be equal:

$$P_t = \exp\left(\frac{-t}{T_{\text{av}}}\right),$$

where, exp – base of natural logarithms; t – time period of public easement boundaries` correction, years; $T_{\text{av}}$ – road`s life cycle, for seasonal roads this factor is equal to 1 year.

In compliance with maintenance of boundaries` stability with the probability of 0.9 - 0.95, it`s necessary to specify them every 0.05 years, that is economically inadvisable. Therefore, buffer zones with a special legal regime should be established to avoid repeated preparation of land management documents.

Due to figure 1, it`s possible to note a divergence of lines, that is connected with specifics of the object of seasonal existence. It can be explained with the changing location of an object depending on soil conditions, topographic features, relief characteristics.

Creation of the image of the average line is carried by method of perpendiculars on points which location is determined by extreme seasonal lines. It`s necessary to receive the first indicator – deviation of each line from the drawn average line.

Knowledge of position of the average line in the presence of communication of indicators of certain physical characteristics with coordinates of each her point can be used for forecasting the position of border lines in the set buffer zone. Each point correspond to certain results of the studied indicators.

We need to determine by carrying out an analytical research of each lines` groups with a binding to values of their deviation. Distribution of indicators of zones is only the assumption so far, his scientific justification is possible by a clustering method. Each seasonal line is characterized by soil conditions. These observations need to be broken into “k” number clusters. Parameters should be considered with average line and should be spread out on clusters (groups). The main aim of statistical research is to find out dependence between deviations on points of lines of different seasons and physical parameters. For future clustering it`s offered to establish the following zones: maximum, average divergence and minimum divergence.

The term "divergence" offers to designate distance between points of extreme axes regardless of a season. There is the result of research for lines of the highway for the season 2011-2012 as an example. It is obvious that such researches have been conducted for each of seasonal line (figure 2).

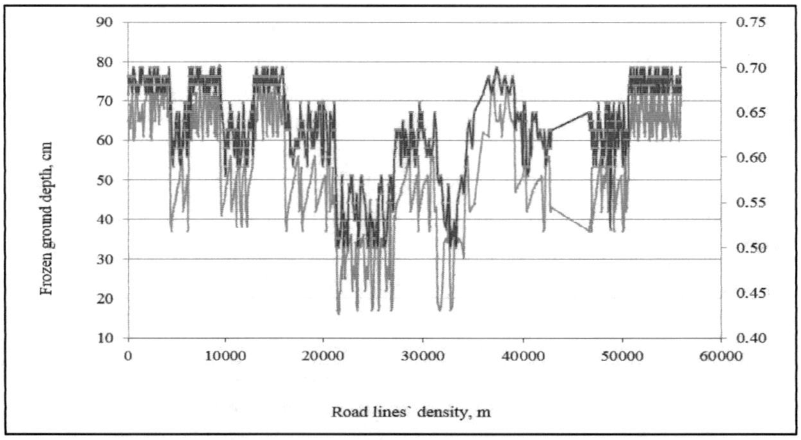

Figure 2: Result of carrying-out of analysis for the road`s line for the season of 2011-2012: with intervals of values of depth of ground freezing.

The distribution of indicators` values is uneven. Certain groups and classes are obviously allocated.

For further calculation, matrix with 9 values is used: 8 numerical indicators, which are characterized soil properties and a distance equal to seasonal road lines` deviation from zero one. As test values have different dimensions, data was previously standardized, that is, the difference between the numerical index of a particular cell and the mean value was determined. A standard deviation separation is performed to eliminate the dimension and align the spread. Scientific justification for indicators distribution is given using clustering method. As a result of spatial analysis of geotechnical maps compiled for design area of the seasonal road route, it was determined that for each previously defined area indicators varied in following ranges (table 1).

Table 1. Physical and mechanical soil properties, which are measured in field conditions during annual seasonal road`s territorial distribution.

№	Soil condition average mark	Units of measurement	Zones of seasonal lines` divergence		
			Minimal divergence (0-23 m)	Average divergence (24-67 m)	Maximal divergence (68-84 and more m)
1	Humidity	%	10	11	11
2	Density	g/cm³	12	7	15
3	Total volume content of ice	%	40	30	32
4	Depth of ground freezing	cm	30	44	29
5	Bearing capacity of soil	kg/cm²	6.1	5.2	4.3
6	Proof of soil stability		4	4	6
7	Average temperature of soil freezing	°C	-0.5	-0.8	-0.3
8	Speed of soil heaving	mm/d	10	12	11
9	Driving resistance	MPa	0.55	0.87	2.86

Using the results of longstanding engineering and geological survey, as well as using the cluster analysis method, we determine the possible value of axis divergence for the maximum recorded indicators of soil characteristics for each previously identified zone of seasonal axis divergence. Thus, the maximum possible divergence indicators have been determined for each zone (table 2), which will determine the boundaries of the buffer zone.

Table 2. Maximal marks of physical and mechanical soil properties, which are measured during longstanding engineer survey, for each divergence zone.

№	Soil condition maximal mark	Units of measurement	Zones of seasonal lines` divergence		
			Minimal divergence	Average divergence	Maximal divergence
1	Humidity	%	12	13	15
2	Density	g/cm³	13	7	19
3	Total volume content of ice	%	45	32	32
4	Depth of ground freezing	cm	30	46	34
5	Bearing capacity of soil	kg/cm²	6.8	5.2	4.4
6	Proof of soil stability		4	4	7
7	Average temperature of soil freezing	°C	-0.1	-0.1	-0.2
8	Speed of soil heaving	mm/d	11	12	12
9	Driving resistance	MPa	0.61	0.89	2.90

As is shown in table 2, it`s possible to calculate the width of the buffer zones and, consequently, to determine the coordinates of their boundaries on the ground by establishing the relationship between the numerical values of the specified soil characteristics and the deviation value. Knowledge of the position of the central baseline in accordance to soil properties can be used to predict the position of seasonal for the next season (table 3).

Table 3. The projected width of public easement`s buffer zones for seasonal road on forestry fund lands.

	Minimal divergence, m	Average divergence, m	Maximal divergence, m
Results of geotechnical maps analysis and cluster analysis method	0-23	24-67	68-84
Adjustment according to maximal marks of soil properties	0-28	28-79	79-92

Analytical study was carried out for each of the seasonal axes, as a result groups of divergence were identified. The result of cluster analysis is clear division of zones of divergence into three groups. Having information about the coordinates of points for each zone, you can indicate them on the territory.

As an illustration of carried out calculations (figure 3), there an example of buffer zone model built for the seasonal road according to the data obtained during the four operating seasons.

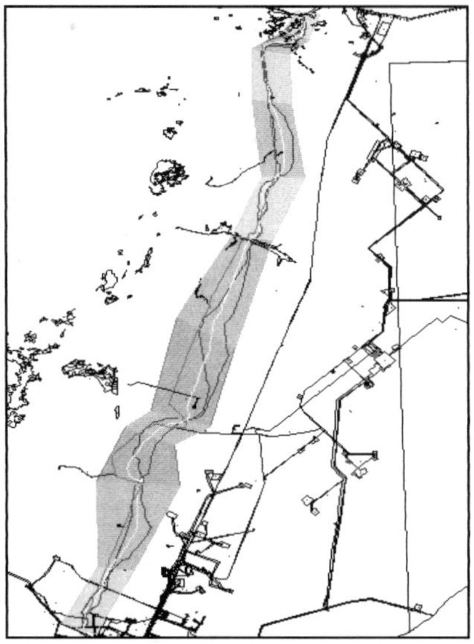

Figure 3. Model for boundaries` setting of public easement for seasonal road on forestry fund lands.

The solution of the problem of the location of seasonal roads on the forest fund sections is possible with the application of the forecast of the change in the location of the boundaries of the investigated objects, as well as with the establishment of the boundary of the public easement as a spatial basis, as shown in figure 3. By this means, establishment of public easement boundaries for study objects, with an appropriate legal status, avoids repeated preparation of land management documents to clarify the boundaries of adjacent forestry fund lands.

4. Conclusion

As a result of research: 1. Particular characteristics of seasonal roads as real estate items is defined; 2. Need to improve the technology of study objects` designing taking into account the established specifics is justified; 3. Methodological approaches to the location of linear road transport objects have been studied, as well as the peculiarities of such approaches for seasonal roads; 4. Physical and mechanical soil properties affecting location changes of the study objects are revealed; 5. Factors determining soil properties are ranked according to the degree of influence on the location of seasonal roads; 6. The model has been developed to determine the bandwidth of the tap depending on the degree of influence of the factor; 7. The technology of spatial basis, defining the limits of seasonal roads` passage has been developed; 8. GIS project has been created to map the boundaries of the buffer zone`s territorial distribution of seasonal roads.

As a result of boundaries` establishment of public easement with for seasonal roads, their lawful location on large area comprising many forestry fund lands is arranged. Registration of public easement implies a simplified procedure for establishing the boundaries of study objects on the territory in spite of seasonal changes of the measured points` coordinates and the presence of registered rights to adjacent land parcels.

The results of research ensure the development of existing technology for seasonal roads` design on forest fund lands. It can become not only conceptual basis for the system of operation and management of road transport objects as real estate items, but it can also be used to improve the legal framework in land and forest laws.

The results are significant for business processes of land management organizations, correct work of territorial bodies of the Federal Service of State Registration, Cadastral and Cartography, as well as for the territorial bodies of the Federal Road Agency of the Russian Federation during seasonal roads` management.

References

[1] Federal law «On introducing amendments to the Land Code of the Russian Federation and certain legislative acts of the Russian Federation regarding the simplification of linear objects placement» (341-FZ from 03.05.2018)
[2] Fedaral law «Forest Code» (200-FZ from 04.12.2006)
[3] Bondarchuk D 2018 *Public easement* [in Russian – Publichnyj servitut] (Moscow: EZH-YUrist). **34**. 1035
[4] Kovyazin V, Romanchikov A and Kitenko A 2019 Classification of lands infrastructure forest fund. *IOP Conference Series: Earth and Environmental Science.* **316**. 012622
[5] Skvorcov A, Pospelov P and Kotov A 2005 *Geomatics in road branch* [in Russian – Geoinformatika v dorozhnoj otrasli] (Moscow: MADI (GTU) p 249
[6] Kostyrchenko V A, Madjarov T M and Merdanov Sh M 2016 *Basic Aspects of the Transport Infrastructure's Development of the Far North. Fundamental Research* (Penza: Akademiya Yestestvoznaniya) p 136
[7] Hinkka V, Mantsinen H, Leviäkangas P and Hautala R 2016 Integrated winter road maintenance management. New directions for cold regions research. *Cold Regions Science and Technology* **121** pp 108-117
[8] Karjalainen T, Leinonen T, Husso M and Karvinen S 2009 *Intensification of Forest Management and Improvement of Wood Harvesting in Northwest Russia* (Helsinki: Working papers of the Finnish forest research institute) **110** p 151

IOP Conf. Series: Materials Science and Engineering **817** (2020) 012020 doi:10.1088/1757-899X/817/1/012020

Substantiation of connection of diagnostic indicators with parameters of technical condition of hydraulic manipulators

B Martynov[1*], N Sidorenkov[2], M Taraban[1]

[1]*St. Petersburg State Forest Technical University*, 5 Institutskiy Lane, St. Petersburg 194021, Russian Federation
[2]JSC «Podyemnye Mashiny», 182112, 8Korniyenko street, Velikiye Luki, Pskov oblast, Russian Federation

*Corresponding email:lesbisnes@mail.ru

Abstract. At the organization of proactive system of maintenance of mechanisms the technique based on determination of relative probability is applied. But in contrast to the technique, when the machine is for maintenance and adjusted the type and frequency of the hydraulic manipulators (HM) it is necessary to carry out constant monitoring of the technical condition of the main elements of the mechanism. The main elements of the HM, which primarily determine its technical condition, are the distribution system, including the pump, oil pressure distributors in the executive bodies of the HM, as well as the column, the extension-grip joint and the rotator. Nominal and permissible pressure values in the distribution system are determined by the technical documentation on the HM. The paper considers the dependence of the vibration signal on the technical condition of the coupling shaft-gear column HM.

1. Introduction

By organizing proactive system maintenance of mechanisms, the technique is applied based on determination of relative probability [1]. However, unlike the technique when the machine is currently on maintenance where the type and frequency are adjusted, for hydraulic grapple it is necessary to control constantly the technical condition of the main elements of the mechanism. The main elements of the grapple, which define primarily its technical condition, are the distribution system, including pumps, valves oil pressure in the Executive apparatus, the hydraulic grapple and the column joint extension tube-grab and rotator. Minor and allowable pressure in the distribution system are defined by technical documentation of the hydraulic grapple. In the work the dependence of the vibration signal index upon technical condition of the finger joint of the grapple column's shaft gear.

The purpose of the studies was to define the correlation of vibro-acoustic signal and the technical condition's parameters of the grapple column's pairing.

2. Methods and Materials

The technical condition of grapple's column will be determined specifically directly by the finger joint of the shaft gear and by the technical condition of the bearing of this pairing.

In the work [2] the matter are reviewed concerning appearance of the errors in finger joint and determination of the force causing oscillation in transmission.

The impact force of finger joint defines by:

$$g_0 = \frac{r_i \omega_1 J_1 J_2}{J_1 \omega_1 + J_2 \omega_2} * \frac{\Delta_i}{i} \quad (1)$$

where, J_1, J_2 – moments of inertia of the drive and driven wheels; ω_1, ω_2 –the angular velocities of the wheels; r_i – radius of pitch circle; $\frac{\Delta i}{i}$ - relative deviation of the transmission ratio.

For coupling a rack pinion of the grapple's mechanism, expression for impact force is presented as following:

$$g_0 = \frac{J_{r1} J_{r2} V}{J_{r1} V + J_{r2} \omega_{r2}} * \frac{\Delta_i}{i} \quad (2)$$

where, J_{r1}, J_{r2} — the moments of inertia of the rack and pinion respectively; V – linear velocity of the rack; ω_{r2} – angular velocity of the pinion.

Considering that angular velocity is connected to linear relationship $V = R\omega$, and solving the moments of inertia, we will get:

$$g_0 = \frac{m_{r1} m_{r2} R_{r1} R_{r2} V}{2(m_{r1} R_{r1} V + 0.5 m_{r2} R_{r2} V_{r2})} * \frac{\Delta_i}{i} \quad (3)$$

where, m_{r1}, m_{r2} – masses of the rack and pinion respectively; R_{r1}, R_{r2} –radiuses of the rack and pinion respectively; V, V_{r2} – linear velocities of the rack and pinion respectively.

A relative deviation of the transmission ratio of the coupling rack pinion will depend on error of the teeth profiles and side clearance of gearing teeth [3]. On one hand, the error profiles of the teeth appears by their cutting, therefore, it will be systematic error of the gear. On the other hand, the error of increase of side clearance appears at the operating conditions. Therefore, the final expression for impact force value will be presented as following:

$$g_0 = \frac{m_{r1} m_{r2} R_{r1} R_{r2} V}{2(m_{r1} R_{r1} V + 0.5 m_{r2} R_{r2} V_{r2})} . (\delta + ch) = \frac{K_1 V}{2K_2 V + K_3 V_{r2}} . (\delta + ch) \quad (4)$$

where, K_1, K_2, K_3 – constant values, depending on geometric size and masses of coupling; δ – error teeth profile; c–average teeth stiffness; h – value of the side clearance.

For defining the amplitude of vibration signal perceived by the sensor, we use the second Newton's law presented in differential form [4].

$$m \frac{d^2 y}{dt^2} + \varepsilon \frac{dy}{dt} + cy = F(t) \quad (5)$$

where, $m = \frac{m_1 m_2}{m_1 + m_2}$ - provided mass of the sensor-object system for diagnosing; ε – coefficient of the tenacious deformation; c–stiffness of sensor binding to object of diagnosing; $F(t)$ – force oscillating transmissions in object of diagnosing.

Assuming that sensor-object system is stiff system where coefficient of the tenacious deformation is almost equal zero and considering that amplitude of the impact force determines by (4), the formula can be modified as following:

$$A = \frac{K_1 V}{2K_2 V + K_3 V_{r2}} . (\delta + ch) + K_4 \quad (6)$$

where, K_4 – coefficient of the stiffness for the sensor-object system; A – amplitude of the vibration signal.

Considering that error teeth profiles is at least an order of magnitude smaller than stiffness value and knowing value of the velocities and stiffness of the binding sensor, the amplitude of the vibro-acceleration can be made a simple formula:

$$A = f(h) \quad (7)$$

Under operating conditions in the process of wear of a rolling bearing, various types of malfunction occur, leading to an increase in the vibration signal [5].

An arising vibration signal is momentary, but it has a wide range of frequencies from 2 Hz to 25 kHz.

We will consider the case of a rolling bearing used on a grapple. A spherical roller bearing is set on the rotary grapple column. In many cases, the technical condition of the bearing affects the reliability and safety of such low-speed mechanisms.

The main malfunctions of spherical bearings are wear of the rolling path of the rings due to combined loads during operation, chipped ring edges, a shell and cracks on the rolling surfaces, installation defects.

All types of defects occur at the low-frequency area, with the exception of installation defects, which occur at medium frequencies. Moreover, the frequency of occurrence of defects is equal to [5]:

$$f_o = \frac{1}{2} f_V (1 - \frac{r_r}{r_V} \cos\alpha) * z \tag{8}$$

where, f_o – frequency of occurrence of outer ring; f_V - shaft speed; $\propto$- contact angle of the rollers with rolling path; z – number of the rolling elements; r_r, r_V – the radii of the rollers and shaft, respectively.

In [6], it was proved that a change in the frequency of rotation of a defective bearing does not affect the change in the frequency of free oscillations of the node in which it is installed. Therefore, the noise in the form of vibration can be neglected from the frequency of oscillation of the structure.

As the result, the dependence of the vibration acceleration pulse from shock loads in the bearing can be written as follows:

$$m\frac{d^2y}{dt^2} + \varepsilon\frac{dy}{dt} + cy = F(t) + Q(t) \tag{9}$$

where, $F(t)$ – restoring force due to contact elasticity [7]; $Q(t)$– radial load.

Since the grapple's operation occurs in the start-stop mode, i.e. the main impact energy is concentrated in the impulse of force arising at the beginning of work, therefore, the restoring force does not have a noticeable effect on the amplitude of vibration due to the duration of the process.

Moreover, the radial load will be characterized by the impulse of force, which is determined from the expression [8]:

$$g_1 = S(1 + l)(\frac{F*m}{r}\sin\frac{\pi}{N})^{0.5} \tag{10}$$

where, g_1 – force impulse; l–coefficient of restoring; F – radial force acting on the bearing; m – bearing mass; S – gap; r – rolling elements radius; N –number of the rolling elements.

Radial load is associated with the clearance correlation [2, 7]:

$$F = Nl^{1.5}.2rS^{1.5} \tag{11}$$

Substituting (11) into (10) in conclusion for the force impulse we will get:

$$g_1 = KS^{1.25}[2N(\sin\frac{\pi}{N})]^{0.5} \tag{12}$$

where, K – constant depending on the type of the bearing.

Because of changing the parameters of the technical state, the value of the disturbing force changes. This is due to the presence of amplitude modulation of the signal:

$$g_1 = KS^{1.25}\,[1 + \xi\sin(\psi t + \alpha)](2N\sin\frac{\pi}{N})^{0.5} \tag{13}$$

where, ξ – modulation coefficient, which generally considers the errors of the rings, rolling elements, etc.; ψ –modulation frequency; α – modulation phase.

Substituting expression (13) into (9), we obtain the dependence of the amplitude of vibration acceleration on bearing errors.

$$m\frac{d^2y}{dt^2} + \varepsilon\frac{dy}{dt} + cy = KS^{1.25}\,[1 + \xi\sin(\psi t + \alpha)](2N\sin\frac{\pi}{N})^{0.5} \tag{14}$$

Since the stiffness condition of the sensor-object system is preserved, equation (14) is converted to:

$$\frac{d^2y}{dt^2} = \frac{1}{m} KS^{1.25} \left[1 + \xi \sin(\psi t + \alpha)\right](2N \sin\frac{\pi}{N})^{0.5} \quad (15)$$

As has been said, all types of bearing defects mainly occur in the low frequency area. Therefore, in this case, it is necessary to measure not the amplitude of vibration acceleration, but the amplitude of vibration velocity, since it more accurately describes the impact energy [5].

3. Results and Discussion

For determining the relationship of the amplitude of the velocity of the vibration signal from the parameters of the technical condition of the bearing, firstly a particular solution of the equation (15) must be found in the form [3]

$$y = b \sin(\omega t + \varphi) \quad (16)$$

where, b – the amplitude of the excitation force; ω – excitation force frequency; φ – excitationforce phase.

After that, we must find the first derivative of the obtained value of the vibration movement. Because of these transformations, we obtain:

$$\frac{dy}{dt} = \frac{2\omega_0^2 \omega g_1}{(\omega_0^2 - \omega^2)} \left[1 + \xi \sin(\psi t + \alpha)\right] \sin(\omega t + \varphi) \quad (17)$$

where, ω_0 – natural system frequency.

Since the natural frequency of the system is at least two orders of magnitude lower than the frequency of the exciting force, and the frequency and the modulation phase are almost the same as the carrier frequency, equation (17) reduces to the following:

$$B = \frac{KS^{1.25}}{\omega m} (1 + \xi) \quad (18)$$

where, B – amplitude of the vibration velocity.

Thus, analyzing the formula (18), we can conclude that the amplitude of the vibration velocity increases with growing radial clearance in the bearing and modulation coefficient.

References

[1] Martynov B G 2005 Substantiation of individual strategy of effective operation of forest machines by results of diagnostics (in Russian – Obosnovanie individualnoi strategi effektivnoi ekspluanaci lesnyh machin porezultatam diagnostirovanija) *Izvestia Sankt-Peterburgskoj lesotehniceskoj akademii* (Saint-Petersburg: SPbFTA) issue 172, pp 62-65

[2] Dynamic processes in mechanisms with gears 1976 (in Russian – Dinamicheskie processy v mehanizmah s zubchatymi peredachami) ed M D Genkin (Moscow: Nauka) 154

[3] Martynov B G 2005 Justification of effective operation of individual machines according to the results of their technical condition (in Russian – Obosnovanie effektivnoi ekspluanacii individualnyh machin po rezultatam ih tehniceskogo sostoianiia) Doctoral thesis. St. Petersburg, 361

[4] Shkalikov V S 1980 Measurement of vibration and shock parameters (in Russian – Izmerenie parametrov vibracii i udara) (Moscow: Publishing house of standards), 280

[5] Barkov AV, Barkova N A and Azovtsev A Yu 2000 Monitoring and diagnostics of rotary machines by vibration (in Russian – Monitoring i diagnostika rotornyh machin po vibracii) (Saint-Petersburg: Spbgmtu), 158

IOP Conf. Series: Materials Science and Engineering **817** (2020) 012020 doi:10.1088/1757-899X/817/1/012020

[6] Kostyukov V N 1985 Rank method of vibroacoustic diagnostics and quality assessment of machines (in Russian – Rangovyii metod vibroakusticheskoii diagnostiki I ocenki kachestva machin) *Hydraulic Drive and control systems of construction, traction and road machines*, (Omsk: OMPI), pp 113-124
[7] Sheftel B T and Shanitsyn A A 1973 Vibration of ball bearing with radial clearance(in Russian – Vibraciia sharikopodshipnika s radialnym zazorom), *Machine science* (Mashinovedenie), Moscow, 4, pp 29-35
[8] Pavlov B V 1971 Acoustical diagnosis of mechanisms (in Russian – Akusticheskaia diagnostika mehanizmov) (Moscow: Mashinostroenie), p 221

IOP Conf. Series: Materials Science and Engineering **817** (2020) 012021 doi:10.1088/1757-899X/817/1/012021

Improvement of timber industry logistics using information systems

S Medvedev[1*], A Mokhirev[2], T Rjabova[1]

[1]Department of economic and natural Sciences, *Lesosibirsk Branch of Reshetnev Siberian State University of Science and Technology*, 29 Pobedy Street, Lesosibirsk 662543, Russian Federation
[2]Department of Technology of Logging and Wood Processing Industries, *Lesosibirsk Branch of Reshetnev Siberian State University of Science and Technology,* 29 Pobedy Street, Lesosibirsk 662543, Russian Federation

*Corresponding email: medvedev_serega@mail.ru

Abstract. The article deals with the problems of using software to solve the problems of timber logistics. Tools for information processing and analysis are proposed. The possibilities of using software products to optimize the activities of enterprises of the forest industry are shown. Further directions on the use of modern information technologies in timber industry logistics are revealed.

1. Introduction

In modern world, the information analysis is an extremely important task and the factor of success of many industrial companies. Information using has a diverse character. The field of the company's activity and the management system faces are essential here.

It can be used to make various decisions and predict the dynamics of various indicators. Currently, there are many different software packages for data analysis and processing. Software products can be used to solve various application tasks. Various industries may also seriously vary the used software. The use of software should not be an end in itself. Ultimately, the use of modern information technology is designed to solve complex engineering problems with a large amount of computational operations [1].

Timber industry complex (TIC) is a complex system with many structural elements that are significantly different from each other. Each of the subfields has a unique specificity and a set of tasks. As for any other field, the companies and specific indicators of their activity are influenced by a range of factors of internal and external environment. Their analysis, identification of patterns, forecasting, assistance in making managerial decisions and a number of other tasks are the set of priorities in the use of modern information technologies and various software in particular [2].

Logging and timber transportation are the most important areas in the development and achievement of efficiency of activity of the entire TIC. The cost, individual qualitative and quantitative characteristics of essential resource of the field (timber and secondary resources) are formed on the initial stage [3-4]. The purpose of the study is to assess the capabilities and determine the most important directions of the use of certain software products for development of the logging industry, including in the context of timber transportation.

IOP Conf. Series: Materials Science and Engineering **817** (2020) 012021 doi:10.1088/1757-899X/817/1/012021

2. Methods and Materials

The process of forest management and information processes at the enterprise of large timber enterprises, including the Krasnoyarsk region were the objects of the study. The following main methods were used: analysis, statistical-mathematical planning, modeling, forecasting, etc. Various application program packages, widespread and potentially used for solving issues of companies in the industry were considered as the tested software products.

3. Results and Discussion

3.1. Results

The studies determined in that the most commonly used software in the companies of logging industry is Microsoft Office. Various companies use various applications starting with the most common Word to, for example, InfoPath. The simplicity of the interface, universalism and popularity determine the priority use of this software for the purpose of solving the applied issues of industry.

The list of tasks for timber industry logistics (regarding harvesting and transportation of timber products) was determined in the study. Various information technologies can be used for solution of this tasks [5-6]:

- optimization of technological process of logging;
- searching of logging sites with the desired characteristics;
- optimization of the storage of timber and wood raw material;
- optimization and improvement of delivery routes;
- selection and optimization of used equipment (both for logging and transportation);
- ensuring unity of transport and warehouse process;
- selection of variants and types of infrastructure facilities;
- selection and justification of technological solutions;
- planning and forecasting company development and various aspects of the activities in timber harvesting and timber transportation, etc.

Ultimately, the activities of logging companies, taking into account timber transport, is designed to solve the essential task of satisfying consumer demand for timber products (not only wood but also secondary wood resources) with minimum costs [7]. The task can be complicated by introducing restrictions on time, product quality, cost of products, related services or the place of delivery. However, the key task is to minimize the cost of delivering the final product to the consumer. Moreover, any optimization decisions should be fully justified. The essential parameters of the justification are the fulfillment of requirements for the compliance of the final product with the expectations (contracts) of consumers and economic efficiency. The latter is directly or indirectly influenced by many parameters - from the used equipment to the change in ambient temperature during the period of timber removal [8].

Based on the tasks described above, the functional diagram of the use of individual information technologies for solution of issues of timber industry was determined (figure 1).

Step 1. Assessment of the current state of the company (if the company is an operating one). The necessary assessment of the current technical, organizational, financial-economic and other conditions. The assessment is to be subject to statistical processing.

Step 2. Making the strategic decision about the necessity of application of information technologies and software for company modifications in order to improve its efficiency and planning peculiarities of implementation of the changes.

Step 3. Identifying potential software and information technologies able to adequately assess and/or contribute to decision making on the choice of possible ways of converting the logging activities and timber transport.

Step 4. "Informatization" (digitizing) of processes. In this case, "Informatization" means the detailing and structuring of all processes of the company, preparing of information cards (reference data) for each of them. In this description ("digitizing") is subject to alternative options of activity, if any. The latter include, for example, alternative technologies of timber processing delivery chains (routes), transport equipment, etc.

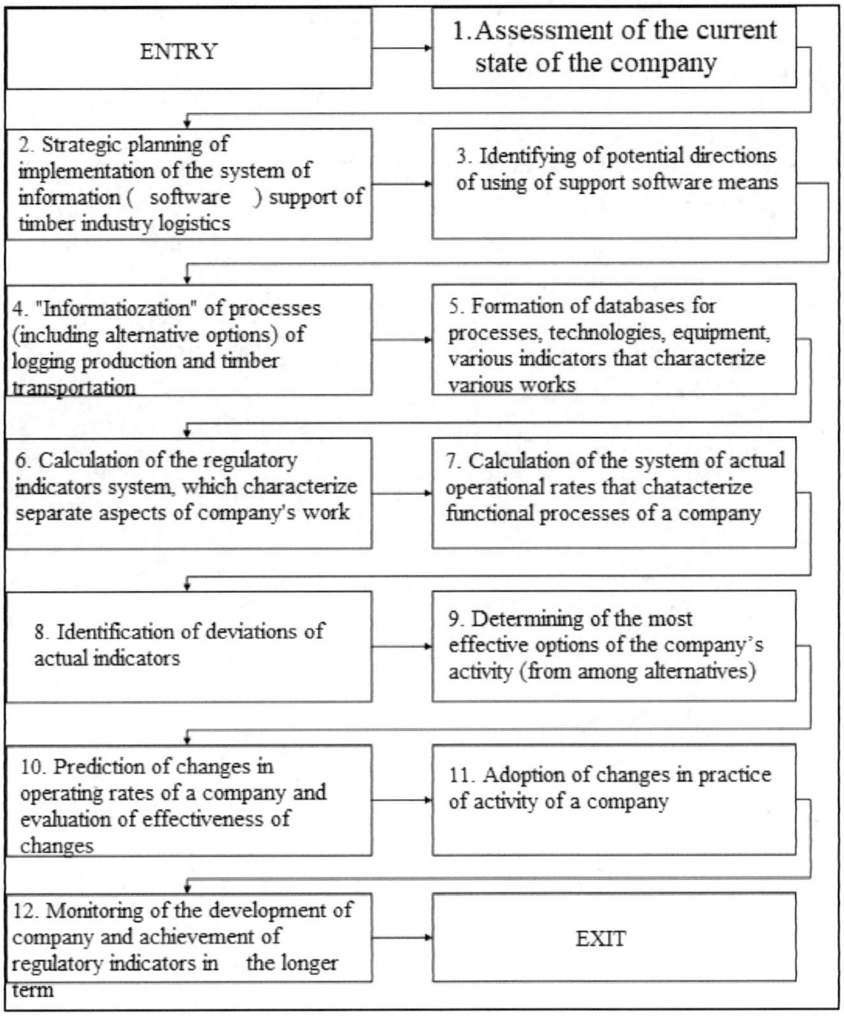

Figure 1. Functional diagram of the management system of logging activities and timber transportation considering optimization using information technologies.

Step 5. Formation and filling in of databases for processes, technologies, equipment, climatic conditions and other aspects of the company and characteristics affecting the company.

Step 6. Calculation of the regulatory indicators system. This step is necessary to facilitate subsequent calculations and form a database for comparison with the actual values of indicators of work of the company.

Step 7. Calculation of the system of actual operational rates of the company. It is necessary to make a complex calculation of various indicators characterizing the efficiency of the individual stages of logging and transport of wood materials.

Step 8. Identification of deviations of actual indicators from the regulatory ones. This step is necessary for identifying the most problematic and underperforming sites, processes, equipment, technologies and reasons for deviations.

Step 9. Determining of the most effective options of the company's activity (from among alternatives), including necessary investments and organizational changes.

IOP Conf. Series: Materials Science and Engineering **817** (2020) 012021 doi:10.1088/1757-899X/817/1/012021

Step 10. Prediction of changes in operating rate and evaluation of effectiveness of changes in accordance with the most efficient solutions, defined in the previous stages.

Step 11. Adoption of changes in practice of activity of the company. After identifying optimal options for the organization, it is necessary either to convert the current activities (alternative routes, equipment, technology, etc.), or to organize the work of the company in the most effective ways.

Step 12. Monitoring of the development involves the evaluation and tracking of location of company within certain regulatory boundaries.

A number of software products that allow to solve the tasks within this functional scheme was identified in the study (table. 1).

Table 1. Analysis of the characteristics of the northern timber industry region of Krasnoyarsk Territory.

Process of implementation of information technology	Software program				
	Microsoft Office Excel	Microsoft Office Project	Microsoft Office Access	Statistica	ArcGIS
Assessment of the current state of the company	+			+	
"Informatization" of the company work processes	+	+	+		+
Creating and filling in of databases	+	+	+	+	+
Calculation of the system of regulatory indicators	+			+	
Calculation of the system of actual indicators	+	+			+
Identification of deviations from regulatory indicators	+	+	+	+	
Determination of the most effective options for the company's activity	+	+			
Prediction of changes in operating rate of the company and evaluation of effectiveness of changes	+			+	
Monitoring of the development of the company and the achievement of regulatory indicators in the long term perspective	+	+			

Table 1 shows the use of software products in the process of implementing information technology in the timber industry and timber transport. The given five types of software allow to implement the basic necessary functional as part of the information technology implementation process. At the same time, all information about the work of the studied companies can and are to be connected within the framework of these software products. It should be clarified that the Statistica software package (developed by StatSoft) is used for statistical, graphical analysis, forecasting and a number of other purposes; ArcGIS (developed by ESRI) is used to build geoinformational systems at any level.

As part of the study, a conceptual model of processing and analyzing information of timber industry logistics in selected software products was developed (figure 2).

The essential software product in this case is Microsoft Office Excel. It allows to organize the exchange of data between all other software products. Also, a large amount of calculations is carried out in this program: from the elementary summation of various indicators, for example, the cost or volume of transported wood products, to the cost-effectiveness analysis of alternative options for the development of the company, for example, various delivery routes [9].

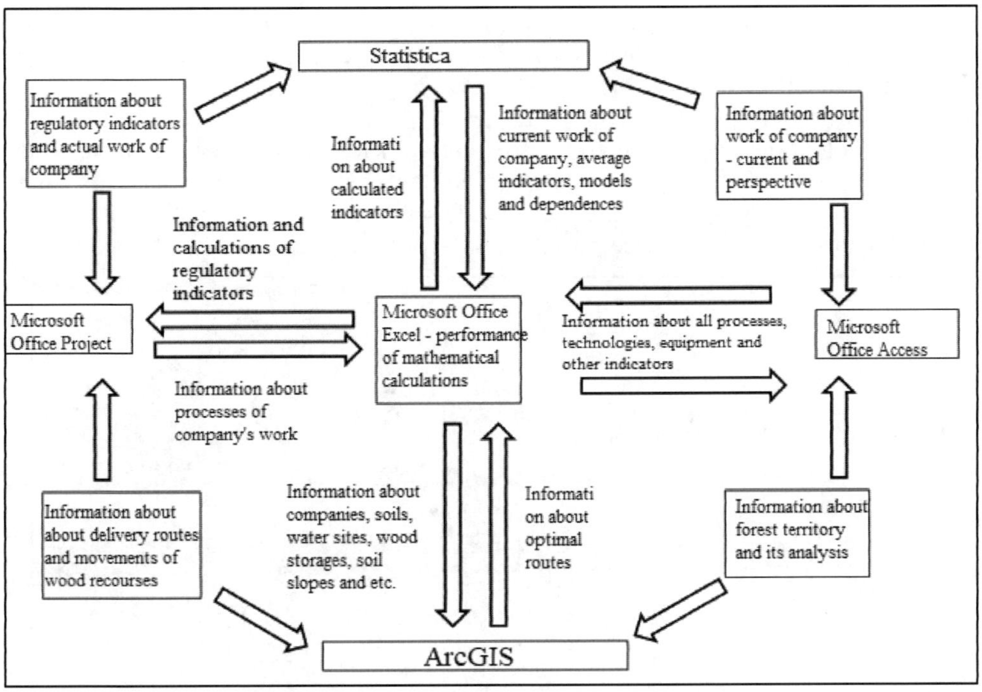

Figure 2. Conceptual model of processing and analysis of information of timber industry logistics in selected software products.

While developing the provisions of the conceptual model, it is necessary to provide information regarding the filling of databases in each of the considered software products (basic information):

1. Microsoft Office Project:

- a list of all the processes of the logging company (timber transport), including alternative options;
- information on the resources involved in the implementation of the processes (with an assignment to specific types of work);
- a time schedule of processes implementation;
- various types of processed (calculated) data regarding the processes of the company (workload of staff, behind schedule, cost of individual stages, etc.), etc.

2. Microsoft Office Excel:

- technological schemes for the implementation of logging and transportation of wood resources;
- characteristics of equipment and machines;
- detailed costing according to various technological schemes;
- natural and climatic characteristics of the forest territory;
- requirements for products from customers;
- regulatory indicators;
- industry average indicators, etc.

3. ArcGIS [10]:

- mapping materials;
- natural and climatic characteristics of the forest territory;
- information on industrial and infrastructure facilities located in the forest territory, etc.

4. Statistica:

- regulatory indicators;
- information on the actual operation of the company;
- planned indicators of future periods;
- industry average indicators, etc.

IOP Conf. Series: Materials Science and Engineering **817** (2020) 012021 doi:10.1088/1757-899X/817/1/012021

5. Microsoft Office Access:

- information for filling in the databases regarding all processes, works and sites that are subject to analysis by the described software products;
- results of calculations and studies using software.

As previously noted, the data connectivity is essential factor in the feasibility of this model. It is important that all used software products allow not only importing data, but also maintaining communication and, if one of the parameters in one of them changes, recalculating all the others. The visualized scheme of work of the applications is shown in figure 3.

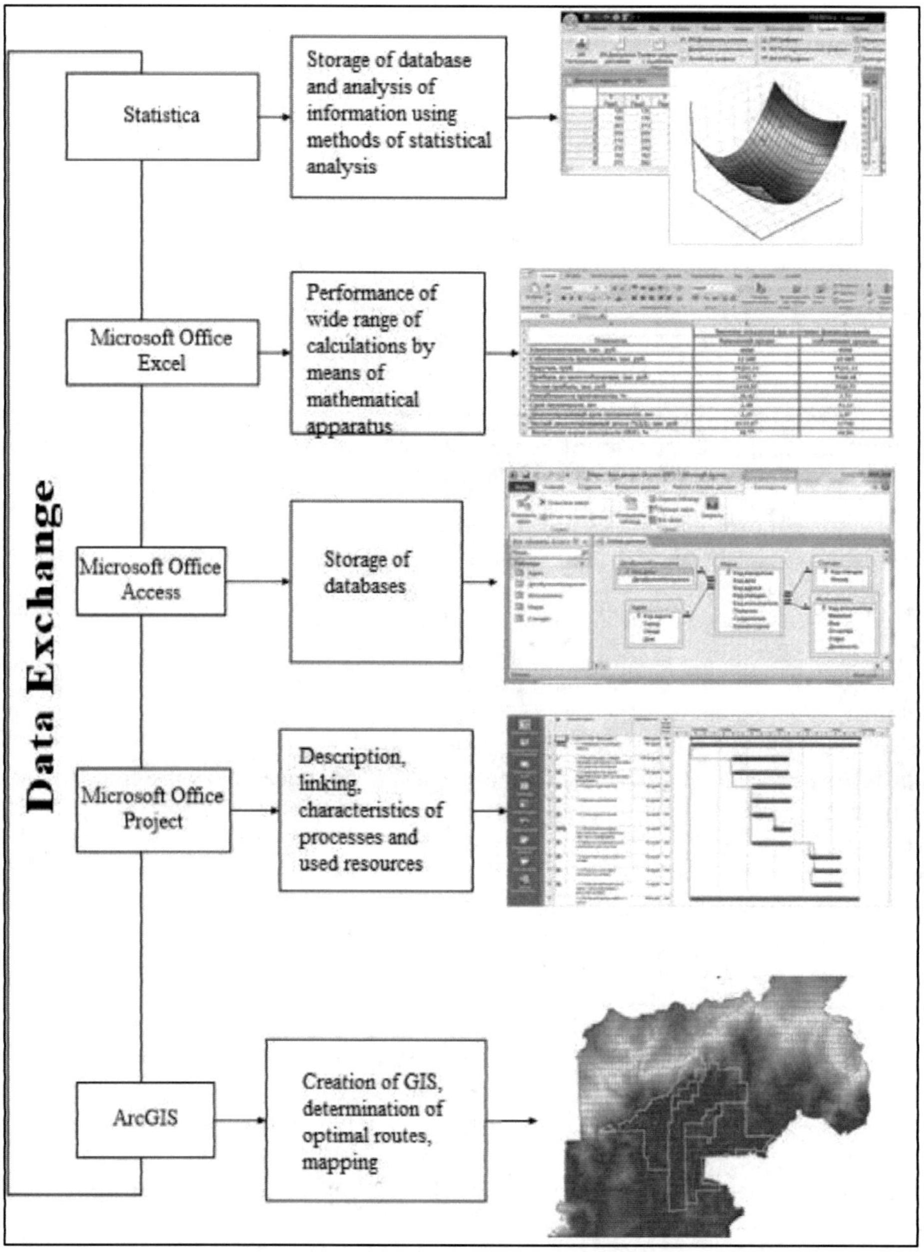

Figure 3. Visualized scheme of operation of software products when optimizing logging company's activity.

It is important to note that it is not difficult to associate data in Microsoft Office applications. For example, when pasting previously copied data, it is necessary to select "Paste Special" \ "Link" command and select the desired insertion option (data or file). Similarly, working with ArcGIS is straightforward [11]. At the same time, it is more difficult to organize the relationship of data with the Statistica program.

This software product allows to associate data with the ability to process only using dynamic data exchange (DDE). DDE communication allows to establish the relationship between the source file (Microsoft Excel spreadsheet) and the STATISTICA data file (client file). At the same time, changes made to the source file automatically appear in the corresponding Data Table (client file).

The use of information technologies must allow the following characteristics of information to function: accuracy, completeness, relevance and clarity. Each of these functions can be implemented through the use of selected software:

- reliability – the use of real data of companies, information on forest areas, standard values, calculated values obtained according to the known and tested formulas and methods;
- completeness – the ability to use a wide (practically not limited) databases, using various software products which allow to evaluate the processes from different perspectives of conducted analysis;
- relevance – the possibility of quick data updating by making adjustments and automatic recalculation of the whole complex of interrelated data;
- clarity – each of the software products allows to demonstrate various aspects of the studied work of the companies – from the presentation in Microsoft Office Project to maps in ArcGIS.

3.2. Discussion

The following priority tasks are determined by the team of authors as promising in the short term for testing the direction of use of data or other software products:

- using the business opportunities to solve issues of the use of secondary wood resources;
- integration of graph theory in applied software products;
- automatic selection of product portfolio for logging enterprises;
- extension of the range of companies that are covered by the survey to all companies of the timber industry complex;
- creation of a standalone product that integrates the capabilities of all the programs represented in the study;
- implementation of possibility of automatic creation on the basis of the conducted calculations of a text document (Microsoft Office Word) with recommendations for changing the activities of the studied company and containing a feasibility study of such changes;
- use of the possibilities of offered software products for development and implementation of a regional policy in logging industry, etc.

4. Conclusion

As the result of the conducted study, the use of a number of software products for the purposes of timber industry logistics was justified. The use of modern information technologies in the field of logging and wood transportation allows to obtain justified decisions on the optimal delivery routes, used technologies and equipment, economic efficiency of individual transformations in the work. The ability to establish a link between the used data is essential for integrated use of various software products. The selected set of software systems can be modified and extended to achieve the specific goals of a company in the industry. Further work of the team of authors is associated with the expansion of the functional use of the suggested software and testing of the suggested solutions.

Acknowledgement

The reported study was funded by a grant from the President of the Russian Federation - for young scientists, candidates of sciences MK-1902.2019.6 and RFBR according to the research project № 18-310-00311.

IOP Conf. Series: Materials Science and Engineering **817** (2020) 012021 doi:10.1088/1757-899X/817/1/012021

References

[1] Chernikov S Yu 2013 The main directions of forestry informatization. *Bulletin of the Voronezh Institute of High Technologies* [In Russian: Vestnik Voronezhskogo instituta vysokih tekhnologij] **10** pp 58-61

[2] Sackov I, Smrecek R and Tucek J 2014 Forest transportation survey based on airborne laser scanning data and GIS analyses. *GIScience&Remote Sensing* **51** 83

[3] Zozulya V V , Romanchenko O V, Zuykov A V, Sergeeva A Y, Medvedev S O and Zozulya I V 2017 Financial stimulation of forest resources deep processing. *Journal of Advanced Research in Law and Economics* **8** pp 306-12

[4] Mokhirev A P, Pozdnyakova M O, Medvedev S O and Mammatov V O 2018 Assessment of availability of wood resources using geographic information and analytical systems (the Krasnoyarsk territory as a case study). *Journal of Applied Engineering Science* **3** 313

[5] Vagizov M R and Sirazetdinova G R 2015 Development of an interactive mapping service: description of the structure and mechanisms of system interaction. *Forest survey and forest management* [In Russian: Lesnaya taksaciya i lesoustrojstvo] **52** 56-61

[6] Zhuravskaya M and Tarasyan V 2014 Forming of the regional core transport network taking into account the allocation of alternative energy sources based on artificial intelligence methods. *Transport Problems* **4** 121-30

[7] Zozulya V V, Sakhanov V V, Medvedev S O, Bezrukikh Y A and Romanchenko O V 2017 The features of industrial modernization management in forest complex. *International Multidisciplinary Scientific GeoConference Surveying Geology and Mining Ecology Management, SGEM 17, Ecology, Economics, Education and Legislation* pp 927-934

[8] Eremeeva L E 2012 Some problems of trucking logistics in activity schemes of timber industry companies. *Forests of Russia and the economy in them* [In Russian: Lesa Rossii i hozyajstvo v nih] **3** 173-6

[9] Porozov P E, Avdeev Yu M, Mokretsov Yu V, Khamitova S M and Lukashevich V M 2018 Development of a forest innovation cluster in the NWFD regions. *Economy and Entrepreneurship* **1** 369-72

[10] Soloviev A A, Zharkikh J I, Krasnoperov R I, Nikolov B P and Agayan S M 2016 GIS-oriented solutions for advanced clustering analysis of geoscience data using ARCGIS platform. *Russian Journal of Earth Sciences* **6** 1-6

[11] Saiger P 2008 Development, implementation and validation of a planetary information system based on ARCGIS. *D L R - Forschungsberichte* **4** 1-129

Formation of the through technological chains within the forest industry cluster

S Medvedev[1*], A Mokhirev[2], T Rjabova[1], M Gerasimova[3]

[1]Department of economic and natural Sciences, *Lesosibirsk Branch of Reshetnev Siberian State University of Science and Technology*, 29 Pobedy Street, Lesosibirsk 662543, Russian Federation
[2]Department of Technology of Logging and Wood Processing Industries, *Lesosibirsk Branch of Reshetnev Siberian State University of Science and Technology,* 29 Pobedy Street, Lesosibirsk 662543, Russian Federation
[3]Department of information and technical systems, *Lesosibirsk Branch of Reshetnev Siberian State University of Science and Technology,* 29 Pobedy Street, Lesosibirsk 662543, Russian Federation

*Corresponding email: medvedev_serega@mail.ru

Abstract. The article is devoted to the problem of formation of technological chains in timber industry clusters. Special attention is paid to the movement of material flows and the method of forming technological chains. As a result, a model of the movement of material flows of wood resources within the timber cluster has been developed.

1. Introduction

The production process of any industrial enterprise is a difficult economic category. Building optimal relationships between the individual elements of a working system is a long-term, continuous process. Moreover, the larger the enterprise, the greater the difficulties in organizing effective activities are. However, the effectiveness of large companies is a sequence higher than the one of small and medium-sized enterprises.

A more difficult task is the organization a sequence higher of relationships between business associations. In this case, in addition to building internal processes, there should be a functioning system of interaction between independent elements that differ significantly from each other in a set of parameters. They can operate according to various standards, in various fields and areas of the economy and production, focus on achieving different results, operate using different levels of automation, etc. Of course, the more participants in such an association, the more difficult their interaction and coordination is in achieving a shared (strategic) goal. The task of forming such the through technological chains is one of the most important within the framework of the activity of clusters, including forest industry ones [1]. Clusters act as a form of association of enterprises within one or several related fields in order to increase efficiency of activity and optimize processes in the conditions of developed infrastructure concentrated in a certain territory. The main goal of the study is to identify the foundations of the formation of the through technological chains within the forest industry cluster.

2. Methods and Materials

Technological chains for the processing of wood resources at forest enterprises were the objects of study. The study was built on the basis of theoretical and practical developments by the team of authors regarding the creation of forest industry clusters and the organization of production processes at the enterprises of the forest industry. Analysis, statistical and mathematical planning, modeling, forecasting, etc. were the main used methods.

3.Results and Discussion

3.1. Results

The feature of the forest industry and, in particular, forest industry clusters is the use of unique raw materials of vegetable origin – the wood. Its feature lies in the possibility of almost full use for obtaining a variety of products. At the same time, the waste received in the production process can be used for release of other goods [2]. The use of recycled wood resources is fully reflected in this industry. The term "waste" is incorrect and applied only to inefficient enterprises and (possibly) regional complexes. It should be noted that products obtained from recoverable wood resources, in some cases (according to the Russian Federal State Statistics Service) can be even more expensive compared to the main product obtained (table 1).

Table 1. The average price for certain types of FIC products in 2018, rubles [3].

Name of goods	Cost, rub.
Coniferous timber, density m^3	2232
Round coniferous timber for sawing and planning, density m^3	2458
Round hardwood for sawing and planning, density m^3	2169
Round hardwood for production of wood pulp (pulpwood), density m^3	1122
Wood fuel, density m^3	725
Softwood lumber, m^3	6616
Hardwood, m^3	6692
Chippings, density m^3	1345
Plywood, m^3	26580
Chipboards and similar boards made of wood or other lignified materials, m^3	11286
Fuel granules (pellets) from wood processing waste, t	4695
Woodpulp and pulp from other fibrous materials, t	41552
Newsprint, rolled and sheets, t	27945
Offset paper, t	49036
Rosin and rosin acids and derivatives, t	72432
Charcoal, t	18530

The range of products of the forest industry and the specifics of the activities of its enterprises are extremely wide [4]. A wide assortment is used in various industries - from construction and decoration to chemical and medical industries. The general (principal) structure of the industry is shown in figure 1.

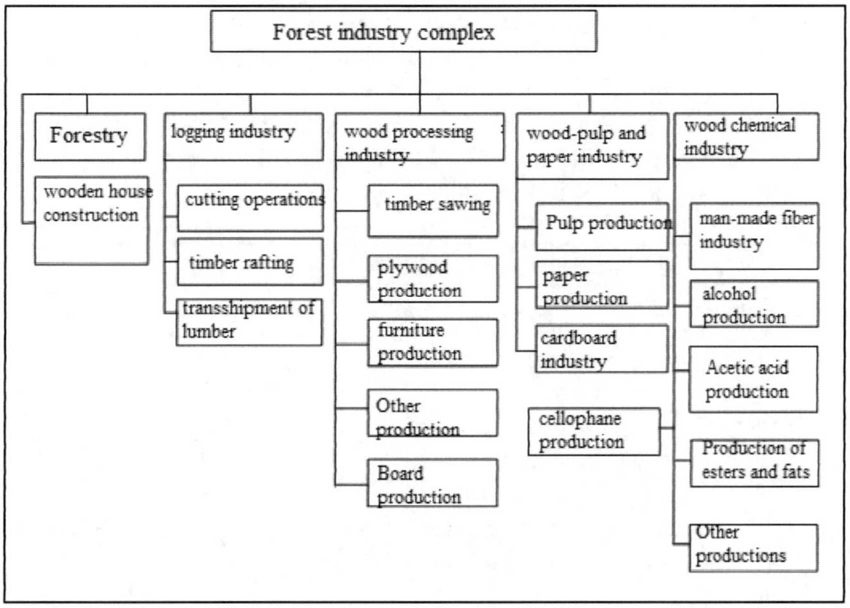

Figure 1. Structure of the forest industry complex (FIC) [5].

Obviously, the forest industry complex includes a wide list of enterprises, where the production of various products was organized. However, any production operates according to a clearly defined algorithm, where the satisfaction of the buyer's needs for a particular product is the key element. The conditions that are given by buyers to the products for sale determine the volume of its production, quality, production technology, etc. The theory of logistics indicates the organization of material flow management in industrial enterprises according to the principles of pulling and pushing systems. In the first case, the crucial role in the functioning of the entire production process was given to the market. The second case is typical, for example, for large businesses that requires continuous production and optimization of the involved production equipment. However, with a sharp change in market demands for products, the transformation of production for new requirements in a short time is practically impossible [6].

The uniqueness of forest industry clusters is including, but not limited to the availability in the organizational association of participants that perform various roles - both large-scale production and small business that can quickly respond to market fluctuations. It is possible to distribute responsibility and resources for the simultaneous operation of two material management systems with the functioning of the Cluster as a single economic system. At the same time, as already noted, the most important resource is recoverable wood raw materials that circulate throughout the cluster formation and serves as its strategic advantage and a connecting element of its participants. The principal diagram of the movement of material flows within the forest industry cluster is shown in figure 2.

In this case, the management system is the coordination body of the Cluster and / or the management (managers) of its individual enterprises. The level of management is determined by the scale and features of product orders by the market. Highlighted key blocks - woodworking, wood processing and deep processing - are the successive stages of involvement of wood raw materials in the production process. If some of the processing stages are not required, the proposed model shall be simplified by eliminating unused production processes.

A separate position in the system of material flows is occupied by logging enterprises and sites. Currently, there is a situation, where the market has an indirect effect on their activities. Almost all regions of the country have a steady demand for logging products (round timber). For this reason and in conditions of the need to develop allotted forest areas, their main task is to minimize the cost of logging and transport operations, sell products at the highest possible price. If the logging is carried out by a division of a large enterprise for its own needs, then in reality all the logged wood used in production is used. Low market dependence is largely due to the production and economic specifics of the forest industry. The period for the return of funds invested in logging can reach one and a half years, provided that the company uses it to produce expensive products.

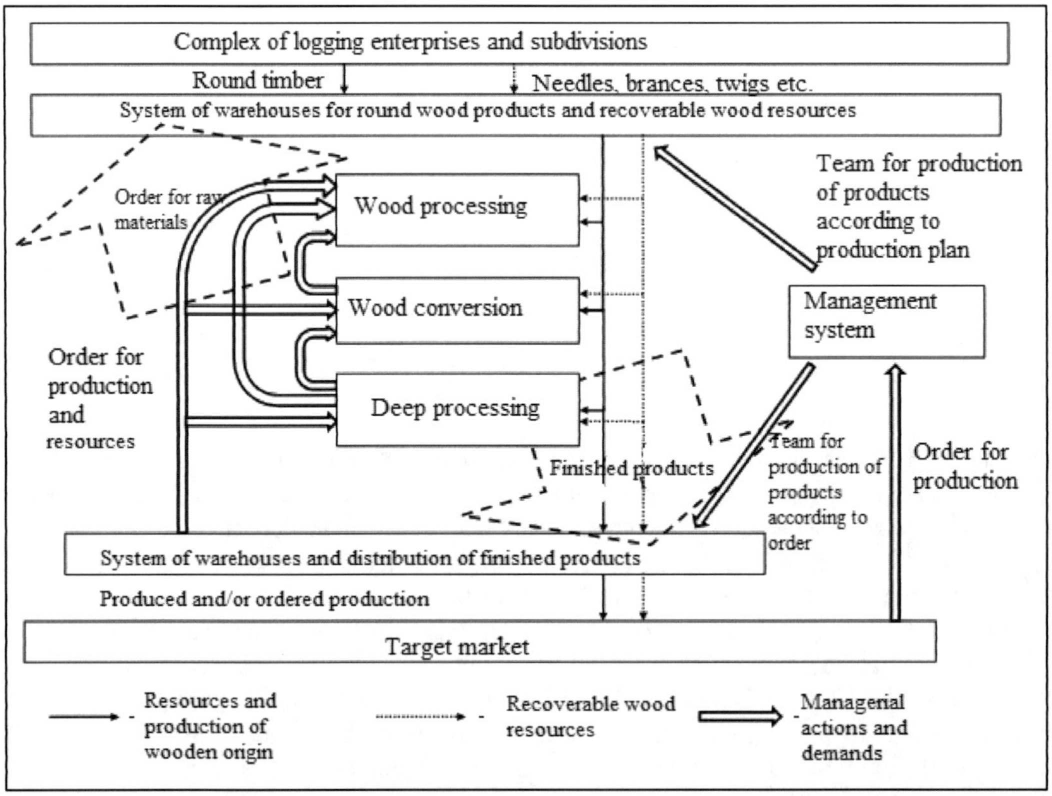

Figure 2. Model of the movement of material flows of wood resourceswithin the forest industry cluster.

It should also be clarified that recoverable wood resources, like any other material (wood) resources that are transported (used) in various processes inside the Cluster can act as a semi-finished product and as a final product. For example, coniferous flour, sawdust, woodchips, lignin, etc. are sold at the market. The market for these products is significantly limited, but exists. At the same time, in practice, in some cases, domestic enterprises not only do not involve them in the production process for profit, but dispose of them, incurring financial losses, or distribute them to the local population (entrepreneurs, state bodies, etc.) for free. In the latter case, obviously, there is a shortfall in revenue.

Analyzing the activities of enterprises within the framework of forest industry clusters, we can distinguish the following key features associated with the formation of technological chains [7]:

- a variety of types of wood resources, differing in quality characteristics, species, cost, directions of use, etc.;
- a wide range of options for using wood resources to produce a variety of products and, as a consequence, the use of a variety of equipment and technologies;
- limited resources, including wood ones;
- disunity and fragmentation of production capacities of enterprises;

- narrow focus of many technologies for processing of wood raw materials;
- remoteness from large sales markets, etc.

The main emphasis in the formation of technological chains should be given to the optimal use of wood resources in order to increase the efficiency of the Cluster. The extent of using of the main resource – (wood) and the extent of effectiveness of the technology determine the success of the functioning of the entire organization in many respects. Obviously, the task of forming of the technological chains should be set at the stage of creating the Cluster, and the process itself is to be strictly regulated. The study proposed a scheme for the formation of the through technological chain of the use of wood resources in the framework of the forest industry cluster (figure 3).

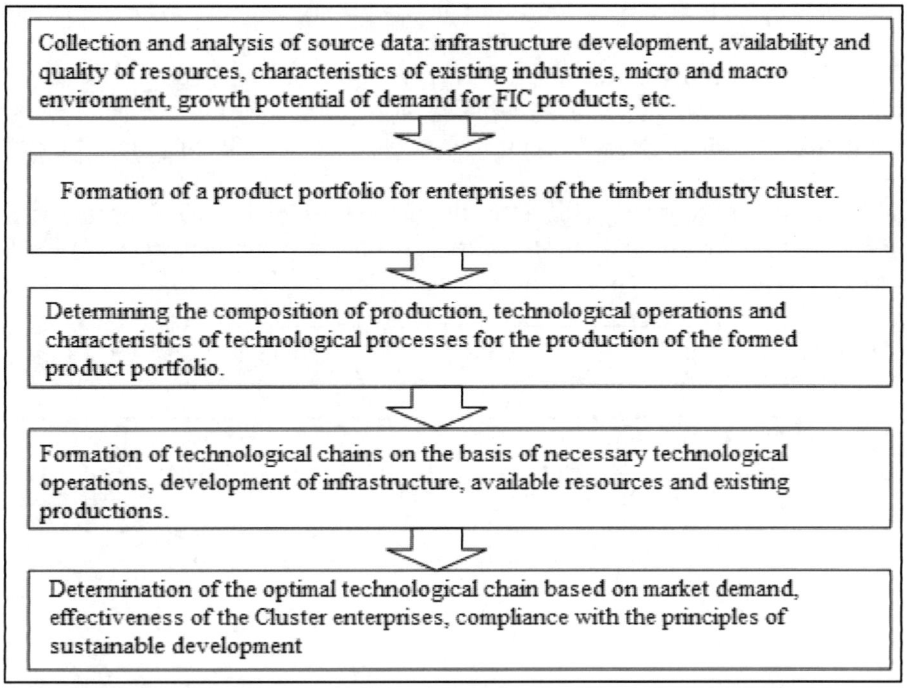

Figure 3. Methodology for the formation of the through technological chain of the use of wood resources within the framework of the forest industry cluster.

According to this technique, an important aspect is the determination of the optimal technological chain based on a number of factors:

1. Market demand, including forecasted one. It determines, ultimately, the volume of products for sale, its quality characteristics, assortment, price and a number of other parameters. The technological chains should be formed considering forecasts of changes in demand. Otherwise, there is a possibility of a lack of sales for the manufactured assortment of goods and a set of related problems. According to the research [Review of the FIC of Russia], the following FIC products are the most promising in terms of demand growth:

- lumber;
- cellulose;
- plywood;
- packaging materials;
- wood pellets;
- OSB;

- MDF and others.

2. The effectiveness of the enterprises of the Cluster. It is important to note that the selection of the optimal technological chain is influenced not only by the effectiveness of technological solutions, but also by the complex of organizational and economic impacts. Situations with an increase in productivity with competent management of enterprises, or losses even of the most technologically competently built industries are possible. Management is the most important component in the work of the Cluster, the functioning of individual technological chains and the activities of individual enterprises [8].

3. Compliance with the principles of sustainable development. Another key component in the functioning of technological chains in the forest industry cluster is the focus on compliance with environmental norms, rules and requirements. The principles of sustainable development should be adopted and put into practice in the Cluster. This provision is necessary both for the sale of products on the world market and for solving a range of issues in activities on the territory of Russia: relations with the local population, state bodies, creating a favorable image, reducing fines for environmental impact, etc.

Based on the results of the work, a model of the through technological chain of the use of wood resources in the framework of the forest industry cluster was formed. It provides the through movement of wood resources on the way from logging to deep processing (figure 4).

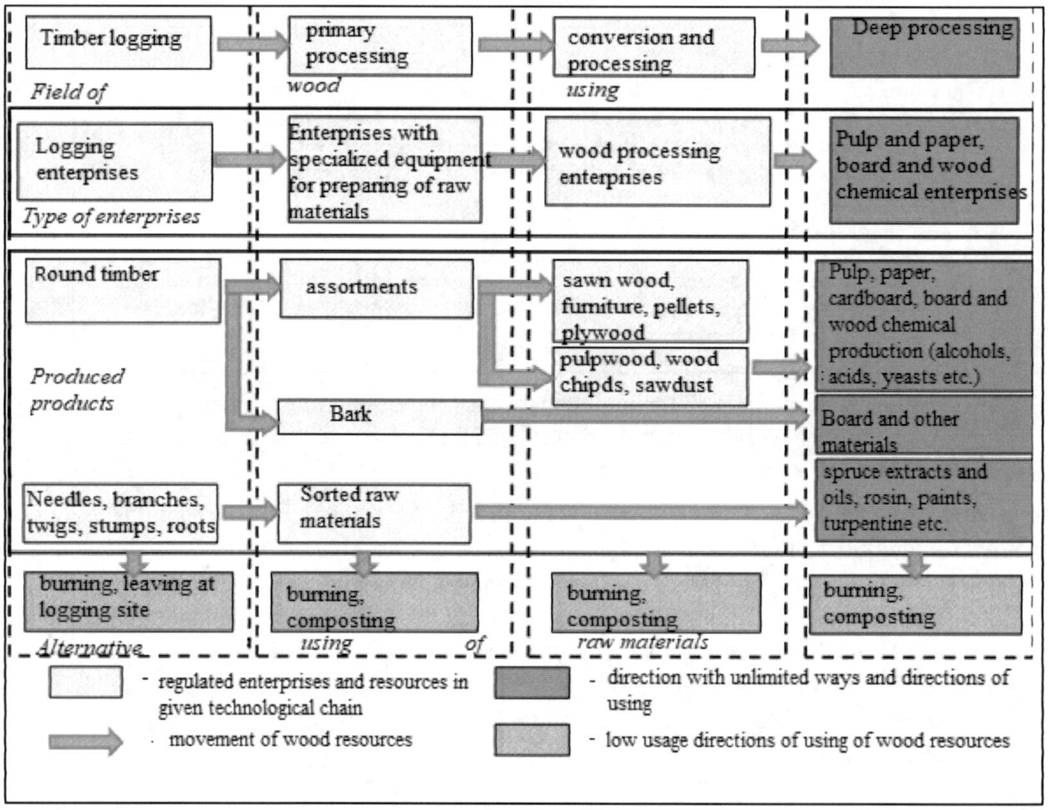

Figure 4. Model of the through technological chain of the use of wood resources in the framework of the forest industry cluster.

Four blocks are provided in this model: the field of use of wood, the type of enterprises, manufactured products and the alternative to using raw materials. The movement of wood resources begins with the stage of logging and goes through the process of conversion into final products up to goods for deep processing. However, it is obvious that not all raw materials are converted to expensive products. Individual Cluster enterprises may specialize in the production and sale of, for example, lumber. Recoverable wood resources - croaker, wood chips, sawdust etc. are to be transferred at domestic prices to other enterprises of the Cluster. However, in some cases, they can be implemented outside. The given alternative options for the use of raw materials - burning, composting, leaving at the cutting area - may also take place in the Cluster technological chain. Their use in some cases is technologically and economically feasible. However, these areas have low economic efficiency and should be avoided.

3.2. Discussion
The following areas can be identified as promising areas of study by the authors:

- economic and mathematical modeling of processes of technological chains of the use of wood resources within the framework of the forest industry cluster;
- the formation of an algorithm for determining the optimal portfolio of products;
- the formation of models for the development of production and economic indicators of enterprises of the forest industry cluster;
- development of a methodology for assessing the effectiveness of technological chains of the use of wood resources in the framework of the forest industry cluster;
- development of methods, algorithms and software for the selection of optimal technological chains of the use of wood resources in the framework of the forest industry cluster;
- development of models for the development of production chains and interconnections of enterprises of the forest industry cluster, taking into account the factors of micro- and macroenvironment;
- development of a methodology for analyzing the impact of the state of infrastructure on the technological chains of the forest industry cluster, etc.

4. Conclusion
As a result of the study, the foundations of the formation of the through technological chains within the forest industry cluster were determined. The model of the movement of material flows of wood resources within the forest industry cluster was developed. Its main difference is the simultaneous use of pulling and pushing principles. This is one of the distinguishing features of clusters. The identified key factors affecting the determination of the optimal technological chain shall also be considered as priorities in the development of individual enterprises and organizations of the forest industry. The development of the Russian forest industry, according to the author, is to be based on the principles of clustering and increasing the efficiency and depth of processing of wood resources. The most important production and economic factors of the development are the applied through technological chains of the use of wood resources.

Acknowledgement
The reported study was funded by a grant from the President of the Russian Federation - for young scientists, candidates of sciences MK-1902.2019.6 and by RFBR, the Government of Krasnoyarsk Territory, Krasnoyarsk Regional Fund and LLC «Krasresurs 24», project number 20-410-242901.

References
[1] Chernikov S Yu 2013 The main directions of forestry informatization. *Bulletin of the Voronezh Institute of High Technologies* [In Russian: Vestnik Voronezhskogo instituta vysokih tekhnologij] **10** pp 58-61.

[2] Sackov I, Smrecek R and Tucek J 2014 Forest transportation survey based on airborne laser scanning data and GIS analyses. *GIScience&Remote Sensing* **51** 83.

IOP Conf. Series: Materials Science and Engineering **817** (2020) 012022 doi:10.1088/1757-899X/817/1/012022

[3] Zozulya V V, Romanchenko O V, Zuykov AV, Sergeeva A Y, Medvedev S O and Zozulya I V 2017 Financial stimulation of forest resources deep processing. *Journal of Advanced Research in Law and Economics* **8** pp 306-12.

[4] Mokhirev A.P., Pozdnyakova M.O., Medvedev S.O. and Mammatov V.O. 2018 Assessment of availability of wood resources using geographic information and analytical systems (the Krasnoyarsk territory as a case study). *Journal of Applied Engineering Science* **3** 313.

[5] Zozulya V V, Sakhanov V V, Medvedev S O, Bezrukikh Y A and Romanchenko O V 2017 The features of industrial modernization management in forest complex. *International Multidisciplinary Scientific GeoConference Surveying Geology and Mining Ecology Management, SGEM 17, Ecology, Economics, Education and Legislation* pp 927-934.

[6] Zhuravskaya M and Tarasyan V 2014 Forming of the regional core transport network taking into account the allocation of alternative energy sources based on artificial intelligence methods. *Transport Problems* **4** 121-30.

[7] Eremeeva L E 2012 Some problems of trucking logistics in activity schemes of timber industry companies. *Forests of Russia and the economy in them* [In Russian: Lesa Rossii i hozyajstvo v nih] **3** 173-6.

[8] Porozov P E, Avdeev Yu M, Mokretsov Yu V, Khamitova S M and Lukashevich V M 2018 Development of a forest innovation cluster in the NWFD regions. *Economy and Entrepreneurship* **1** 369-72.

The forecast of forestry tractor speed with mechanical transmission

O Mikhailov*, G Taradin, M Taraban, S Cozmin

St. Petersburg State Forest Technical University, 5 Institutskiy Lane, St. Petersburg 194021, Russian Federation

*Corresponding e-mail: kerro07@mail.ru

Abstract. One of the most energy-intensive logging processes is skidding of trees. The correct choice of engine and transmission system parameters significantly affects the efficiency of forest machine use. The article suggests the method of determining the optimal range for mechanical transmission of skidders, which are widely used in skidding of log the logging area to the upper warehouse. This technique allows considering properties of fiber (rolling resistance, drag part of the forest pack and the slope of the fiber), truckload and parameters of the motor installation. It should be also noted that the coefficient of adaptability of the diesel engine (the coefficient of engine torque loading) and the engine loaded at crankshaft speed are taken into account. Results of experimental researches that have shown good convergence with theoretical calculations are presented.

1. Introduction

Currently about 40% of the world's timber is logged using whole tree technology, and these operations are mainly carried out with the use of machines with a classic mechanical transmission. These transmissions have a number of advantages over hydrodynamic and hydrostatic. They are easy to maintain and operate especially at low temperatures, do not require expensive synthetic hydraulic fluids, they have higher efficiency (no double conversion of energy), and therefore lower fuel consumption.

It should also be noted that there are conditions at the logging site where it is not feasible, and sometimes even impossible, to use modern logging complexes with hydrostatic transmissions. Classic skidder tractors for forestry, reforestation and firefighting operations are indispensable in the forest.

Therefore, increasing the efficiency of the forestry tractor with mechanical transmissions is a task of current interest.

The aim of the research is to increase the traction and velocity properties of the skidder and to reduce the energy consumption of the skidder process by determining the optimal parameters of the engine and transmission unit.

2. Methods and Materials

The proposed method allows to evaluate the parameters of powerpack unit at the design stage. The formula (10) shows that the accuracy of the results is influenced by the engine torque factor (K_{lo}^{i}), which in its turn is significantly influenced by the adaptability coefficient of the diesel engine torque (K_{lo}) [1, 2]. This is very important due to the increased use of Common Rail diesel engines, which have increased torque reserves.

The results of the study can also be used in the development of test programs and the calculation of the strength of components.

3. Results and Discussion

One of the main criteria for evaluating the efficiency of work is the transport capacity of the skidder system consisting of a skidder and a trawl bundle of wood. This criterion essentially depends on the time of T_i operation of the system at the i front and on the number of switching operations.

Each gear ratio i determines the lower and upper limits of the velocities used at $\upsilon_i^{\min}$ and at $\upsilon_i^{\max}$. The velocity of $\upsilon_i^{\max}$ is determined by the operation of the engine on the velocity characteristic regulator branch and $\upsilon_i^{\min}$ by the crankshaft velocity n_{en} corresponding to the maximum engine torque.

Time T_i of tractor operation at i gear is determined by the formula:

$$T_i = T \int_{v_i^{\min}}^{v_i^{\max}} f_V(v)dv, \qquad (1)$$

where, the T – full time skidding, $f_V(v)$ – the density of the random velocity distribution of the tractor V.

When skidding, the skidding velocity of the skidding system V is a function of the skidding resistance coefficient (f_c), the skidding slope (α) and the radius R. These values of the trail properties can be considered as random values determined by the movement of the tractor and the slip of the packs (X_1), the slope of the trail (X_2) and the slew resistance of (X_3), and besides

$$X_k = A_k V^{-1} + B_k \qquad (2)$$

where, the coefficients A_k and B_k $(k = 1, 2, 3)$ are determined by the parameters of the engine and the skidding system [3-7].

In the given work, the technique of direct determination of density $f_V(v)$ is offered that simplifies calculation of time T_i of system work on separate transfers.

At the same time, no special assumptions are made regarding the type of distribution of resistance forces X_k, which is quite important, since the laws of distribution X_2 and X_3 significantly differ from normal. The developed scheme is based on the possibility to rearrange the operations of addition of independent random values of one dimension and calculation of their distribution densities.

Suppose there is a functional dependence of $Y = t(X)$ between random variables X and Y. The distribution densities $f_X(x)$ and $f_Y(y)$ of these random variables are related by the ratio [6, 7]

$$f_Y(y) = \int_{-\infty}^{+\infty} f_X(x)\,\delta[y - t(x)]dx, \qquad (3)$$

where, δ is a Dirac function belonging to many broad functions.

If $t(x)$ has no multiple roots, the generalized function $[y - t(x)]$ can be represented as a linear combination of broad functions $\delta(x - x_m)$ [5]:

$$\delta\big(y - \varphi(x)\big) = \sum_m |\tau'_m(y)|\delta[x - \tau_m(y)]. \qquad (4)$$

here, $\tau_m(y) = x_m$, where x_m form a set of equation solutions

$$y = t(x) \qquad (5)$$

Therefore, the non-negative number m depends on y, i.e. $m = m(y)$. Let us denote by Φ set of all $t(x)$ values at $-\infty < x < +\infty$. If $y \overline{\in} \Phi$, then $m(y) = 0$, equation (5) has no solution and sum (4) turns to zero. We will further assume that (5) is a monotone function. Then there is an inverse function with respect to (5) $x = \tau(y)$ defined for all as $y \overline{\in} \Phi$. Hence $m(y) = 0$ if $y \overline{\in} \Phi$ and $m(y) = 1$ if $y \in \Phi$ and according to (4) we have:

$$\delta\big(y - t(x)\big) = \begin{cases} 0 \; if \; y \overline{\in} \Phi; \\ |\tau'(y)|\delta[x - \tau(y)] \; if \; y \in \Phi. \end{cases} \quad (6)$$

By virtue of (3) and (6) we find:

$$f_Y(y) = \begin{cases} 0 \; if \; y \overline{\in} \Phi; \\ |\tau'(y)|f_X[\tau(y)] \; if \; y \overline{\in} \Phi. \end{cases} \quad (7)$$

Formula (7) allows to pass from the distribution density $f_X(x)$ of a random variable X to the distribution density $f_Y(y)$of random variable $Y = t(X)$ if the function $\tau(y)$ inverse to the monotonous function (5) is known.

If $f_\psi(x)$ is the density of the generalized law of distribution of the properties of the skidding wire corresponding to a random value:

$$\psi = f_c + i + f_t, \quad (8)$$

where,f_c, i and f_t are dimensionless analogues of resistance forces X_1, X_2 and X_3 then according to (2) we obtain:

$$\psi = AV^{-1}(A = A_1 + A_2 + A_3), \quad (9)$$

because $B_1 + B_2 + B_3 = 0$. Coefficient A is found by the formula [6, 7]

$$A = 0.377 M_{en} K_{lo}^i \eta_{tr} n_{en} K_n^i (G + Q)^{-1}, \quad (10)$$

where, M_{en} is the motor torque at rated crankshaft speed; K_{lo}^i is an coefficient of the engine torque load; η_{tr} is a transmission efficiency;n_{en} is a nominal engine speed; K_n^i- speed utilization factor at i-th gear; G is a tractor weight; Q is a weight of the forest pack.

According to (7) by replacing (9) the following formula is obtained

$$f_V(v) = \begin{cases} 0 \text{ if } v \leq 0; \\ Av^{-2} f_\psi(Av^{-1}) \text{ if } v > 0. \end{cases} \quad (11)$$

May $MX_k = \alpha_k$ (mathematical expectations); $DX_k = \sigma_k^2$ (dispersion) and α_k (asymmetry coefficients) are parameters of the random components X_k. In real situations, the asymmetry coefficient α_1 is statically insignificant and therefore we will assume $\alpha_1 = 0$. Using the transition formulas (11) we can show that the distribution density $f_V(v)$is set in the form:

$$\begin{cases} f_V(v) = \dfrac{A}{\sigma v^2} \exp\left(-\dfrac{t^2}{2}\right)\left[1 + \dfrac{\alpha}{6} H_3(t) + \dfrac{\beta}{720} H_6(t)\right]; \\ t = \dfrac{Av^{-1} - \alpha}{\sigma}; \alpha = \alpha_1 + \alpha_2 + \alpha_3; \sigma^2 = \sigma_1^2 + \sigma_2^2 + \sigma_3^2; \\ \alpha = (\alpha_2 \sigma_2^3 + \alpha_3 \sigma_3^3)\sigma^{-3}; \; \beta = 20\alpha_2\alpha_3(\sigma_2\sigma_3\sigma^{-2})^3, \end{cases} \quad (12)$$

where,

$$H_3(t) = t^3 - 3t; \; H_6(t) = t^6 - 15t^4 + 45t^2 - 15 \quad (13)$$

- Chebyshev-Ermita polynomials [8, 9].

Besults and discussion concrete calculations will be carried out with reference to the tractor TLT-100M with rated power of the engine $N_{en} = 72$kw. Let us consider two variants of the travel loads $Q = 5.0 \text{ m}^3$ and $Q = 6.5 \text{ m}^3$ at different values of the adaptability coefficient of the diesel engine K_{lo}. For characteristic operating conditions of tractors OTZ in the region of Karelia these loads correspond to the total factors of resistance to movement $\psi_1 = 0.221$, $\psi_2 = 0.230$ and average square deviation $\sigma_1 = 0.032$, $\sigma_2 = 0.034$.

Table 1. Use of tractor transmission time TLT-100M.

Q, m³	Transmissionn umber	i_{tr}	$\upsilon_i^{min} - \upsilon_i^{max}$, km/h	$T_i T^{-1}$, %			
				K_{lo}=1.15	K_{lo}=1.30	K_{lo}=1.40	K_{lo}=1.50
with five-speed serial gearbox							
5.0	I	2.88	1.98–2.71	0	0	0	0
	II	2.14	2.66–3.65	8.1	0	0	0
	III	1.64	3.47–4.76	79.4	16.5	10.5	4.3
	IV	1.06	5.38–7.37	12.5	82.6	88.1	94.1
	V	0.65	8.77–12.02	0	0.9	1.3	1.6
6.5	I	2.88	1.98–2.71	0	0	0	0
	II	2.14	2.66–3.65	20.6	0.6	0	0
	III	1.64	3.47–4.76	75.4	42.7	3.4	9.7
	IV	1.06	5.38–7.37	4.0	56.4	65.8	89.1
	V	0.65	8.77–12.02	0	0.3	0.8	1.2
with four-speed reversing gearbox							
5.0	I	2.27	2.47–3.39	2.4	0	0	0
	II	1.57	3.57–4.89	85.7	24.9	16.2	7.8
	III	1.03	5.44–7.46	11.9	73.1	80.9	88.9
	IV	0.69	8.12–11.13	0	2.0	2.9	3.3
6.5	I	2.27	2.47–3.39	8.9	0.1	0	0
	II	1.57	3.57–4.89	86.4	59.4	42.5	15.1
	III	1.03	5.44–7.46	4.7	39.9	56.1	82.5
	IV	0.69	8.12–11.13	0	0.6	1.4	2.4

Figure 1 shows the results of experimental studies of TLT-100M tractor transmissions with DPM (K_{lo} = 1.40) at the Onega Tractor Plant test site. Results of researches show good convergence with theoretical calculations (table 1).

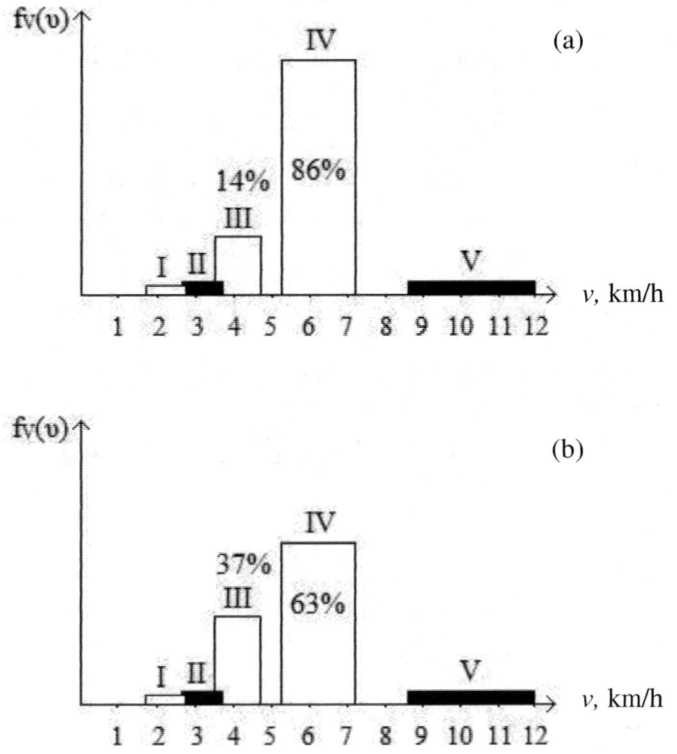

Figure 1. Use of tractor's TLT-100M gearboxes with CDA (K_{lo} = 1.40) and five-speed gearbox on the FFW polygon. a) $Q = 5.0$ m^3, b) $Q = 6.5$ m^3

References

1. Ivanov G A 1982 Influence of the engine adaptability factor on the choice of the number of gears in the tractor transmission (in Russian – Vliyanie koeffizienta prisposoblyaemosti dvigatelya na vybor chisla peredach v transmissii traktora)/ Tractors and agricultural machines, **8**, 12-13
2. Izvekov V S 1980 About efficiency of use on the skidding tractor of the engine with the increased correcting factor of a reserve of torque (in Russian – Ob effektivnosti ispolzovaniya na trelevochnom traktore dvigatelya s povyschennym korrekturnym koeffizientom zapasa krutyaschtgo momenta)/ Collection of data from Moscow State Forestry Engineering Institute. Issue 126, 104-109
3. Anisimov G M 1975 Operating conditions and load of the skidder transmission (in Russian - Usloviya ekspluatacii i nagrugennosti transmissii trelevochnogo traktora) (Moscow: Forestry industry) 165
4. Anisimov G M and Pustoshny P A 1981 Prediction of the skidder gear use (in Russian – Prognozirovanie ispolzovanya peredach trelevochnogo traktora) - Ex. higher education institutions. Forestry Magazine (Lesnoy Zhurnal), **2**, 55-58
5. Antipin V P, Durmanov M Yand Mikhaylov O A 2019 Choosing Transmission Gearset for Agricultural Aggregates Based on Energy Consumption eds. Radionov A, Kravchenko O, Guzeev V and Rozhdestvenskiy Y Proceedings of the 4th International Conference on Industrial Engineering. ICIE 2018. Lecture Notes in Mechanical Engineering. Springer,

IOP Conf. Series: Materials Science and Engineering **817** (2020) 012023 doi:10.1088/1757-899X/817/1/012023

Cham, pp 1591-1600. https://doi.org/10.1007/978-3-319-95630-5_170

6. Mikhailov O A, Spiridonov S V, Taradin G S and Durmanov M Ya 2018 Prediction of the time of use of transmission transmission numbers taking into account the random nature of the weight of the pack» (in Russian– Prognozirovanie vremeni ispolzovaniya peredatochnyh chisel transmissii s uchetom sluchainogo haraktera vesa pachki) / Forests of Russia: politics, industry, science, education / materials of the third international scientific and technical conference. Volume 2 / ed V M Gedjo (St. Petersburg: SPbFTU) 195-197
7. Mikhailov O A, Martynov BG , Durmanov M Ya and Taradin G S 2018 Efficiency of the use of diesel engines with high adaptability on the harvesting machines (in Russian – Effektivnosti ispolzovaniya dizelei s vysokoi prisposoblyaemostyu na lesosechnyh maschinah)/ Forests of Russia: politics, industry, science, education / materials of the third international scientific and technical conference. Volume 2 / ed by V M Gedjo (St. Petersburg: SPbFTU) 197-199
8. Gelfand I M and Shilov G E 1958 Generalized functions (in Russian – Obobschennye funkzii) Issue. 1 (Moscow: Science) 439
9. Pugachev V S 1968 Introduction to probability theory (in Russian – Vvedenie v teoriyu veroyatnostei) (Moscow: Science) 368

Finding an effective technological chain of the logging process in the natural and production conditions of the Krasnoyarsk region

A Mokhirev[1]*, K Rukomojnikov[2], M Gerasimova[1], S Medvedev[1]

[1]*Lesosibirsk Branch of Reshetnev Siberian State University of Science and Technology,* 29 Pobedy Street, Lesosibirsk 662543, Russian Federation
[2]*Volga State University of Technology,* 3 Lenin square, Yoshkar-Ola 424000, Mari El Republic, Russian Federation

*Corresponding email: ale-mokhirev@yandex.ru

Abstract. In the logging process, it is possible to build a technological chain in different ways. When transporting wood from the cutting area, water or land transport can be used. Using of lumberyards and processing of wood on them also increase the multivariance of the technological process. Execution of operations in various natural and industrial conditions differs in productivity and material costs. The efficiency of production depends on the choice of logging technology in dynamic natural production conditions. The authors presented a graphic-analytical model for solving the given task in the publication. The article is devoted to the graphic-analytical model for solving the task. A numerical example is solved for this purpose. The numerical example showed the efficiency of the method of finding a rational technological chain of logging operations in dynamic natural and production conditions.

1. Introduction

In recent years the removal of wood from the cutting area to the consumer is characterized by a multivariate technological process with the use of forest warehouses. Most of them are used for wood storage between transport seasons [1]. Finding the optimal variant of the technological chain is an actual task.

Logging operations require adequate and careful planning in order to rationally select the technological chain and effectively solve the problems of forest management. In addition, logging operations must be coordinated with the transportation of harvested wood to the consumer, since transport costs account for the majority of the total logging costs [2]. Also, when choosing technological chains, there is a question of the wood processing degree, the type of commodity products obtained from it. After all, this affects the consumer ability of the product and its price. Therefore, searching for optimal solutions that take into account both transport and procurement costs, as well as the cost of products sold is important for building reliable wood supply chains and maintaining the profitability of the enterprise.

Planning the movement of marketable products to the consumer in a dynamic environment is probably the most important decision at the operational level and is critical to maintaining the wood supply chain. These operations should be carried out ideally with minimal costs, while meeting the requirements of the consumer to the product. Besides, logging and transport operations account for the bulk of the costs, and disruption of wood supplies can lead to high costs in processing industries.

To solve such problems graphic-analytical models may be successfully used, envisaging a construction of links between operations, allowing to perceive the technological chain for certain types of work in a logical sequence, eliminating the duplication of certain technological elements, giving the opportunity to evaluate different systems of machines and the structure of production flows in a variety of industrial processes of forest works [3].

Methods of substantiation of finding the shortest paths between the vertices of the graph are described in Moore's [4], Floyd's [5], Dijkstra's [6], Bellman's researches [7], etc.

They allow to carry out the analysis of static elements of the enterprises' work and serve as a basis for realization of stationary-dynamic tasks of stream programming at rationalization of flows in transport systems. [8-10].

The simplex method of linear programming can be used to solve such problems [11].

The above methods and algorithms can be used in the process of considering the operations of the logging technological process in dynamics when presenting them in the form of time-stretched graphs. These studies demonstrate in detail the basic scheme and analysis capabilities of the operating network. However, they characterize the possibility of passing through the arcs of each time interval only one flow option and do not take into account the specifics of the logging industry, envisaging the need for a comprehensive solution to the problems of developing several cutting areas within one time range when justifying the schedule of the logging technological process. In view of this, they cannot always be used in solving the issues of substantiating the technological chain of work at timber enterprises.

In the conditions of the need to implement such a comprehensive analysis, it becomes obvious that the use of machines and mechanisms involved in the performance of work on one of the analyzed areas of the logging enterprise reduces the resources of their possible application in the course of performing the same operations on the territory of another site in the time interval adopted for the analysis. The mentioned researches, stipulating theoretical possibility of passing several flows on the eponymous operations of technological process of each analyzed period provide the researcher with the possibility of the analysis only the graphs with independent from each other carrying capacities of separate parallel arcs of analyzed time intervals.

In the paper [12] the authors propose detailed graphical models of wood transportation from the cutting area to the consumer, loading, unloading and processing operations taking place at the intermediate and lower lumber yard. The research [13] suggests a methodology for solving this problem. The proposed mathematical dependences allow to carry out search of the maximum flow of the minimum cost in dynamic structure of technological process of works performance at the enterprise. In addition to the dynamic component, the methodology differs by taking into account the income of the enterprise from the sale of marketable products. The purpose of this study is to use this technique to improve the technological chain of transport, loading, unloading and processing operations of the logging process in dynamic natural-production conditions using the developed graphic-analytical model.

The initial data for the implementation of the numerical example were formed from the real data of logging enterprises of the Krasnoyarsk territory of the Russian Federation.

It is advisable to divide the natural and climatic conditions of the logging area into 5 periods: winter (124 days), winter-spring (31 days), spring (72 days), summer (62 days), autumn (71 days). In each period, the performance of machines on operations is different. Logging is carried out on two sites, from which it is possible to transport wood directly to the consumer or through the lower coastal warehouse (figure 1). The roads are partly temporary logging roads or year-round roads. The schematic diagram of the logging site is shown in figure 1. On two sites timber can be harvested in the form of whips or sortiments of up to 20 thousand m^3 in the first area and up to 50 thousand m^3 in the second area. It is possible to transport the wood by rafting along the river from the lower coastal warehouse, rafts can be formed for this purpose. Timber storage between periods is carried out in stacks in the lower warehouse, for this purpose a land area is rented and personnel are hired for security. The consumer is ready to accept wood in any period in the amount of 70 thousand m^3. To solve the problem, the cost of performing technological operations *C*, the productivity of *П* machines and equipment for their implementation, as well as labor costs *f* for performing operations are determined.

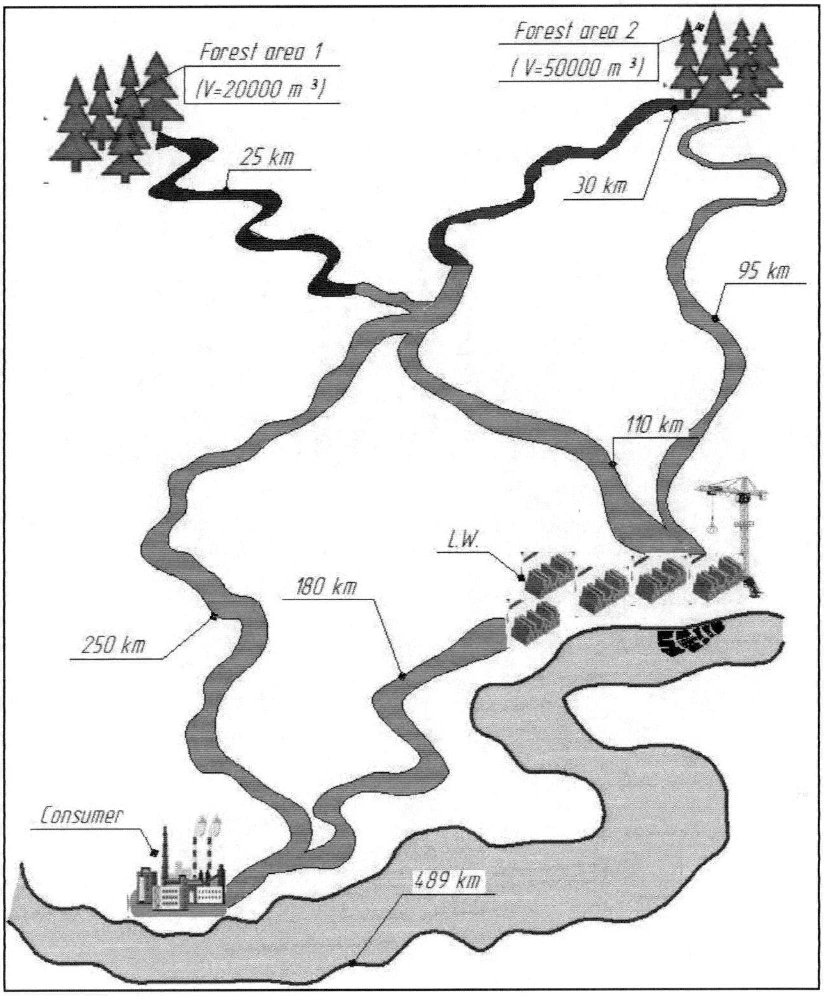

Figure 1. Schematic diagram of the logging site.

It is required to determine the technological chain of operations for primary processing and transportation of harvested wood from two forest areas to the consumer, characterized by a minimum cost and providing maximum integrated production when performing of all types of work, taking into account the initial data.

2. Results and Discussion

Analyzing the initial data, it can be assumed that the most effective periods of transportation are winter (I) and summer (IV). To simplify the solution of the numerical example, according to the method [13], make a graph for these periods, taking into account only the time parameters of the remaining periods (figure 2).

IOP Conf. Series: Materials Science and Engineering **817** (2020) 012024 doi:10.1088/1757-899X/817/1/012024

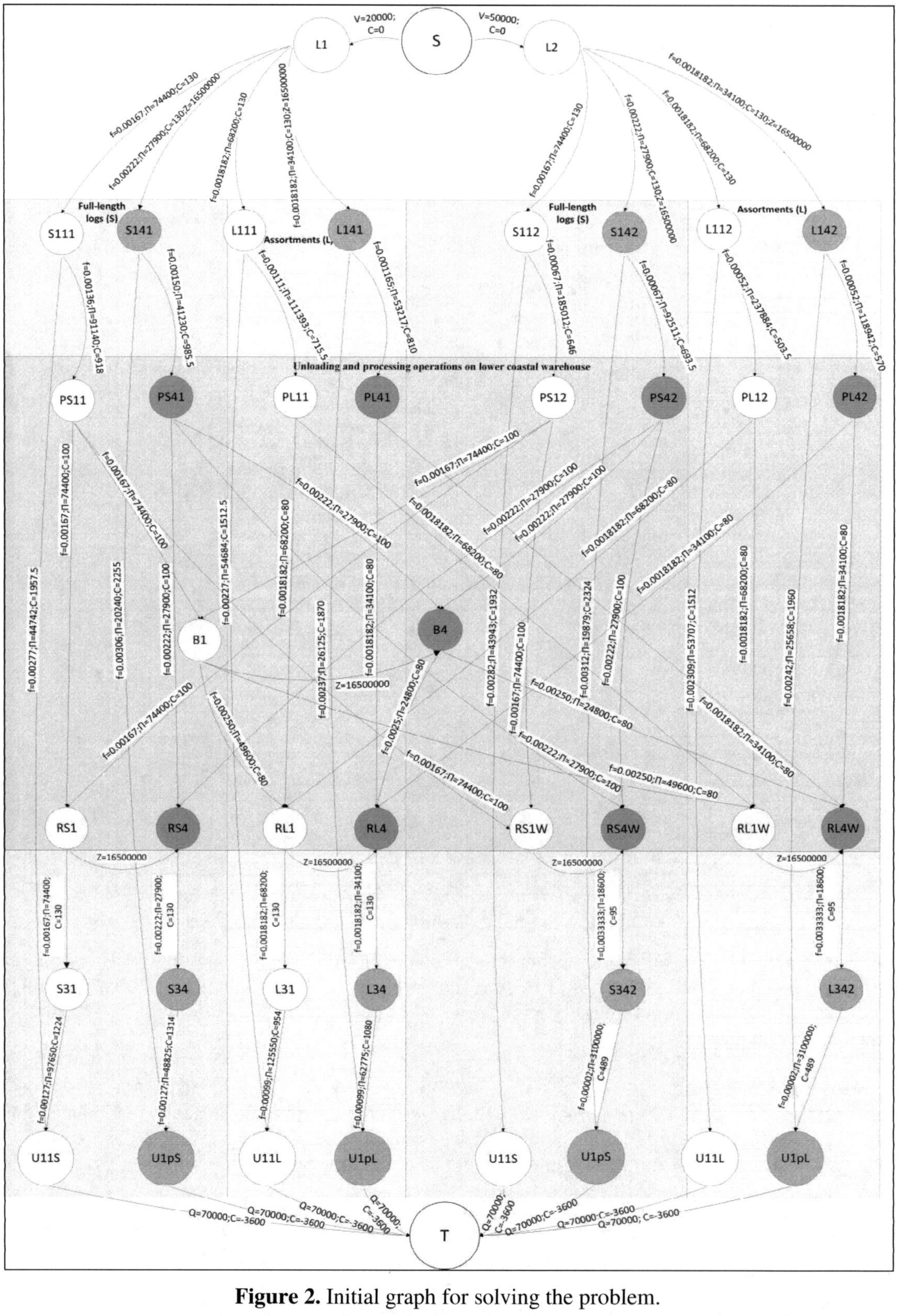

Figure 2. Initial graph for solving the problem.

This graphical model represents the "stretched in time" dynamic graph and allows to carry out graphic-analytical approach to the substantiation of logging process' technological chain of forest warehouses, raids of loading and unloading operations, processing operations, modes of transport (water, land), the choice of the consumer and type of final commodity products in dynamic natural production environmental conditions.

The vertex *S* (figure 2) represents a fictitious source, the vertex *T* is a fictitious final point. From the fictitious source, the arcs enter the vertices L, denoting forest areas. These arcs are characterized by capacity, which is determined by the volume of wood removal from the corresponding forest area *V*. Arcs are also characterized by the costs *C* with which the wood resource is included in the technological chain. This may be the costs of acquiring wood, performing technological operations preceding the considered technological chain. The entire technological chain is represented on the graph by intermediate vertices located between the vertices denoting the forest areas *L* and consumers *U*. From vertices *L* arcs are coming in vertices $Mh\theta N$ (*M* - name of the object of labor during the operation (whips - *S*,sortiments - *L*); *h* - stage of transportation; θ- period number; *N* - forest area number). At the stage, after the lower warehouse, the designation *W* can appear in the characteristic of the vertices, which symbolizes transportation by water. Arcs entering the vertex determine the loading operation, leaving the vertex - the transportation of wood. Arcs are characterized by capacity, limited by performance *П*. Also, arcs are characterized by labor costs *f*, variable *C* and fixed *Z* costs. Fixed costs are taken into account when storing wood between periods. In the arcs outgoing the fictitious vertex *S* to the arcs characterizing the cutting areas, the characteristic $V_{M\theta N}$ indicates the volume of the analyzed object of labor in a certain period θ, moved from the cutting area *N*. In the arcs incoming in the fictitious vertex *T* from the vertices of consumers, there is a characteristic Q_{AM}, symbolizing the maximum volume of production *M*, possible for the acquisition of consumer *A*. Also, this arc is characterized by the cost of products sold *C*.

The solution to this problem is based on the Basaker-Gowen's algorithm for determining the minimum cost flow in the transport network. It is modified taking into account the fact that when the flow passes through any of the arcs, the capacity of arcs reflecting similar operations in the same period decreases [14]. Thus, the decrease in productivity of machines and equipment involved in different technological areas is considering. The given minimum cost flow found in this way is optimal if there are no negative cycles in the network with modified arc values.

This graphical model allows to carry out the analytical approach to the substantiation of logging process' technological chain of forest warehouses, raids of loading and unloading operations, processing operations, modes of transport (water, land), the choice of the consumer and type of final commodity products in dynamic natural production environmental conditions.

Four iterations were required for a complete solution of the problem. The calculation result is shown in the graph of figure 3. Double line sindicate the flowpath.

Thus, the timber harvested in the first forest area in the volume of 20000 m^3 will be exported to the consumer by road in first period. From the second forest area, the volume of harvested wood equal to 33770 m^3 will be exported to the consumer by road in the 1st period, the remaining volume of 16230 m^3 will be transported by road to the lower warehouse and then by water in the 4th period.

The costs of the proposed technological chain, taking into account the sale of wood will be - 125061940 rub.

Characteristics of the technological chain of the entire route of wood delivery from the forest areas to the consumer are summarized in table 1.

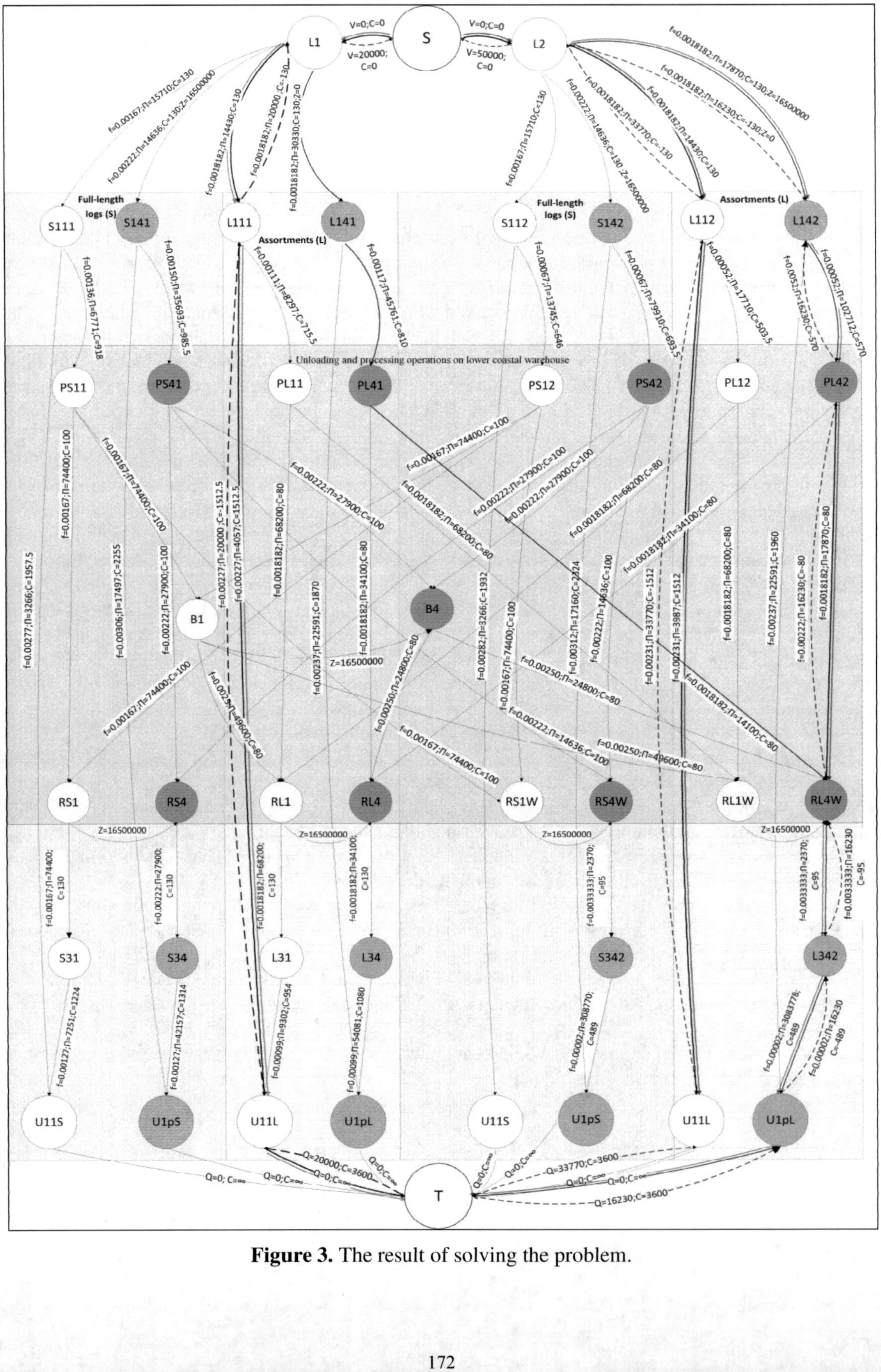

Figure 3. The result of solving the problem.

Table 1. Characteristic of the technological chain of the entire route of wood delivery from the forest areas to the consumer.

Characteristic	Route		
	S, L1, L111, U11L, T	S, L2, L112, U11L, T	S, L2, L142, PL42, RL4W, L342, U11L, T
Technological chain	L, T	L, T	L, T, U, R, WT
Type of products sold	sortiments	sortiments	sortiments
Forest area number	first	second	second
Transportation (implementation) period	first	first	forth
Transportation volume, m^3	20000	33770	16230
Distance of wood transportation by cars, km	275	280	95
Distance of wood transportation by rafting, km	-	-	589
Variable costs, rub	32 850 000	55 450 340	22 137 720
Fixed costs, rub	-	-	16 500 000
Sales income, rub	72 000 000	121 572 000	58 428 000
Profit, rub	39 150 000	66 121 660	19 790 280

where, L - loading, T - transportation by road, U - unloading, R - raft, WT - water transport.

3. Conclusion

The proposed graphical model allows to carry out an analytical approach to justifying the sequence of transporting timber from different forest areas, the use of lumber yards in the logging process, the raids of application loading and unloading works, processing operations, modes of transport (water, land), the choice of the consumer and type of final commodity products in a dynamic natural production environmental conditions.

A numerical example of the problem solution showed the efficiency of the methodology of finding a rational technological chain of logging operations in dynamic natural production conditions.

Acknowledgement

The reported study was funded by Russian Foundation for Basic Research, Government of Krasnoyarsk Territory, Krasnoyarsk Regional Fund of Science, to the research project: « Development of the fundamental principles of forest infrastructure design as a dynamically changing system in the conditions of logging production», grant № 19-410-240005; The project «Development and implementation of effective technology for integrated processing of logging waste» was funded by Krasnoyarsk Regional Fund of Science.

References

[1] Rukomojnikov K P 2014 Graphic-analytical modeling of technology of forest harvesting in forest compartment with fuzzy dynamic natural production conditions *Modern problems of science and education* **6** Available at: http://www.science-education.ru/120-16417

[2] Ralevic P, Ryans M and Cormier D 2010 Assessing forest biomass for bioenergy: operational challenges and cost 339 considerations. *The Forestry Chronicle*, **86(1)**, 43-50. Available at: https://www.researchgate.net/publication/269745366 DOI: 10.5558 / tfc86043-1

[3] Sazonova E A 2005 Research and justification of technological processes of logging works on the basis of the transparent energy analysis: *Ph. D. thesis*. Ekaterinburg, p 145

[4] Moore E F The shortest path through a maze *Proc. Int. Symp. On the Theory of Switching*, Part II, p 285

[5] Floyd R W 1962 Algorithm 97 – Shortest path *Comm. Of ACM*, **5**, p 345

[6] Dijkstra E W A note on two problems in connection with graphs *Numerische Mathematik*, **1**, p 269

[7] Bellman R 1958 On a routing problem *Quart. of Applied Mathematics*, **16**, p 87

[8] Belyakov C L, Belyakova M L, Bozhenyuk A V and Savelyeva M N 2014 Optimization of flows in transport systems *Izvestiya S Fed U. Engineering sciences*, **5(154)**, pp 161–7

[9] Bozheniuk A V and Gerasimonko E M 2013 Development of algorithm of finding the maximum flow of minimum cost on fuzzy dynamic transport network. *Don Engineering bulletin* [in Russian -*Inzhenernyy vestnik Dona*], Vol 24, **1 (24)**, pp 104. Available at: http://www.ivdon.ru/magazine/archive/n1y2013/1583

[10] Mokhirev A, Gerasimova M and Pozdnyakova M 2019 Finding the optimal route of wood transportation *IOP Conf. Series: Earth and Environmental Science*, Vol 226, conference 1. Available at: iopscience.iop.org/article/10.1088/1755-1315/226/1/012053 DOI:10.1088/1755-1315/226/1/012053

[11] Kumar A and Kaur M 2010 A Fuzzy Linear Programming Approach to Solve Fuzzy Maximal Flow Problems *IJPMS*. Vol 1, **1** p 6-12

[12] Mokhirev A and Rukomojnikov K 2019 Graphic-analytical modelling of technological chain of logging operations in dynamic natural and production conditions. *IOP Conf. Series: Earth and Environmental Science* **316** 012039 DOI:10.1088/1755-1315/316/1/012039

[13] Rukomojnikov K P and Mokhirev A P 2019 Validation of the Logging Operations Scheme through the Creation of Dynamical Model of the Enterprise Functioning. *Russian Forestry Journal* [Lesnoy Zhurnal – in Russian], **4**, pp 94–107. DOI: 10.17238/issn0536-1036.2019.4.94

[14] Rukomojnikov K P 2015 Justification of methods and parameters of rational technologies of forest areas quarterly development *Abstract of Doctor's degree dissertation* Volga State University of Technology, Yoshkar-Ola, p 328

IOP Conf. Series: Materials Science and Engineering **817** (2020) 012025 doi:10.1088/1757-899X/817/1/012025

Finding an effective technological chain of logging operations

A Mokhirev[1]*, K Rukomojnikov[2], M Gerasimova[1], S Medvedev[1]

[1]*Lesosibirsk Branch of Reshetnev Siberian State University of Science and Technology,* 29 Pobedy Street, Lesosibirsk 662543, Russian Federation
[2]*Volga State University of Technology,* 3, Lenin square, 424000, Yoshkar-Ola, Mari El Republic 424000, Russian Federation

*Corresponding email: ale-mokhirev@yandex.ru

Abstract. The choice of technological process' parameters certainly affects the result of its implementation. When performing logging operations, there are many options for their sequence, the use of machines at different stages of performance and periods of the year. Choosing the technological chain, enterprise management is faced with various possible solutions. In most cases, they are taken speculatively - based on personal experience. The given study is a continuation of the group of authors work on the complex problem of algorithmization and optimization of logging operations, taking into account the analysis of various moving and processing operations of wood resources. In previous publications, the authors presented a graphic-analytical model for solving the task. This article presents a description of the algorithm developed on the basis of the presented mathematical and graphic-analytical models for finding an effective technological chain of transport, loading and unloading and processing operations of the logging process in dynamic natural production conditions. The algorithm was successfully tested on the example of data as close as possible to the conditions of the Krasnoyarsk territory.

1. Introduction

The scientific community has not yet formed a unified universal methodology that can be used to increase the availability of wood resources located in the forest territory. Such a task requires the use of two key types of tools:

1. Economic and mathematical apparatus used to assess the resource potential of a certain territory;
2. Geographic information system, which allows to solve a wide class of problems, with reference to the geographical location of the investigated objects and the use of cartographic material processed with the use of computer technology.

The necessary difference of such a technique should be in the study of processes and the development of appropriate solutions in a changing environment. The latter circumstance is often overlooked, but it is quite obvious that the economic conditions of industrial enterprises are dynamic, which is influenced primarily by the natural environment.

The development of such a methodology is designed to solve the applied problems of the forest, in particular, the logging industry. At the same time, its development is a fundamental task, the solution of which involves bringing the proper theoretical and methodological justification. This paper presents the separate results of the study of the author's team in this area.

Currently, the remoteness of the territories in which logging enterprises operate from processing facilities is rapidly increasing. At the same time, there is an increase in the cost price of harvested raw

IOP Conf. Series: Materials Science and Engineering **817** (2020) 012025 doi:10.1088/1757-899X/817/1/012025

materials, which reduces the economic efficiency of logging as such. This situation, as well as a number of technogenic and natural-climatic factors, leads to the fact that the cost of individual forest resources is higher than the sale price. Such wood is considered to be economically inaccessible. Alongside this, the use of the existing mathematical apparatus in conjunction with the economic justification allows to identify the most effective forest infrastructure. Its implementation in practice should address a key challenge beyond the economy - the organization of continuous and sustainable use of forests. At the same time, the problem of increasing the availability of wood resources is solved by determining the optimal routes of delivery and dispersal of loggers in the territories. The latter is understood not only economic but also transport, environmental, etc. [1].

The key task of researches in this direction is to solve a complex of engineering and economic problems. The first of them consists, first of all, in choosing not only effective, efficient, but also optimal technical solutions for the delivery of wood resources from the forest area to the consumer. The second - in the economic justification of the selected technologies, machines and equipment, as well as the product portfolio and the direct consumer of products.

There are studies devoted to the justification of through flows of harvesting, transportation and processing of wood based on mathematical models, performed by Shegelman I R[2, 3]. The use of graphic-analytical models providing the construction of relationships between operations, allows to perceive technological chain for certain types of work in a logical sequence, eliminates duplication of certain technological elements, provides an opportunity to assess the various systems of machines and the structure of production flows in various technological processes of forestry [4].

However, these studies do not provide for the possibility of their use for modeling the technological chain in dynamic conditions, characterized by the need to take into account the natural and production characteristics when transporting wood from the cutting area to the consumer.

The methods of substantiation of finding the shortest paths between the vertices of the graph are described in Moore's [5], Floyd's [6], Dijkstra's [7], Bellman's [8] and other researches. They allow to carry out the analysis of static elements of work of the enterprises and serve as a basis for realization of stationary-dynamic tasks of stream programming at rationalization of flows in transport systems [9-12]. The simplex method of linear programming can be used to solve such problems [13].

The above methods and algorithms can be used in the process of considering the operations of the technological process of logging operations in dynamics when presenting them in the form of time-stretched graphs. These researches demonstrate in detail the basic schemes and analysis capabilities of the operating network. However, they characterize the possibility of passing through the arcs of each time interval only one flow option and do not take into account the specifics of the logging industry, envisaging at justification of the schedule of the logging technological process the need for a comprehensive solution to the problems of developing several cutting areas within one-time range. In view of this, they cannot always be used in solving the issues of justification of the technological chain of work at timber enterprises.

In the conditions of the need to implement such a comprehensive analysis, it becomes obvious that the use of machines and mechanisms involved at performance of works on one of the analyzed sites of the logging enterprise reduces the resources of their possible application during the execution of the same operations in the territory of another site in the time interval accepted for analysis. The noted researches stipulating theoretical possibility of passing several flows on the same operations of the technological process of each analyzed period provide the researcher with the possibility of the analysis only the graphs with independent from each other throughput capacity of separate parallel arcs of analyzed time intervals.

The purpose of the study is to develop an algorithm for the technological chain of transport, loading and unloading and processing operations of the logging process in dynamic natural-production conditions using the developed graphic-analytical model.

IOP Conf. Series: Materials Science and Engineering **817** (2020) 012025 doi:10.1088/1757-899X/817/1/012025

2. Materials and methods

The object of the study is the process of wood transportation (delivery) from the cutting area to the consumer. Analysis was used as a key research method. Meanwhile, the developed algorithm was based on the results of the work of the author's team and using the Bellman-Ford algorithm. Approbation of the results was carried out on the example of logging enterprises of the Krasnoyarsk territory.

3. Results and Discussion

This study is a continuation of the work of the author's team on the complex problem of algorithmization and optimization of logging operations, taking into account the analysis of various moving and processing operations of wood resources. In the paper Mokhirev A and Rukomojnikov K [14], the authors propose detailed graphical models of wood transportation from the cutting area to the consumer, loading and unloading and processing operations taking place at the intermediate and lower forest warehouse. In the study RukomojnikovKP, Mokhirev AP [15], on the basis of a graphic-analytical model, a technique for solving this problem is proposed. Suggested mathematical dependencies allow to carry out search of the maximum flow of the minimum cost in dynamic structure of technological process of works' performance at the enterprise. In addition to the dynamic component, the method differs by taking into account the income of the enterprise from the sale of marketable products.

The next stage of the researches is a description of the algorithm developed on the basis of the presented mathematical and graphic-analytical models for finding an effective technological chain of transport, loading, unloading and processing operations of the logging process in dynamic natural and production conditions.

The developed algorithm includes the following steps:

1. Take for calculation the vertices and arcs of the graph from the first period of the technological process $\theta = 1$, using a variant of the time-stretched graph G_p, relating to a particular situation of the technological process of timber removal from logging sites.

2. Taking into account the values of the flow $\xi\left(x_i, x_j, \theta, \theta + \tau_{ij}(\theta)\right)$ and $\xi(x_i, x_j, \theta, \theta)$moving along arcs of a graph G_p, build a residual network $G_p^\mu = (X_p^\mu; A_p^\mu)$. At the same time, each arc of the new network connecting the «vertex-time» pair (x_i, θ) with the «vertex-time» pair (x_j, θ), along which a flow of any value is started up at the first stage of calculation, has an inverse arc connecting (x_j, θ) with (x_i, θ) with the residual throughput $V^\mu(x_j, x_i, \theta, \theta) = \xi(x_i, x_j, \theta, \theta)$ and cost $C^\mu(x_j, x_i, \theta, \theta) = -C(x_i, x_j, \theta, \theta)$. If the value of the transported flow $\xi(x_i, x_j, \theta, \theta)$ is equal to the throughput of the arc, so $C^\mu(x_j, x_i, \theta, \theta) = \infty$. The movement of the reverse flow along any of the reverse arcs of the residual network leads to the possibility of increasing the throughput of any of the arcs characterizing the same operation of the technological process in the analyzed time interval by the value of:

$$V^N_{N_h(i=b)j}(\theta) \leq \frac{\sum_{i\in[1;b)\cup(b;g]} \sum_{x_j \in X} f_{ij}^N(\theta) \cdot \xi_{ij}^N(\theta)}{f_{(i=b)j}^N(\theta)} \quad (1)$$

Each arc connecting a «vertex-time» pair (x_i, θ) with a «vertex-time» pair $\left(x_j, \theta + \tau_{ij}(\theta)\right)$ has a residual bandwidth $V^\mu\left(x_i, x_j, \theta, \theta + \tau_{ij}(\theta)\right) = V\left(x_i, x_j, \theta, \theta + \tau_{ij}(\theta)\right) - \xi\left(x_i, x_j, \theta, \theta + \tau_{ij}(\theta)\right) = \infty$, and a reverse arc with a residual bandwidth $V^\mu(x_j, x_i, \theta + \tau_{ij}(\theta), \theta) = \xi\left(x_i, x_j, \theta, \theta + \tau_{ij}(\theta)\right)$ and cost $Z^\mu(x_j, x_i, \theta + \tau_{ij}(\theta), \theta) = -Z\left(x_i, x_j, \theta, \theta + \tau_{ij}(\theta)\right)$.

The residual network initially coincides with the original graph.

Determine the path P_p^μ of the minimum cost in the constructed residual network using the Bellman-Ford algorithm.

IOP Conf. Series: Materials Science and Engineering **817** (2020) 012025 doi:10.1088/1757-899X/817/1/012025

If there is no such path and the analysis of all analyzed p periods is performed, then it is possible to draw a conclusion about the correspondence of the previously found path variant to the optimal variant of the given flow movement and proceed to step 9.

If there is no such path only within the analyzed periods (at the moment of time (θ)) and it is possible to make a transition to the next period $\left(\theta + \tau_{js}(\theta)\right)$, then add the vertices and arcs of the next period to the graph. Since the throughput capacities of the arcs of each of the periods in the areas associated with the dummy source and dummy effluent depend on the flow started up along similar arcs of the previous periods, then carry out the calculation of the throughput capacities of its arcs connecting a «vertex-time» pair $\left(x_S, \theta + \tau_{ij}(\theta)\right)$ with the «vertex-time» pair $\left(x_j^{\mu}, \theta + \tau_{ij}(\theta)\right)$, and «vertex-time» pair $\left(x_i^{\mu}, \theta + \tau_{ij}(\theta)\right)$ with the «vertex-time» pair $\left(x_T, \theta + \tau_{ij}(\theta)\right)$, according to the formulas:

$$V_{L_N}\left(\theta + \tau_{ij}(\theta)\right) = V_N - \sum_{\theta=1}^{\theta=p} \xi_{L_N}(\theta), \tag{2}$$

$$V^y\left(\theta + \tau_{ij}(\theta)\right) = Q_U^y - \sum_{\theta=1}^{\theta=p} \xi^y(\theta). \tag{3}$$

If the path is found, then proceed to the next step.

3. Determine the maximum throughput of the identified minimum cost path in the residual dynamic network.

$$\delta_p^\mu = \min \begin{cases} \begin{pmatrix} V^\mu\left(x_S, x_j^h\right) \\ \text{where } \left(x_S, x_j^h\right) \in P^\mu_{p\,(S\to T)} \end{pmatrix}; \\[2ex] \min\begin{pmatrix} \underset{\substack{\text{where } \left(x_S, x_i^h\right) \in P^\mu_{p\,(S\to T)} \\ \theta\in[0,p]}}{V^\mu\left(x_S, x_i^h, \theta, \theta\right)} + \dfrac{1}{f^*}\cdot \underset{\substack{\text{where } \left(x_i^h, x_S\right) \in P^\mu_{p\,(T\to T)} \\ \theta\in[0,p]}}{V^\mu\left(x_i^h, x_S, \theta, \theta\right)} \end{pmatrix}; \\[2ex] \min\begin{pmatrix} V^\mu\left(x_i^h, x_j^h, \theta, \theta+\tau_{ij}(\theta)\right) \\ \text{where } \left(x_i^h, x_j^h\right) \in P^\mu_{p\,(S\to T)} \\ \theta\in[0,p] \end{pmatrix}; \\[2ex] \min\begin{pmatrix} \underset{\substack{\text{where } \left(x_i^h, x_j^{h+1}\right) \in P^\mu_{p\,(S\to T)} \\ \theta\in[0,p]}}{V^\mu\left(x_i^h, x_j^{h+1}, \theta, \theta\right)} + \dfrac{1}{f^*}\cdot \underset{\substack{\text{where } \left(x_j^{h+1}, x_i^h\right) \in P^\mu_{p\,(T\to T)} \\ \theta\in[0,p]}}{V^\mu\left(x_j^{h+1}, x_i^h, \theta, \theta\right)} \end{pmatrix}; \\[2ex] \min\begin{pmatrix} \dfrac{1}{f^*}\cdot V^\mu\left(x_i^h, x_j^{h+1}, \theta+\tau_{ij}(\theta), \theta+\tau_{ij}(\theta)\right) \\ \text{where } \left(x_i^h, x_j^{h+1}\right) \in P^\mu_{p\,(T\to T)} \\ \theta\in[0,p] \end{pmatrix}; \\[2ex] \min\begin{pmatrix} \dfrac{1}{f^*}\cdot V^\mu\left(x_j^h, x_i^h, \theta+\tau_{ij}(\theta), \theta\right); \\ \dfrac{1}{f^*}\cdot V^\mu\left(x_i^h, x_j^h, \theta, \theta+\tau_{ij}(\theta)\right) \\ \text{where } \left(x_i^h, x_j^h\right) \in P^\mu_{p\,(T\to T)} \\ \theta\in[0,p] \end{pmatrix} \\[2ex] \begin{pmatrix} V^\mu\left(x_i^h, x_T\right) \\ \text{where } \left(x_i^h, x_T\right) \in P^\mu_{p\,(S\to T)} \end{pmatrix}; \\[2ex] \min\begin{pmatrix} \underset{\substack{\text{where } \left(x_i^h, x_T\right) \in P^\mu_{p\,(S\to T)} \\ \theta\in[0,p]}}{V^\mu\left(x_i^h, x_T, \theta, \theta\right)} + \dfrac{1}{f^*}\cdot \underset{\substack{\text{where } \left(x_T, x_i^h\right) \in P^\mu_{p\,(T\to T)} \\ \theta\in[0,p]}}{V^\mu\left(x_T, x_i^h, \theta, \theta\right)} \end{pmatrix}; \end{cases}; \qquad (4)$$

$$f^* = \frac{f^*_{h,h+1}}{f^*_{h+1,h}}, \qquad (5)$$

where $f^*_{h,h+1}$ - labor costs characterizing the analyzed saturated arc of a narrow production site as part of the selected path, directed at a time point (θ) to a dummy effluent, machine-shifts; $f^*_{h+1,h}$ - labor costs characterizing the response to the analyzed saturated arc of a narrow production area, the reverse arc at the time point (θ), machine-shifts;

4. Update the values of flows along the arcs of the graph G_p.

a) for arcs connecting the «vertex-time» pairs $\left(x_i^\mu, \theta + \tau_{ij}(\theta)\right)$with $\left(x_j^\mu, \theta\right)$in the graph G_p^μ with cost $C\left(x_i, x_j, \theta, \theta + \tau_{ij}(\theta)\right) \leq 0$, replace the flow $\xi\left(x_j, x_i, \theta, \theta + \tau_{ij}(\theta)\right)$ along similar arcs of the graph G_p, directed from (x_j,θ) to $(x_i,\theta+\tau_{ij}(\theta))$, by the value from $\xi\left(x_j,x_i,\theta,\theta+\tau_{ij}(\theta)\right)$ to $\xi\left(x_j,x_i,\theta,\theta+\tau_{ij}(\theta)\right)-\delta_p^\mu$;

b) for arcs connecting the «vertex-time» pairs (x_i^μ,θ) with (x_j^μ,θ)in the graph G_p^μ, with cost $C(x_i,x_j,\theta,\theta)\le 0$, replace the flow $\xi(x_j,x_i,\theta,\theta)$ along similar arcs of the graph G_p, directed from (x_j,θ) *to* (x_i,θ), by the value from $\xi(x_j,x_i,\theta,\theta)$ to $\xi(x_j,x_i,\theta,\theta)-\delta_p^\mu$;

c) for arcs connecting the «vertex-time» pairs (x_i^μ,θ) with $(x_j^\mu,\theta+\tau_{ij}(\theta))$in the graph G_p^μ, with cost $C\left(x_i,x_j,\theta,\theta+\tau_{ij}(\theta)\right)\ge 0$, replace the flow $\xi\left(x_i,x_j,\theta,\theta+\tau_{ij}(\theta)\right)$ along similar arcs of the graph G_p, directed from (x_i,θ) to $(x_j,\theta+\tau_{ij}(\theta))$ by the value from $\xi\left(x_i,x_j,\theta,\theta+\tau_{ij}(\theta)\right)$ to $\xi\left(x_i,x_j,\theta,\theta+\tau_{ij}(\theta)\right)+\delta_p^\mu$;

d) for arcs connecting the «vertex-time» pairs (x_i^μ,θ) with (x_j^μ,θ)in the graph G_p^μ, with cost $C(x_i,x_j,\theta,\theta)\ge 0$, replace the flow $\xi(x_i,x_j,\theta,\theta)$ along similar arcs of the graph $\mathrm{G_p}$, directed from (x_i,θ) to (x_j,θ) by the value from $\xi(x_i,x_j,\theta,\theta)$ to $\xi(x_i,x_j,\theta,\theta)+\delta_p^\mu$;

e) update the values of the flows along the arcs connecting:

- «vertex-time» pair (x_j^μ,θ) with «vertex-time» pair (x_S,θ), by the value from $\xi(x_j,x_S,\theta,\theta)$ to $\xi(x_j,x_S,\theta,\theta)-\delta_p^\mu$;

- «vertex-time» pair (x_T,θ) with «vertex-time» pair (x_i^μ,θ), by the value from $\xi(x_T,x_i,\theta,\theta)$ to $\xi(x_T,x_i,\theta,\theta)-\delta_p^\mu$.

5. Determine the cost of the accepted path

$$\sum C=\sum_{(x_i,x_j)\in\tilde{P}^\mu_{p(S\to T)}} C_{(x_i,x_j)}+\left(\sum_{(x_j,x_i)\in P^\mu_{p(T\to T)}} C_{(x_j,x_i)}+\sum_{(x_i,x_j)\in P^\mu_{p(T\to T)}} C_{(x_i,x_j)}\right)\cdot f^*, \tag{6}$$

where, $P^\mu_{p(S\to T)}$ - a section of a path consisting of straight arcs going in the direction from a dummy source to a dummy effluent; $P^\mu_{p(T\to T)}$ - a section of the path including forward and reverse arcs, directed from the dummy effluent and returning back, creating a cycle; $\sum_{(x_i,x_j)\in P^\mu_{p(S\to T)}} C_{(x_i,x_j)}$ - total variable costs on the section of the path from a dummy source to a dummy effluent along straight arcs, m.u.; $\sum_{(x_j,x_i)\in P^\mu_{p(T\to T)}} C_{(x_j,x_i)}$ - total variable costs on a cyclic section of the path along the reverse arcs, m.u.; $\sum_{(x_i,x_j)\in P^\mu_{p(T\to T)}} C_{(x_i,x_j)}$ - total variable costs on a cyclic section of the path along the straight arcs, m.u.;

6. Calculate the time $m^*(\theta)$ remaining until the end of the period.

7. Replace the bandwidth and weight of the arcs connecting:

- «vertex-time» pair (x_S,θ) with «vertex-time» pair (x_j^μ,θ), by the value from $V_{L_N}(x_S,x_j,\theta,\theta)$ to $V_{L_N}(x_S,x_j,\theta,\theta)-\delta_p^\mu$;

- «vertex-time» pair (x_i^μ,θ) with «vertex-time» pair (x_T,θ), by the value from $V^y(x_i,x_T,\theta,\theta)$ to $V^y(x_i,x_T,\theta,\theta)-\delta_p^\mu$;

The weight characteristics of the reverse arcs obtained when the flow passes through the arcs of the graph are equal to $\mathrm{C}_{ji}^{\mathrm{X}}=-\mathrm{C}_{ij}^{\mathrm{X}}$.

Carry out replacement of capacities on the arcs corresponding to the same technological operations:

$$\Pi_{ij}^{\mathrm{X}}(\theta)=\frac{m^*(\theta)}{f_{ij}^{\mathrm{X}}(\theta)}. \tag{7}$$

IOP Conf. Series: Materials Science and Engineering **817** (2020) 012025 doi:10.1088/1757-899X/817/1/012025

8. Build a residual network. Proceed to step 2.

9. When identifying the maximum flow $v(p)$ of the minimum cost in the graph G_p, move to the initial dynamic graph, discarding the dummy vertices S and T.

To solve this problem, the algorithm was formed in Excel.

4. Results and Discussion

The implementation of any algorithm in practice is a matter of discussion. The operation of an industrial enterprise in difficult natural-climatic and production conditions often diverges from theoretical developments. Nevertheless, it is the approbation that checks the adequacy and viability of algorithms related to the transportation of goods, in particular wood resources.

The initial data for the implementation of the numerical example were formed from real data of logging enterprises of the Krasnoyarsk territory of the Russian Federation.

The natural and climatic conditions of the logging area should be divided into 5 periods: winter (124 days), winter-spring (31 days), spring (72 days), summer (62 days), autumn (71 days). In each period the performance of machines on operations is different. Logging is carried out on two sites, from which it is possible to transport wood directly to the consumer or through the lower coastal warehouse. The roads are partly temporary logging roads or year-round roads. On two sites timber can be harvested in the form of whips or sortiments of up to 20000 m^3 in the first area and up to 50000 m^3 in the second area. It is possible to transport the wood by rafting along the river from the lower coastal warehouse, rafts can be formed for this purpose. Timber storage between periods is carried out in stacks in the lower warehouse, for this purpose a land area is rented and personnel are hired for security. The consumer is ready to accept wood in any period in the amount of 70000 m^3. To solve the problem, the cost of performing technological operations C, the productivity of Π machines and equipment for their implementation, as well as labor costs f for performing operations are determined.

According to the proposed algorithm and data, a rational technological chain is found. To compare the indicators, an analysis of other variants of the technological chains of transport and moving works was carried out. In the Krasnoyarsk territory, under similar conditions, the most commonly used options are:

1) Transportation in the summer to the lower warehouse with further rafting of the maximum possible volume of wood, the rest of the volume during the harvesting period (winter) by road to the consumer. In this case, the volume of wood equal to 18600 m^3 is transported from the second forest site in the summer to the lower warehouse with further water transport to the consumer. The rest of the volume of wood of the second forest area (31400 m^3) in winter by road to the consumer. The entire volume (20000 m^3) from the first forest area is transported in winter.

2) Transportation by motor transport of the maximum possible volume in the winter to the consumer, the remaining volume-by road in the summer to the consumer. In this case, the transportation of the entire volume (50000 m^3) from the second site will be delivered to the consumer in the winter, part of the volume (3770 m^3) from the first site is also transported in the winter, the rest of the volume (16230 m^3) of the first forest site in the summer.

A brief comparative description of the routes is presented in table 1.

Table 1. Comparative characteristics of technological chain options.

Characteristic	Option		
	first	second	calculated
Volume of transportation, м^3	70 000	70 000	70 000
Costs, RUB.	130 361 200	137 252 225	126 938 060
Salesincome, RUB	252 000 000	252 000 000	252 000 000
Profit, RUB.	121 638 800	114 747 775	125061940

Thus, as a result of using the proposed methodology in solving the problem:

- the rational sequence of development of forest areas is established;

- the rational volume of timber, which can be exported to the lower coastal warehouse for further transportation by water transport at the minimum cost of labor, has been established;
- the possibility of reducing the total costs for the implementation of the work plan for transport and relocation operations by 3% compared to other possible variants of technological chains was revealed. The economic effect reaches 49 rubles / m^3.

5. Conclusion

The proposed algorithm allows to carry out an analytical approach to justify the sequence of timber transportation from different forest areas, to substantiate the use of forest warehouses in the logging process, raids of application loading and unloading operations, the type of transport (water, land), the choice of the consumer and the type of final commercial product in a dynamic natural production environment.

A numerical example of solving the problem has shown the efficiency of the method of finding a rational technological chain of logging operations in dynamic natural-production conditions.

Acknowledgement
The reported study was funded by Russian Foundation for Basic Research, Government of Krasnoyarsk Territory, Krasnoyarsk Regional Fund of Science, to the research project: «Development of the fundamental principles of forest infrastructure design as a dynamically changing system in the conditions of logging production», grant № 19-410-240005 and grant from the President of the Russian Federation - for young scientists, candidates of sciences MK-1902.2019.6.

References

[1] Mokhirev A P, Pozdnyakova M O, Medvedev S O and Mammatov V O 2018 Assessment of availability of wood resources using geographic information and analytical systems (the Krasnoyarsk territory as a case study). *JAES* **3** 313
[2] Shegelman I R, Skrypnik V I and Kuznecov A V 2010 Analysis of performance indicators and evaluation of the effectiveness of logging machines in various natural and industrial conditions *Scientific notes of Petr.stU*, **6 (109)** pp 13-23
[3] Shegelman I R, Shchegoleva L V and Ponomarev A YU 2005 Mathematical model of the choice of through-flow of procurement, transportation and processing of wood raw materials *News of the St. Petersburg Forestry Academy*. **172** pp 32-7
[4] Sushkov S I, Burmistrova O N and Pil'nik YU N 2015 Optimization of Transport Process Parameters at Enterprises of the Timber Industry Complex *Fundamental Research*. **11. 2** pp 237-41, available at: http://fundamental-research.ru/ru/article/view?id=39317
[5] Moore E F The shortest path through a maze *Proc. Int. Symp. On the Theory of Switching*, Part II, p.285
[6] Floyd R W 1962 Algorithm 97 – Shortest path *Comm. Of ACM*, 5, p345
[7] Dijkstra E W A note on two problems in connection with graphs *Numerische Mathematik*, **1**, p 269
[8] Bellman R 1958On a routing problem *Quart. of Applied Mathematics*, 16 p87
[9] Bershtejn L S, Belyakov S L and Bozhenyuk A V 2012 Using fuzzy temporal graphs for modeling in GIS *News of the S Fed U. Technical science*, **126**pp 121-127
[10] Bozhenyuk A V and Rozenberg I N 2007 Placement of service centers in GIS based on interval graphs *Placement of service centers in GIS based on interval graphs*, **14(4)** pp 682-3
[11] Bozheniuk A V and Gerasimonko E M 2013 Development of algorithm of finding the maximum flow of minimum cost on fuzzy dynamic transport network. *Don Engineering bulletin*, **1 (24)** pp 104. Available at: http://www.ivdon.ru/magazine/archive/n1y2013/1583.
[12] Mokhirev A, Gerasimova M and Pozdnyakova M 2019 Finding the optimal route of wood transportation *IOP Conf. Series: Earth and Environmental Science*, **226, 1.** Available at:

iopscience.iop.org/article/10.1088/1755-1315/226/1/012053 Doi:10.1088/1755-1315/226/1/012053

[13] Kumar A and Kaur M 2010 A Fuzzy Linear Programming Approach to Solve Fuzzy Maximal Flow Problems *IJPMS*. **1** pp6-12

[14] Mokhirev A and Rukomojnikov K 2019 Graphic-analytical modelling of technological chain of logging operations in dynamic natural and production conditions. *IOP Conf. Series: Earth and Environmental Science* **316** 012039 doi:10.1088/1755-1315/316/1/012039

[15] Rukomojnikov K P and Mokhirev A P 2019 Validation of the Logging Operations Scheme through the Creation of Dynamical Model of the Enterprise Functioning *Russian Forestry Journal* [Lesnoy Zhurnal – in Russian]**4** pp 94–107. DOI: 10.17238/issn0536-1036.2019.4.94

IOP Conf. Series: Materials Science and Engineering **817** (2020) 012026 doi:10.1088/1757-899X/817/1/012026

On the question of the mechanism of dust cloud formation on forest roads

B P Motovilov, O V Zubova, A A Borozna*, G K Motovilov, D A Skachek

St. Petersburg State Forest Technical University, 5 Institutskiy Lane, St. Petersburg 194021, Russian Federation

*Corresponding email: 3349280@mail.ru

Abstract. The justification of the model of dust cloud formation during the passage of a forest train on a dry road surface is given. Given the previous studies on sedimentation of single particles and a dust cloud, a formula is obtained that allows one to calculate the initial and average sedimentation velocity of a dust cloud taking into account changes in the concentration of dust in the cloud.

1. Introduction

The main types of coatings of forest roads are gravel, crushed stone and soil. Operation of forest roads with such coatings in dry weather is accompanied by increased wear and tear, and intense dust formation. The movement of the same type of road trains with the same speeds contributes to the emergence of "waves" on the surface. As a result, the road surface disintegrates, the speed and safety of traffic decline, the service life of engines shortens, and the consumption of fuel and lubricants increases. In addition, dust has a harmful effect on humans and the environment.

2. Materials and Methods

Taking into account the theoretical principles of the formation of a dust cloud on roads with gravel and soil surface proposed in [1, 2], the mechanism of dust cloud formation on forest roads with such surface appears to be as follows.

When a timber-carrying road train travels on a dry, untreated with dust-suppressing reagents sand-gravel surface, both fine-grained (dusty-clay particles and fine-grained sand) and larger fractions become detached and get transported. This is due to the fact that a compacted mass of air is formed in front of the road train that moves some of the dust. The bulk of the dust along with loose material is captured and rises into the air by the tread of the tires of the wheels of the road train. Separation of this material from the wheels occurs under the influence of the centrifugal force:

$$C = \frac{M_v{}^2}{R}, \quad (1)$$

where, C is the centrifugal force; M is the mass of particles of the material pinched between the protrusions of the tires; v is the peripheral speed of the treadmill, equal to the speed of the road train; R is the radius of the wheel.

The dust raised into the air is carried away by turbulent air flows, which are formed both between the lower parts of the road train (bottom) and the carriageway, and behind the passing road train. When dust particles come off the coating surface, humidity has a decisive effect. The water meniscus formed at the point of contact between the spherical dust particle and the flat particle of the surface material holds the particle on the surface with a force [3] according to formula (2):

$$F_k = 4\pi * r * \delta * \cos\theta \tag{2}$$

where, F_k is the magnitude of the capillary force; r is the particle radius; θ is the wetting angle; δ is the surface tension of water.

From equation (2), it can be seen that capillary forces depend on the particle size, the surface tension of the liquid, and the ability of the contacting bodies to wet, i.e., with an increase in the contact angle (θ), the adhesion forces decrease. Therefore, for dust suppression of gravel coatings, it is advisable to use reagents with a small wetting angle.

A moving forest train causes the formation of turbulent vortices (figure 1). The width of the coverage zone of the turbulent vortices during the passage of a road train can be determined from the expression:

$$B = G_{\text{sh}} + b_v \tag{3}$$

and the height

$$A = G_{\text{v}} + b_v \tag{4}$$

where, G_{sh} and G_v are the width and height of the road train, respectively; b is a coefficient depending on the streamlining of a car or road train. According to [4], for timber-carrying road trains loaded with "b" whole-lengths, it is 0.28-0.30, and for single vehicles, 0.06 - 0.075; v is the speed of the train;

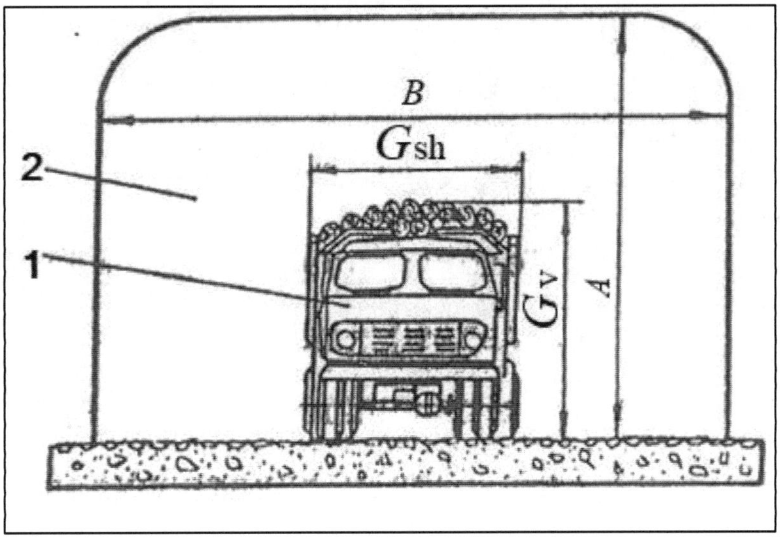

Figure 1. A moving forest train (1) forms a zone of turbulent vortices (2).

Let us determine the height of the perturbed air flow according to expression (4) for a KAMAZ-6426 timber transport train moving with a pack of whole-lengths at a speed of 40 km/h (11.1 m/s), having a height of 3.3 m, a frontal section of 8.25 m^2 and a streamlining coefficient 0.3. Then we get that the height of the perturbed stream "A", and therefore the height of the dust cloud (during calm periods) will be 15.3 m, that is, "A" is approximately equal to the two frontal sections of the road train. Formulas (3) and (4) are in good agreement with experimental studies [5], which established that trucks moving at a speed of 12-17 m/s cause the formation of turbulent vortices in the vertical and transverse directions, the coverage area of which is from two to five frontal sections of the car. Naturally, with a decrease in vehicle speed, the coverage area of the turbulent vortices will decrease. The length of a dust plume depends on the total length of the forest truck and its speed, the height of the dust particles and the speed of their sedimentation.

IOP Conf. Series: Materials Science and Engineering **817** (2020) 012026 doi:10.1088/1757-899X/817/1/012026

The length of the dust plume L is proposed to be determined by the formula:

$$L = L_{\text{a.n}} - L_a + L_c \tag{5}$$

where, $L_{\text{a.n}}$ is the total length of the forest train; L_a is the length of the car (tractor); L_c is the sedimentation path of dust particles.

The sedimentation velocity of dust particles (v_{c1}, m/s) isolated from each other according to [6] can be determined by the formula:

$$v_{c1} = \frac{d^2}{18\mu_b} * \rho g \tag{6}$$

where, d is the particle diameter, m; ρ is the particle density, kg /m^3; μ_b is the dynamic viscosity of air, Pa*s.

From equation (6) it follows that the sedimentation rate of spherical dust particles with a diameter of 50 μm, a density of 2.65 g/cm^3 at an ambient temperature of 20 ^{0}C will be:

$$v_{c1} = \frac{(50*10^{-6})^2*2650*9.81}{18*17.75*10^{-6}} = 0.19 \text{ m/s}, \tag{7}$$

The sedimentation time of such a particle from a height "*A*" equal to 15 m will be (15: 0.19) = 79 s. In this case, the length of the dust plume at a train speed of 11.1 m/s will be:

$$L = 25 - 6 + (11.1 * 79) = 900 \text{ m}; \tag{8}$$

However, our observations established that the settling of a dust cloud after passing a forest road train occurs in 10-60 s or at a speed of 0.25 - 1.5 m/s, that is, much faster than it follows from formula (6) for single sedimentation particles.

The faster settling of the dust cloud compared to particles isolated from each other is explained by the following. It is known [6, 7] that a characteristic feature of nonspherical particles is their tendency to take a position at which the medium resistance would be maximum; lamellar dust particles are arranged so that their more developed faces and longer edges appear perpendicular to the direction of motion [7]. This orientation increases as the Reynolds criterion increases, while the angle between the direction of particle fall and the direction of the drag force increases. Therefore, the trajectory of the settling particles deviates from the vertical. For sufficiently large Reynolds numbers, the motion of particles becomes spiral or zigzag. With slow settling of small particles (with a radius of less than 10 μm), this phenomenon is not observed [6, 7].

A zigzag drop of dust particles leads to their clustering and coalescence. Such larger aggregates settle at a higher speed, additionally capturing fine particles along the way, resulting in a decrease in the concentration of dust in the cloud. In addition, a falling large particle or aggregate carries air particles with it, so that the air resistance to the movement of other particles falling in the immediate vicinity decreases, and they acquire additional speed.

Thus, the sedimentation of particles isolated from each other sharply differs from the sedimentation of a system of particles, since each of the particles moving with respect to the medium excites flows in it that act on other particles. If the number of particles in the aerosol cloud is very small, and they do not affect each other during movement, then the sedimentation rate of such a cloud will be the same as in the case of sedimentation of single particles (6). With a large concentration of dust particles, the cloud moves as a unit and is surrounded by a counter flow of air. Under the influence of gravity, falling particles carry the medium along, so that the resistance of the medium to the movement of particles decreases. In this case, the cloud sedimentation rate can be much higher than the speed of individual particles and is expressed by formula (9):

$$v = \left(\frac{32*R*n*g*\rho_n*\pi*r^3}{g*\psi*\rho_B}\right)^{0.5} \quad (9)$$

where, v is the sedimentation rate of the dust cloud; R is the radius of the dust cloud; r is the radius of the dust particles; n is the number of dust particles per unit volume; ρ_n is the density of dust particles; g is the gravity acceleration, ψ is the drag coefficient; ρ_B is the air density.

From formula (9) it can be seen that it is suitable for use only when a dust cloud consists of uniform dust particles having the same size. In the case when a dust cloud consists of polydisperse particles, which is typical for road dust, the practical application of formula (7) presents significant difficulties. However, one can replace $r^3\rho_n n = 3Q/4\pi$, where Q is the mass of dust per unit volume. With this in mind, formula (9) is transformed into an expression that allows us to determine the initial sedimentation velocity (v_H) of a road dust cloud from its known (or given) concentration in a unit volume of air:

$$v_H = \left(\frac{8R*Q*g}{3*\psi*\rho_B}\right) \quad (10)$$

or

$$v_H = 5.115\left(\frac{RQ}{\psi*\rho_B}\right)^{0.5}, \quad (11)$$

It should be noted that during the cloud's existence, the value of Q varies from a certain maximum (at the beginning of its formation at a given point) to zero.

The results of calculating the sedimentation rates of a dust cloud depending on its size and dust concentration are presented in figure 2. In the calculations, the sedimentation rate of a dust cloud with a reduced radius of 10, 20 and 30 m was determined at a dust concentration in the air from 0.1 to 20 g/m^3. The air density at 20 ^{0}C is assumed to be 1.205 kg/m^3. The drag coefficient of the medium to the movement of a dust cloud according to [6, 7] in our case is close to 1.0.

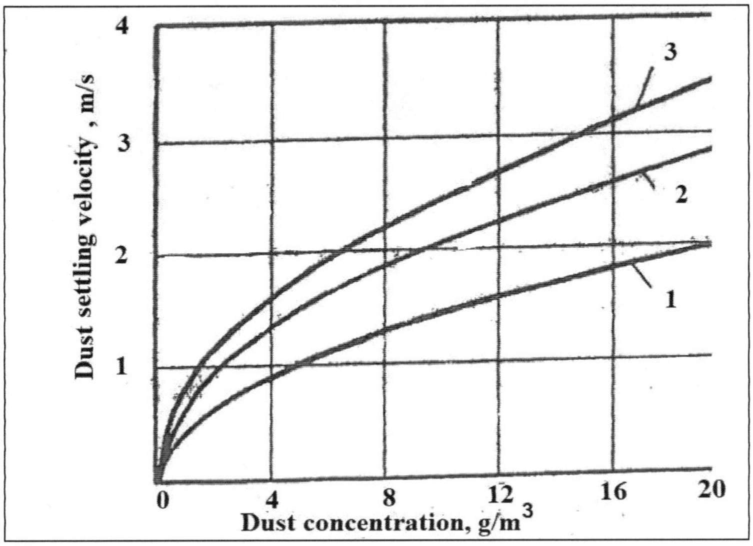

Figure 2. Change in the sedimentation rate of a dust cloud from dust concentration and the cloud radius: 1 - 10 m; 2 - 20 m; 3 - 30 m.

Based on figure 2, the sedimentation rate of a dust cloud increases with increasing cloud size and dust concentration in it. However, the sedimentation rate of the dust cloud, calculated by the formula (10) and (11), and shown in figure 2 turned out to be about two times higher than it was observed by us in the field (0.25 – 1.5 m/s). This can be explained by the fact that over time the concentration of dust in the cloud decreases as indicated above. As a result, the density of the dust cloud decreases, and it no longer settles as a single unit. Therefore, the sedimentation rate of a dust cloud at the end of the sedimentation process will be close to the sedimentation rate of single dust particles. With this in mind, the average value of the velocity v_{av} sedimentation of a dust cloud can by approximated by the half of the initial sedimentation velocity of a dust cloud:

$$v_{av} = 0.5 * v_{н} \tag{12}$$

or

$$v_{av} = 2.558 * \left(\frac{RQ}{\psi * \rho_{в}}\right)^{0.5} \tag{13}$$

3. Conclusions

A design scheme (model) for the formation of a dust cloud during the passage of a forest train on a dry, untreated, dust suppressing sand and gravel surface is proposed. Based on the well-known studies of sedimentation of individual dust particles and a dust cloud, a formula is obtained that allows one to determine the initial and average sedimentation speed of a road dust cloud depending on the concentration of dust in the cloud.

References

[1] Pershin M N, Platonov A L, Markov L A and Rozov Y N 1993 *Dust removal on roads and airfields* (Moscow: Transport)
[2] Rozov Y N 1972 Dust formation and its influence on transport and operational indicators of gravel coatings *Proceedings of TiproDor NLD* **4** pp 33 – 41
[3] Zimon A D 1976 *Adhesion of dust and powders* (Moscow: Chemistry)
[4] Surikov V T 1962 *Study of some issues of the main resistance to the movement of timber road trains*, PhD thesis cand. tech. sciences (Leningrad: LTA)
[5] Preobrazhensky L U, Dergunov A Y, Alexandrov V V and Pelevina N N 1981 An experimental study of the influence of vehicles on the structure of the air flow on the highway *Proceedings Ch. Geophysis. Observatory* (Leningrad: Gidrometeoizdat) pp 132 – 138
[6] Pirumov A I 1981 *Dust removal from air* (Moscow: Stroyizdat)
[7] Fuchs N A 1955 *The mechanics of aerosols* (Moscow: Academy of Sciences of the USSR)

IOP Conf. Series: Materials Science and Engineering **817** (2020) 012027 doi:10.1088/1757-899X/817/1/012027

Procedural framework of the resources cost effective use of cutting works technologies

O Mushkarova[1], M Mikheeva[1], F Chepic[2], S Vasiliev[2]

[1]Department of Economics, Accounting and Analysis of Economic Activities, *St. Petersburg State Forest Technical University*, 5 Institutskiy Lane, St. Petersburg 194021, Russian Federation
[2]Department of Botany and Dendrology, *St. Petersburg State Forest Technical University*, 5 Institutskiy Lane, St. Petersburg 194021, Russian Federation

*Corresponding email: mihchepmar@yandex.ru

Abstract. With a variety of technologies for the cutting works, the choice of them should be determined by a combination of various factors that allow us to estimate the possible level of resource savings at each stage of such work. At the moment, the ecological effect extended for a certain period of time cannot be calculated precisely. It must be taken into account when evaluating the economic effect of resource savings at the enterprises of the timber complex. Using the structural-logical diagram of methodological approaches in order to establish factors and indicators of the resource-saving effect, a system analysis is carried out, which allows us to draw up an algorithm for generating an estimate of the resource-saving effect of various cutting works.

1. Introduction

An intensive environmental management model is supposed to preserve non-renewable mineral resources, the needs of which are covered by the use of waste products, the volume of which in the function of the recyclables is due to innovative technological methods of production.

The evaluation of economic effect requires considering the economic and environmental consequences of the natural resources removal. The factor of exhaustion and non-renewability of the majority of natural resources determines the search for alternatives to traditional technologies in acquisition of a variety of benefits provided by nature.

Technological methods of the industrial stock and timber processing are characterized by a variety of indexes that evaluate the level of resource saving at each stage of cutting works.

The ecological effect should be considered as a significant but accompanying characteristic of the economic effect. The ecological effect is evaluated as a result of the efficient use of resources extracted out of a biota aimed at producing benefits that provide a synergetic effect for the consumer of these benefits.

The limit of the industrial resources does not fully allow meeting the requirements for absolute environmental safety against the footprint. "Necessity" and "opportunity" are the conditions to establish the economic efficiency of the resource saving at the enterprise and they should identify the priority of various indexes of economic effect, which will act as follows: the main ones are economic, the related ones are environmental and the adverse ones are image-building.

IOP Conf. Series: Materials Science and Engineering **817** (2020) 012027 doi:10.1088/1757-899X/817/1/012027

According to the GOST 30166-2014 “Resources saving. Basic principles” the recourse saving are ”activities (organizational, economic, technical, scientific, practical, informational), methods, processes, a set of organizational and technical measures and actions that accompany all stages of the product life cycle, stages of the waste technological cycle and aimed at rational and economical use of resources ”[1].

Resources are "exploitable sources to satisfy the needs of society" [1]. The economic aspects of the resources exploitation should be correlated to the ecological requirements to support the balance of production needs and opportunities of natural potentials. Hence, the evaluation of the effect should be built on the ratio of the volumes of resources involved into the production and the amount of benefits provided by the level of maximum use of resources.

A high level of competition in home and international markets requires the company to constantly increase the image of environmentally friendly production, which is ensured by the involving the significant investments, and hereby requires an adjustment of the criterion of their effectiveness.

The indexes of the resource-saving effect shown in table 1 should be adopted as framework rates, specification of which depends on the nature of the resources, technological methods for their processing, volumes and characteristics of waste production.

Table 1. Classification of the resource saving effect indexes at the enterprise.

ECONOMIC	ECOLOGICAL	IMAGE BUILDING
Saving on waste recycling Resource saving per unit of output Reducing pollution charges Savings on ensuring the environmental risks Decrease of the losses caused by the occupational diseases	Reducing polluting emissions Decrease in waste disposal areas CO_2 reduction in the air Saving on volumes of recoverable resources Reduction of morbidity of the population	Availability of Certificates of Conformity with international standards of the product quality and technologies State support for the environmental projects of the company\ preferential loans “Green” economy status

Primary resources accumulated in bio-natural (natural) conditions are exhaustible and, in most cases, non-renewable. Secondary resources created in anthropogenic (man made) conditions of economic activity are close in their volume to the primary ones. The generated and accumulated secondary material resources are the resources of secondary raw materials.

Timber, as a primary resource, during processing generates significant amounts of wood waste, recycling of which ensures a significant economic effect in the national economy. However, in the face of the aggravation of climate change problem due to increased anthropogenic pressure, the role of trees plantations as a resource to recompense carbon emissions into the atmosphere is increasing.

Related to species and structural parts of forest biocoenoses and their edaphic conditions determine the direction of reforestation processes such as seminal or vegetative, change of species and storeyed structure. Knowledge of the biological and environmental properties of woody plants allows us to ensure the natural reforestation by means of using the certain types of felling.

The timber industry enterprises using resource-saving technologies contribute to the conservation of the forests, which leads to savings in the costs of cultivation the new forests, since forests on the preserved lands of the forest fund do not require reforestation.

Preserved forests provide the constant ability of forests to absorb the carbon dioxide and hereby, reduce the level of greenhouse effect. Thus, the economic effect is complemented by the environmental effect, namely by decreasing the concentration of carbon dioxide in the atmosphere. The use of wood storage and processing waste as fuel, compared to the mineral one which has a higher calorific value, reduces carbon dioxide emissions. In addition, wood fuel waste can replace the exhaustible types of mineral fuel resources, providing a significant level of efficiency of public production costs while reducing the greenhouse effect. Although carbon dioxide is emitted during the burning up of these wastes, it has a beneficial effect on the climate, given that the logging wastes are being decomposed anyway and thus, release carbon and other greenhouse gases [2-6].

Timber harvesting, a variety of machine systems and work methods significantly affect the types and volumes of secondary resources, and determine the degree of concentration of their formation and gathering patterns, and courses for their use as well. The specifics of the conditions of the timber waste origin at all stages of timber harvesting should be characterized by a framework of indicators in resource saving.

2. Materials and methods

The system-wide approach allows us to build up an algorithm to form an evaluation of the resource saving effect of various cutting works. Figure 1 shows the structural-logical diagram of methodological approaches to establish the factors and indexes of the resource-saving effect. A graphical representation of the problem in the form of Ishikawa [7] diagrams allows you to visualize and evaluate the cause-effect relationships of technological, economic and environmental markers to estimate the effect of resource saving.

A general evaluation of the level of resource saving of the cutting works is determined by the matrix method, which allows systematizing private estimates of various indexes. Evaluation of indexes is carried out by the expert method. Expert methods are related to the non-formalized techniques for describing economic problems and are based on intuition and past experience [8].

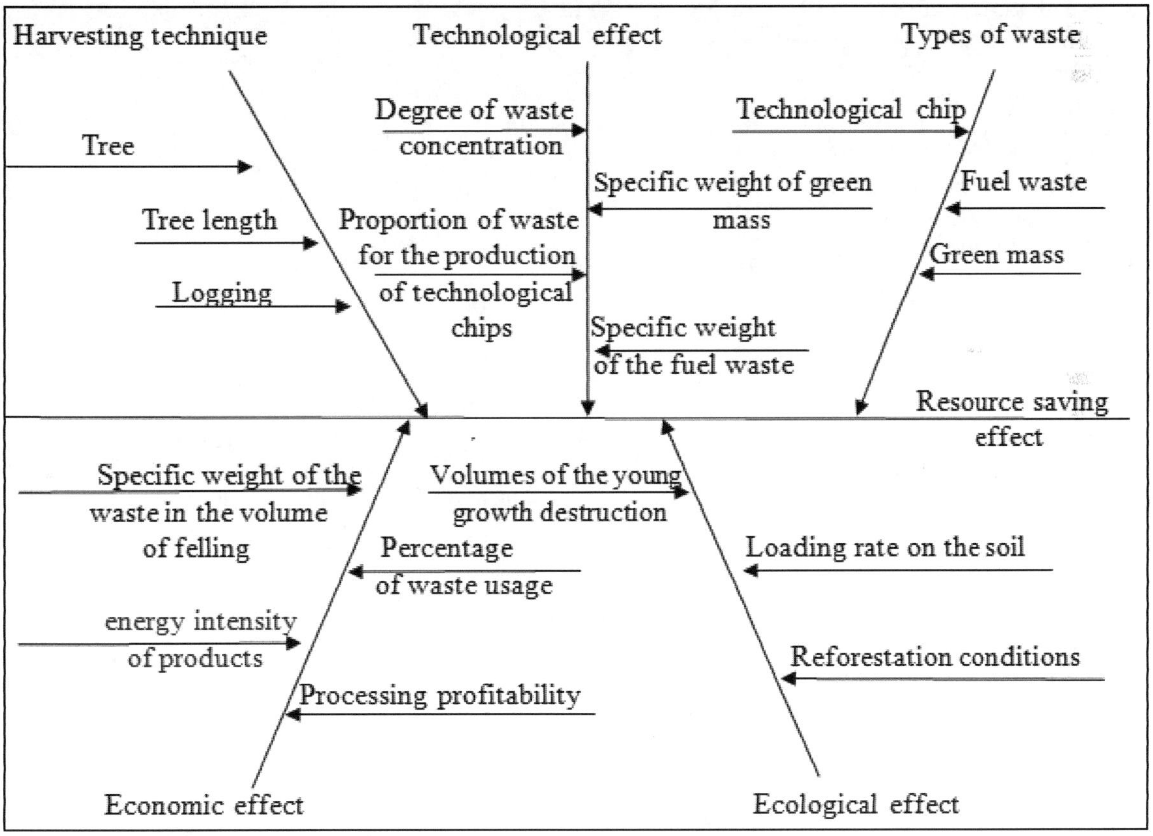

Figure 1. Structural - logical diagram of the resource saving effect indexes making of alternative technologies in the cutting works.

3. Results and Discussion

The enquiry of the experts is carried out by the Delphi method, by sending out a questionnaire (table 2) to evaluate the level of resource saving of technologies that is used in Russian and foreign practice

IOP Conf. Series: Materials Science and Engineering **817** (2020) 012027 doi:10.1088/1757-899X/817/1/012027

of cutting (table 3), and allowing to identify the independent opinion of each expert, in which experts from the forest industry should be involved and designers of logging equipment.

Table 2. Matrix of evaluation of the resource saving effect in diverse technologies in cutting works.

Resource saving effect indexes	Range of the resource saving effect of cutting works technologies									
	№1	№2	№3	№4	№5	№6	№7	№8	№9	№10
TECHNOLOGICAL - degree of waste concentration -proportion of waste for the production of technological chips - specific weight of the fuel waste - specific weight of the green mass										
ECOLOGICAL - loading rate on the soil - volumes of the young growth destruction - conditions of reforestation										
ECONOMIC - energy intensity of products - percentage of waste usage - profitability of waste processing - specific weight of the waste in the volume of felling										

Table 3. Technological processes of the cutting works [9].

Groups of technological works	№ of tech.process	Works in the cutting area	Type of timber to be trailed	Works in the upper timber landing or loading ramp	Type of logged materials
Tree length system	1	Tf - B	T	L	T
	2	Tf - B	T	Tdl - L	S
	3	Tf – Tdl - B	S	L	S
CTL technology	4	Tf – Tdl - B	S	Ccr - L	P_2
	5	Tf - B	T	Tdl - Ccr - L	P_2
	6	Tf - Tdl – Ccr - B	P_2	L	P_2
	7	Tf - Tdl - Ccr – B - L	-		P_2
Added-value wood processing	8	Tf - B	T	Tdl – Ccr – Cr - L	P_1
	9	Tf - Tdl – Ccr - B	P_2	Cr - L	P_1
	10	Tf - Tdl – Ccr – B; Chw - L	P_2	Tdb – Ch - L	P_4 P_3

where, Tf – tree felling; Tdb – debarking; Tdl – tree delimbing; L – loading; Cr – rip-cutting; Ccr – cross cutting; Ch – chipping; Chw – felling waste chipping; B–bundle formation; T – trees; P_1 – timber; P_2 – assortment; S – stems; P_3 – technological chips; P_4 – fuel chips.

The enquiry of the experts is carried out by the Delphi method, which is used along with others, such as SAW, the Preference Ranking Organization Method for Enrichment Evaluations and the Analytical Hierarchy Process and allows you to identify the independent opinion of each expert [10]. The latter should be recruited from the forestry experts and designers of timber cutting equipment. Experts assign a significance coefficient $k\_(i, p) \in [-1; 1]$ to each factor called i. Based on the data obtained by the formula, the total assessment of the timber harvesting technology Q is calculated:

$$Q = \sum_{i=1}^{n} \sum_{p=1}^{m} k_{i,p} , \quad (1)$$

where, Q – total evaluation of the timber harvesting technology; n – numbers of evaluation criteria, in this case, see table 1, $n = 9$; m – number of experts called p, in compliance of the Delphi method not less than 20 people; $k_{i,p}$ – significance factor coefficient called i, assigned by the p expert $k_{i,p} \in [-1; 1]$.

The maximum value of the total rating will determine the preferred timber cutting technology that can be considered as resource-saving from the standpoint of economic, ecological and technological effect.

An evaluation of the technology's weaknesses and strengths will allow further research to justify the most promising ones from the point of view of the most economical and environmentally acceptable technologies for timber harvesting in the context of the development of the timber industry complex which ensure a high level of use of various forest benefits, both production and bioprotective.

References

[1] GOST 30166-2014 Resources saving. Basic prineiples.
[2] IEA. Tracking Clean Energy Progress 2017 (IEA, Paris, 2017).
[3] Searchinger T D, Beringer T, Holtsmark B., Kammen D M, Lambin E F, Lucht W, Raven P and van Ypersele J-P Europe's renewable energy directive poised to harm global forests. *Nat Commun* **9,** 3741 (2018) doi:10.1038/s41467-018-06175-4
[4] Machacova, K, Vainio E, Urban O, Pihlatie M Seasonal dynamics of stem N_2O exchange follow the physiological activity of boreal trees. *Nat Commun* **10,** 4989 (2019) doi:10.1038/s41467-019-12976-y
[5] Energy conservation in the mechanical forest industries. 1990 FAO FORESTRY PAPER 93. Rome: FAO., p 118
[6] Bacherikov I V 2013 Preparation technology for waste to be burnt in Heating Electrical Stations within Pulp and paper industry. Latest problems of timber complex: International scientific research conference materials (Vologda: VoSTU) pp 104-107
[7] Tiutin V V 1999 Development of management solutions: author's lecture course for students of the Faculty of Economy Management (FEU) of all forms of training in the categories 06.08. (Saint-Petersburg: SPbFTU) p 68
[8] Chernov V A 1998 Commercial risk analysis (Moscow: Finance and Statistics) p 128
[9] Grigoriev I V 2013 Technology and equipment in the timber cutting works: study guide for the course design (Saint-Petersburg: SPbFTU), p 132
[10] Leśniak A and Radziejowska A 2017 *Supporting bidding decision using multi-criteria analysis methods* Procedia Engineering pp 208 76-81 DOI:10.1016/j.proeng.2017.11.023.

IOP Conf. Series: Materials Science and Engineering **817** (2020) 012028 doi:10.1088/1757-899X/817/1/012028

E-procurement in developing the outsourcing of transport services

A Nikolaeva*, S Ivshin

Federal State Budgetary Educational Institution of Higher Education «Saint Petersburg State University of Industrial Technologies and Design», 18 Bolshaya Morskaya street, St. Petersburg 191186, Russian Federation

*Corresponding e-mail: nik.anna77@mail.ru

Abstract. The article deals with the development of information technologies in the field of electronic procurement, which are an efficient logistics tool. Modern Internet technologies make it possible to gain competitive advantage in the implementation of business processes through taking the best decision when choosing alternative options related to the outsourcing of transport services. Global trends, which are aimed at digitalising logistics activities, are adapted and applied to the realities of the Russian economy. In the course of the study a comparative analysis of electronic trading platforms of the Russian Federation was carried out and their major characteristics were outlined. The paper touches upon the advantages of logistics outsourcing. One of the major players of e-commerce services in the Russian market was taken as an example to conduct a research covering the issues of e-procurement of transport services.

1. Introduction

E-commerce is a complex and multi-purpose tool of developing economic relations between economic entities, it is also a powerful driver of economic growth at the present stage of informatisation and globalisation of the world economic cooperation [1, 2]. Currently the ongoing processes of globalisation and informatisation have a significant impact not only on the content of economic activity of an economic entity, its structure, relations with external objects, politics, etc., but also evolve into an entirely new digital form. The transfer of economic interactions, such as search and exchange of information, tendering and trading, from a real-world to a virtual space is an important factor in improving trade and procurement activities requiring a global analysis and application of new conceptual approaches [3]. Being a powerful modern means of international trade, e-commerce blurs geopolitical, social and geographical boundaries and forms fundamentally new approaches to the interaction between suppliers and consumers. The number of e-commerce operations carried out with the help of Internet technologies is growing at an incredible rate and has a considerable effect on the world economy [2, 4, 5]. In the modern world, e-commerce is an essential prerequisite for the successful operation of the majority of companies; it opens up new vistas for business development, directly affecting the advancement of the Russian economy [6].

The increasingly deeper immersion of business processes into the infosphere emphasises the need to transform, modify and adapt logistics to modern digital technologies. E-commerce is a fundamentally new set of business processes that requires a completely different approach to management principles, as well as indicators and criteria for evaluating the effectiveness of business activity. Execution of an order via Internet-based technologies reduces the cost of placing an order by one third, with its execution time reduced threefold [7]. It becomes obvious that in the logistics sphere of any organisation, including procurement, transformations and qualitative changes are necessary, and they are already taking place in the document flow, order placement and document handling procedures. When using Internet-based technologies in procurement, organisations can perform many logistics functions such as the selection of suppliers, calculation of the cost of alternative procurement options, electronic documents transfer and tracking the movement of purchased products at a fundamentally different level [8]. A Malyshev in his reasoning about the modern trends in logistics attributes the key role to digitalisation [9]. He pays attention to the current trend which indicates the growth of logistics functions the customer companies delegate to specialized organisations that provide the best tasks execution in the field of logistics.

Digitalisation of modern business generates a new phenomenon known as electronic procurement portals [10]. The use of electronic trading platforms (hereinafter referred to as ETPs) significantly simplifies the customers' activity of ordering products and services due to automation, and has a positive impact on the competition development among supplier companies in producing and selling their goods [11, 12].

2. Methods and materials

When selecting suppliers, the following methods are traditionally used: the rating method, the cost estimation method, the method of dominant characteristics, the method of preference categories [13]. Nowadays, the methods based on information technologies are increasingly used. There is a large number of different electronic trading platforms in the Russian Federation, which makes it possible to search for partners and contractors, including outsourcers of logistics services. Ten out of ninety-six electronic trading platforms of the Russian Federation, which are the leaders in terms of total worth according to the published notices in 2018, are shown in figure 1 [14].

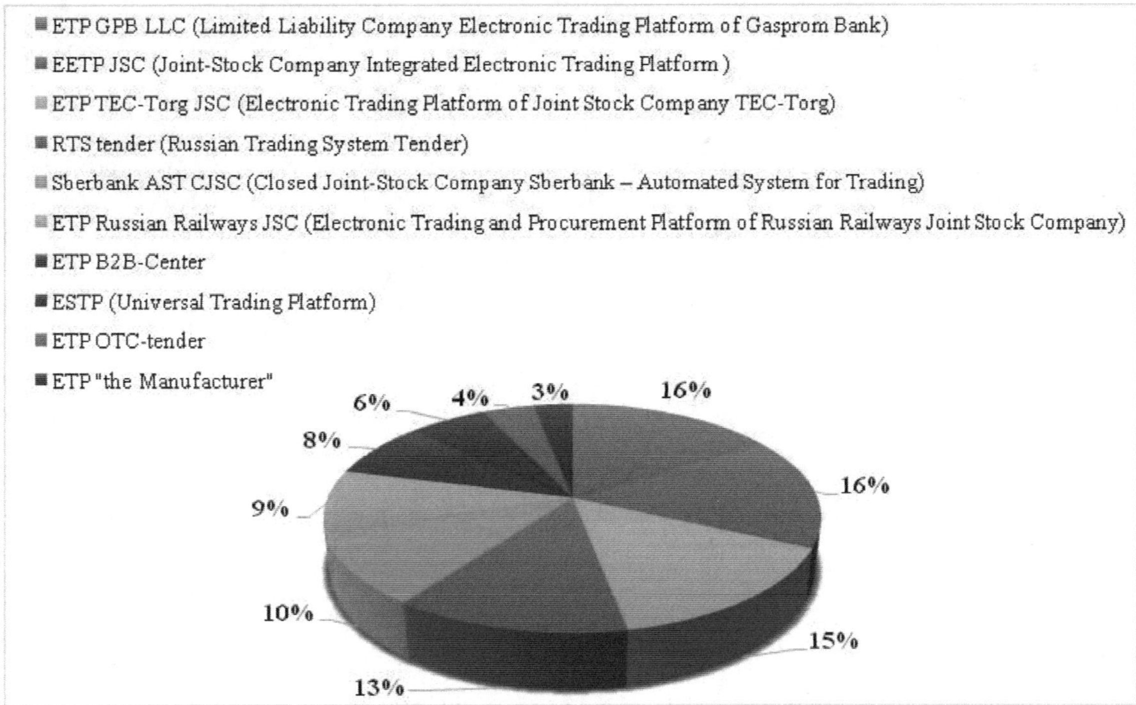

Figure 1. The largest electronic trading platforms in the Russian Federation.

IOP Conf. Series: Materials Science and Engineering **817** (2020) 012028 doi:10.1088/1757-899X/817/1/012028

The study below presents the main characteristics of Russian ETPs. Electronic platforms are divided both by industry (electric power industry, oil and gas sector, metallurgical industry, aircraft construction, agro-industrial sector, insurance, pulp and paper industry, construction, etc.) and by territory, i.e. geographically. Some large holdings have their own ETPs designed to place their own orders.

Apart from the interface, all electronic trading platforms differ in the possibilities of operation. Thus, for example, the search can be carried out by the trade procedure number assigned by the ETP. It is also possible to find purchases in question by whole words or by parts of a word (morphological search), the morphological search takes into account the form of a given word (case, gender, number, declination, etc.). Many platforms can be searched by abbreviations ("CAW"- Construction and Assembly Works, "FEED"- Front End Engineering and Design), word combinations and parts of a word. Other common search criteria for ETPs include the names of organisations and their individual tax number, OKVED (Russian National Classifier of Types of Economic Activity) and OKDP (All-Russian Classifier of Types of Economic Activity, Products and Services).

Some platforms, on the contrary, have the function of excluding search results by certain words. It is the price of a lot which is another additional search filter at most platforms. Many ETPs provide for the search of procedures using a time range (indicating the starting and the ending date of the bidding period). Thus, some of them can be used to select the procedures published in the system within a certain period of time or ongoing procedures, where the submission deadline for the trade procedure comes before a specified date (if the date is not specified in the field "before", all the relevant procedures are displayed).

The most common "filters" refer to the search by the "Purchase" or "Sell" criteria, which display the procedures of the corresponding categories. Additional filtering of purchases or offers by certain types of procedures is also possible.

Another popular option at the ETP is to search only for those procedures that are conducted in accordance with the requirements of Federal Laws № 223-FZ "On procurement of goods, works, services by certain types of legal entities" and/or № 44-FZ "On the contract system of the Federal and municipal procurement of goods, works, services". At more advanced platforms, the search criteria can be narrowed down to procedures that require a system operators to confirm the status of small and medium-sized enterprises (SMEs) as suppliers (under 223-FZ , SME status expands the opportunities for participation in procurement, and this status is also necessary for the participation in procedures available only to SMEs) [15, 16]. Some ETPs offer a search for procurement for the nuclear industry, where it is possible to search only for those procedures whose suppliers were accredited by ROSATOM.

Another distinctive feature of electronic trading platforms is the preservation of the search query in templates. Not all systems have such a function. All ETPs provide for electronic document exchange systems using electronic signatures.

Electronic trading platforms differ in the registration procedure of procurement participants. The platforms have an accreditation system of their own, which implies a set of actions to be effected in accordance with the rules established by a particular organiser and aimed at verifying whether a participant in the system is eligible to participate in the procurement of the organiser. Such verification is based on the analysis of the documents provided by the participant to the organiser, and it results in organiser's granting a special status to the participant of the system, confirming the access to the participation in the procurement of the organiser.

The ETPs make provision for various methods of protection against an unfair counterparty. Secured transaction services guarantee protection for unknown buyers and sellers during reciprocal payments. The guarantor of the fulfillment of bilateral obligations is a third party (escrow agent), which is a company that accepts money for the storage on its account before the parties fulfill their obligations. Therefore, each platform provides for the possibility of making_security payments. Thus, within this framework, a special account for tender security, as well as a sub-account for escrow funds and free collateral are provided. These accounts ensure security for the procurement procedures participation application. For this purpose, the systems may have an electronic safe deposit box for the organiser. An independent bank guarantee can be used at the ETP which is one of the ways of ensuring the fulfillment of obligations of the system participant to the organiser of the procedure, whereby at the principal's request the guarantor issues a commitment letter to pay the beneficiary a sum of money when he or she submits a request for its payment. All reliable systems have their own stop-lists and the corresponding regulatory instructions aimed at improving the level of bona fide participants.

The resources of various procurement systems can be used for the monitoring of the price market, which involves a sequence of actions taken by the organiser in accordance with the rules and regulations established by the organiser of the ETP. As a result, information on the product costs is collected in order to form the total cost of the subsequent procurement procedure.

In addition to the above mentioned characteristics, the ETFs differ in offering various methods of competitive procurement that allow for competition among the participants of the procedures.

Competitive tendering is one of the methods of competitive procurement which implies a compulsory signing_of a contract with the winner (if any). In the course of such tender the organiser informs suppliers about the need for products in advance_and invites them to submit bids; the tender committee appointed by the organiser beforehand considers them and determines the winner as an entity which meets the requirements of the documentation and offers the best conditions for the execution of the contract on the basis of a set of criteria established by the documentation.

Other common methods include the seller's auction, which may provide for a compulsory conclusion of the contract with the winner or the buyer's auction, with or without a compulsory conclusion of the contract with the winner, the Dutch auction, which does not provide for the compulsory conclusion of the contract, requests for proposals, requests for quotations without the need of signing a contract as well.

The majority of ETPs offer such procurement method as competitive negotiations, where the organiser informs suppliers about the need for products or services in advance, invites them to submit bids, then considers them and, based on the results of negotiations with the participants, determines the winner as an entity which meets the requirements of the documentation and offers the best conditions for the execution of the contract on the basis of a set of criteria established by the documentation. As a result of competitive negotiations, the customer has no obligation to determine the winner and sign a contract with him.

In addition to the methods mentioned above, some ETPs offer an opportunity for a multi-lot procurement procedure, which makes it possible for the organiser to pursue the procurement of numerous individual lots and conclude a contract with the winner (winners) for individual lots to obtain the greatest economic effect. In this case, the participants can submit a bid for individual lots of the procedure.

Most electronic trading platforms make it possible to conduct non-public procurement procedure with the participation of only full-fledged members specially invited by the organiser. The procedures providing for pre-qualification of participants can be both independent, when the organiser forms a list of participants in future procurement/sales procedures in accordance with the requirements and criteria established by the documentation, and serve as an additional element of the procurement/sales procedure, which is the admission of the participants in the procedure in accordance with the requirements and criteria established by the documentation prior to the submission of applications with technical and commercial proposals. The participants whose submission of technical and commercial proposals was approved as a results of pre-qualification are admitted to further participation in a tender.

There is also a multi-criteria way of conducting the procedure at the ETP. It allows the organiser to use the following formalised values parameters to assess the received applications while selecting participants. These parameters are:

- economic, technical, organisational, financial and legal attractiveness of the proposal in terms of meeting the needs of the customer (seller);

IOP Conf. Series: Materials Science and Engineering **817** (2020) 012028 doi:10.1088/1757-899X/817/1/012028

• reliability of the participant, based on work experience, business reputation, availability of facilities, production resources, human resources, financial assets and information resources.

Some systems have the possibility of using a multistage procedure. Such procedures are conducted in two or more stages. Based on the results of each stage (except for the last one), the organiser is supposed to clarify the terms of documentation for the procedure (including negotiations with the participants of the procedure) and, accordingly, to submit the refined proposals of the participants of the procedure to the next stage.

One of the additional elements of the procurement procedure is rebidding. It consists in voluntarily increasing the participants' bids preference within a specially organised procedure, whereby the participants can decrease/increase the price of their initial bids in the procurement and, if it is provided for in the documentation to the procedure, improve other conditions of the bid.

3. Results and Discussion

The given features of ETPs can be used by the customer in the selection of the supplier, including an outsourcing contract. The essence of logistics outsourcing is to optimise the cost of transportation of commodities and materials through the partial or full transfer of logistics functions to a specialised organisation that renders high-quality logistics services [17]. In addition to reducing their cost, the profiling of rendered services in outsourcing improves the quality of services provided and potentially increases the market stability of a company in a competitive environment [18].

The capabilities of electronic platforms make it possible to take a well-grounded decision about the transition to outsourcing of transport services, for example. Further maintenance of the services quality provided by the third parties is performed based on regular monitoring of offers in the electronic environment and sustaining healthy competition among suppliers using the possibilities of e-commerce. As a rule, delivery companies, due to high competition in the market, are more motivated and work more efficiently than structural delivery units of the company [17].

Statistical data on the selection of an outsourcing company which provides product transportation services were analyzed using the example of Sberbank AST, one of the largest ETPs operating in the Russian Federation, ranked fifth in terms of the total worth of published notices in 2018. Table 1 presents the procurement of transportation services carried out at Sberbank AST over a five-year period starting from 2014 [19]. It is worth noting that when tendering, different lot names may be used which are actually synonyms and denote the same concepts. Therefore, in order to obtain a complete picture when collecting data such lot names as "transport services", "transportation services", "supply services" and "haulage services" were used.

Table 1. Transport service procurement at ETP «Sberbank AST» during the 2014-2019 period.

Name of the procedure	Number of lots		Initial price		Number of tender organisers		Average price of one lot, million roubles	Average amount for one organiser, million roubles
	number	%	Million roubles	%	number	%		
223-FZ								
Auction	922	2.8	22189.17	19.8	192	2.5	24.07	115.57
Competitive tender	108	0.3	1715.41	1.5	23	0.3	15.88	74.58
Request for proposals	687	2.1	749.20	0.7	87	1.1	1.09	8.61
Request for quotations	472	1.4	923.44	0.8	199	2.6	1.96	4.64
Total under 223-FZ:	**2189**	**6.6**	**25577.22**	**22.9**	**501**	**6.5**	**11.68**	**51.05**

				44-FZ				
Auction	30149	91.1	85414.68	76.4	6944	89.8	2.83	12.30
Competitive tender	10	0.0	75.22	0.1	8	0.1	7.52	9.40
Request for proposals	8	0.0	26.63	0.0	7	0.1	3.33	3.80
Request for quotations	461	1.4	79.43	0.1	240	3.1	0.17	0.33
Total under 44-FZ:	**30628**	**92.6**	**85595.97**	**76.6**	**7199**	**93.1**	**2.79**	**11.89**
				Commercial procurement				
Auction	15	0.0	52.93	0.0	4	0.1	3.53	13.23
Competititve tender	2	0.0	20.00	0.0	1	0.0	10.00	20.00
Request for proposals	31	0.1	108.08	0.1	15	0.2	3.49	7.21
Request for prices and quotations	74	0.2	426.24	0.4	7	0.1	5.76	60.89
Electronic trade	138	0.4	14.25	0.0	6	0.1	0.10	2.37
Total commercial procurement:	**260**	**0.8**	**621.50**	**0.6**	**33.00**	**0.00**	**2.39**	**18.83**
TOTAL:	**33077**	**100.0**	**111794.68**	**100.0**	**7733.00**	**100.0**	**3.38**	**14.46**

As it can be seen from the table, trading on this electronic platform is performed by three main categories of customers, namely:

1. Organisations in whose authorised capital the per cent owned by the Russian Federation, or by a constituent region of the Russian Federation or a municipal entity in total exceeds fifty percent. These are client-organisations, which are obliged to carry out procurement activities in accordance with the Federal Law № 223-FZ "On the procurement of goods, works, services of certain types of legal entities"[15];

2. State clients or local government customers, state-funded institutions, state or municipal unitary enterprises, i.e. organisations that are obliged to perform procurement activities in accordance with the Federal Law No. 44-FZ "On the contract system of the Federal and municipal procurement of goods, works, services"[16];

3. Commercial organisations engaged in procurement activities on the basis of economic feasibility.

As it can be seen from the data presented, the maximum volume of purchases was accounted for by 44-FZ [16], so the volume of purchased transport services (at the maximum initial price) amounted to 85 595.97 million rubles, which is 76.6% of the total volume of placed procedures; the total of 30 628 procedures were carried out, which is 92.6% of the total number of placed orders; the number of organisers of auctions is 7199, which is 93.1% of all customers who held auctions on the electronic platform. At the same time, almost all purchases were implemented by means of electronic auction, where the only criterion for choosing a supplier was the price for the service provided. Thus, the participating companies gradually reduce the initial contract price, and the winner is the participant who offered the lowest cost of transport services.

The second place in the initial maximum contract price and the number of placed procedures is occupied by procurement made in accordance with 223-FZ [15]. Since 2014, 2189 procedures have been placed on the electronic platform in accordance with the Federal Law, which is 6.6% of the total number of tenders for the procurement of transport services; the total initial maximum contract price amounts to 25 577.22 million rubles, which is 22.9% of the total amount of purchases at this ETP. At the same time, 501 companies were the organisers of the tender, which amounts to 6.5% of the total number of customers of transportation services. The electronic auction accounted for the largest volume of purchases at the initial price, which amounted to 19.8% of the total number of performed procedures. However, when assessing the number of lots, it can be seen that the auction accounted for 2.8%, competitive tendering - 0.3%, request for proposals - 2.1%, and request for quotations - 1.4% respectively of the total number of performed procedures for the purchase of transport services.

The share of commercial procurements placed at this ETP is absolutely insignificant and makes less than 1% in both quantitative and monetary terms of the total number of procedures carried out to select a company providing transport services. This is the evidence of the fact, that in public procurement, due to the necessity of staying within the framework of the law, the transition to e-commerce has been much more intensive and has found a wide application not only among state customers, but also among suppliers willing to work in the e-commerce environment.

4. Conclusion

The results of the research make it possible to conclude about the increasing development of e-commerce on a global scale. E-trading serves as a powerful tool in increasing the competitiveness of organisations and business development, it is also a strong driver of economic growth of different countries. Along with the informatisation of logistics and the transition of logistics processes into a virtual information field, organisations are faced with the task of optimising their activities by reducing logistics costs as well, which can be achieved through the outsourcing of logistics functions, using the advantages and tools of e-commerce. On the basis of the study of procurement of an outsourcing company for transport services carried out at one of the largest ETPs of the Russian Federation, it is evident that this area of logistics activities is not widely used among commercial organisations, procurement being carried out mainly by state customers. The research shows that there is a large number of suppliers in the market who provide outsourcing services and are ready to participate in electronic bidding at the corresponding platforms. All this indicates the great potential of commercial companies of the Russian Federation in terms of using electronic auctions for the procurement of outsourcing of transport services, which may become the strongest factor allowing companies to achieve a competitive advantage through the development of e-business.

References

[1] Nazarova I B and Dianova T V 2012 E-commerce: advantages, problems and a "sliding effect" when decreasing transaction costs. *MGIMO University Bulletin*. [in Russian - Vestnik MGIMO] **4 (25)** pp 173-178

[2] Yadova N E 2019 Modern state of the world e-commerce market and its development prospects *Bulletin of Moscow State Regional University. Series: Economics* [in Russian - Vestnik Moskovskogo Gosudarstvennogo Oblastnogo Universiteta. Seriya: Ekonomika] **2** pp 116-125

[3] Zabolotskikh L Yu 2016 Internet-trade as a direction of trade policy of a company *Trade and Economic Journal* [in Russian - Torgovoekonomicheskij Zhurnal] **3(3)** pp 205–222

[4] Zheltenkov A V and Zheltenkov P A 2012 *Innovative Mechanism of Industrial Organisation management Development* [in Russian - Innovacionnyj Mekhanizm Razvitiya Upravleniya Promyshlennoj Organizaciej] (Moscow: IIU MGOU) p 123

[5] Zheltenkov A V and Fedotova M A 2012 Development of strategic management systems in industrial organisations *Bulletin of Moscow State Regional University. Series: Economics* [in Russian -Vestnik Moskovskogo Gosudarstvennogo Oblastnogo Universiteta. Seriya: Ekonomika] **3** pp 77–81

[6] Avdeeva E A and Bochkova E V 2015 E-commerce in Russia: state and problems in the transition to the information-network economy *Economics and Management of the Control*

Systems [in Russian - Ekonomika i Menedzhment Sistem Upravleniya] **3.2** pp 216–222

[7] Kizim A A and Ignatova V O 2012 Virtual logistics as a tool for integration of international business entities *Theory and Practice of Social Development* [in Russian -Teoriya i Praktika Obshhestvennogo Razvitiya] **12** pp. 443-448

[8] Zaburdaev V Yu *E-commerce Logistics* [in Russian - Logistika Elektronnoj], available at: http://logistik03.narod.ru/A_2.htm (date of reference: 17.11.2019)

[9] Malyshev A 2006 On the world market of logistics services *Electronic Journal of Practical Logistics "Warehouse&Technology"* [in Russian - Elektronnyj Zhurnal Prakticheskoj Logistiki "Sklad&tekhnika"] **3**, available at: https://sitmag.ru/article/10399-na-mirovom-rynke-logisticheskih-uslug

[10] Ivanov D 2008 Virtual enterprises: operational cooperation in logistics *Electronic Journal "Corporate Management"* [in Russian - Elektronnyj Zhurnal "Korporativnyj Menedzhment"], available at: https://www.cfin.ru/management/manufact/virtual.shtml

[11] Andreev N Yu and Kordysh F S 2015 Use of electronic auctions in the modern contract system *Property Relations in the Russian Federation* [in Russian - Imushhestvennyje Otnosheniya v Rossijskoj Federaczii] **7 (166)** pp 6-11

[12] Khramkin A A, Kosheleva V V, Kornilov P N, Evstashenkov A N, Vorobieva O M 2015 *Handbook of the State Customer* [[in Russian - Nastolnaya Kniga Goszakazchika] ed A A Khramkin (Moscow: Jurisprudence)

[13] Degtyareva N M and Yakovlev R 2015 Methodical approaches to the selection and assessment of suppliers of the enterprise *Bulletin of Volzhsky Uiversity n.a. V. N. Tatishev* [in Russian - Vestnik Volzhskogo Universiteta imeni V N Tatishcheva] **1 (33)** pp 100-105

[14] Official website of the Ministry of Finance of the Russian Federation [in Russian - Oficzialnyj sajt Ministerstva Finansov Rossijskoj Federaczii] URL: https://www.minfin.ru/ru/document/?id_4=126698 (date of reference: 17.11.2019)

[15] Federal Laws № 223-FZ "On procurement of goods, works, services by certain types of legal entities"

[16] Federal Laws № 44-FZ "On the contract system of the Federal and municipal procurement of goods, works, services"

[17] Krivova A A 2015 Evolution and current state of the logistics outsourcing market in Russia *Economics and Business: Theory and Practice* [in Russian - Ekonomika i Biznes: Teoriya i Praktika] **8** pp 61-64

[18] Bondarenko V A and Guzenko N V 2015 Outsourcing of logistics services: realities and trends *Financial Research* [in Russian - Finansovyje Issledovaniya] **48** pp 176-181

[19] Electronic Trading Platform of Sberbank-AST CJSC, URL: http://www.sberbank-ast.ru (date of reference: 17.11.2019)

IOP Conf. Series: Materials Science and Engineering **817** (2020) 012029 doi:10.1088/1757-899X/817/1/012029

The method of using the harvester for logging in cutting areas with soft terrain

K Rukomojnikov

Volga State University of Technology, 3, Lenin square, 424000, Yoshkar-Ola, Mari El Republic 424000, Russian Federation

Corresponding email: RukomojnikovKP@volgatech.net

Abstract. Article materials relate to the forest industry and can be used in logging operations on cutting sites with soft terrain. The result of the research is a developed technological scheme for the movement of the harvester and forwarder in the cutting area It allows you to reduce the number of skidding trails for moving the forwarder on logging sites with soft terrain, which are able to provide a one-time movement of a harvester, but require additional costs for strengthening of skidding trails for multiple movements of a forwarder. The forwarder works only on skidding trails that are located at distances from each other equal to ten distances of movement of harvester manipulator. This helps to reduce the cost of transport operations on cutting areas with soft terrain. The proposed technology can be used for clear cutting in cutting areas with soft terrain.

1. Introduction

Currently, in forests on cutting areas with soft terrain, a large potential stock of wood is concentrated. In these forest areas, various cutting techniques and various forestry equipment can be used, in particular a set of machines consisting of a harvester and forwarder [1-4].

Among the various options for reducing the cost of transport operations when working on cutting areas with soft terrain include options for the strengthening of skidding trail. However, studies of different options for increasing the bearing power of soils [5] show significant costs for strengthening skidding trails.

The marked options do not exclude the possibility of simultaneously increasing of the width of the forest swathes. This is especially true on soils that are able to provide a single movement of harvester, but requiring significant additional costs for strengthening the skidding trails for multiple movements of forwarder.

2. Methods and Materials

Currently, in the forest industry, there are various options for work technology that can increase the width of forest swathes [6-8]. Among them are variants of technological schemes for the operation of the harvester and forwarder in cutting areas with wide forest swathes with one or two intermediate trails between the skidding trails for moving the harvester along them [9]. The main part of damage to soils, undergrowth and reserved trees is carried out as a result of movement of the forwarder. On the cutting areas with one intermediate trail in a forest swathes, the forwarder works only on skidding trails, which are located at distances from each other equal to 3-3.5 distance of extension of the harvester manipulator. On the cutting areas with two intermediate trails in the forest swathes, the forwarder works only on skidding trails, which are located at distances from each other equal to 4-5 distance of extension of the harvester manipulator. The harvester performs felling, limbing, cutting into lengths when moving along the intermediate trail, as well as along the skidding trails. In both cases, the harvester stacks the bundles of logs so that the logs can be reached by the manipulator of the forwarder when it moving along the skidding trails. This ensures the availability of logs harvested by the harvester moving along the intermediate trail for the forwarder moving along the skidding trails. Using these methods allows to reduce the total length of skidding trails on the cutting area and increase the width of the forest swathes. However, when using these methods, the width of the forest swathes does not exceed 4-5 effective distances of movement of the manipulator of the harvester and can be further increased when developing new technological schemes. An additional increase in the width of forest swathes can increase the efficiency of work on cutting areas with soft terrain.

The closest in technical essence is the method of logging with obtaining logs by harvesters and forwarders using intermediate trail in forest swathes when performing voluntary-selective cutting, even gradual cutting, two-pass cutting system [10].

Similarly to the methods described earlier, this method, in order to reduce the degree of impact of forestry machines on the forest environment and reduce the area of the technological network of skidding trails in the cutting area, involves placing additional intermediate trails on the forest swathes between skidding trails for the movement of the harvester. The width of forest swathes can be up to 40-50 meters. The harvester, moving along intermediate trails, felling trees and laid trees in the direction of the nearest skidding trail, perpendicular to it. The harvester moves along the intermediate trail only once. Further, the harvester moves to the skidding trail and moves along it. The harvester performs felling, limbing, cutting into lengths, and also it additionally limbing and cuts to length of the trees which lying on the ground. The harvester pulls the trees lying on the ground, grabbing them by the top, into the processing zone. The harvester moves the harvester head to the basis of tree and cuts into lengths. These trees were harvested earlier at a time when the harvester was moving along the intermediate trail. Sawed logs are stacked on the right and left near the skidding trail. Felling residues are stacked on the skidding trail. The forwarder begins skidding on the skidding trail after the harvester moves to another swathes. The machines operate at a safe distance from each other.

The disadvantages of this method include the fact that this method, like the previous methods described, does not make it possible to make apiary widths of more than 4 effective distances of movement of the manipulator of the harvester. Consequently, it is possible to create a new method that allows even more to reduce the damage to the soil cover on the cutting area with soft terrain by the forwarder.

The result of this study, proposed on the basis of the search for equivalent technological solutions, is the reduction in the number of skidders trails for the movement of the forwarder on logging sites with soft terrain, which are able to provide a one-time movement of a harvester, but require additional costs for strengthening of skidding trails for multiple movements of a forwarder.

3. Results and discussions

The technical result of the technological scheme of work in the cutting area with soft terrain proposed in this article is achieved by the fact that the forest swath is divided into five parts. On the forest swath on each side of the skidding trail, there are two additional trails for the movement of the harvester. The width of the swath reaches ten effective distances of movement of the manipulator of the harvester.

The development scheme of one swath in the cutting area is presented in figure 1.

The method provides for the separation of the cutting area into forest swathes, the placement of the skidding trail 1 in the middle of the forest swathes for the movement of the forwarder 2 and harvester 3 in the process of work on the cutting area.

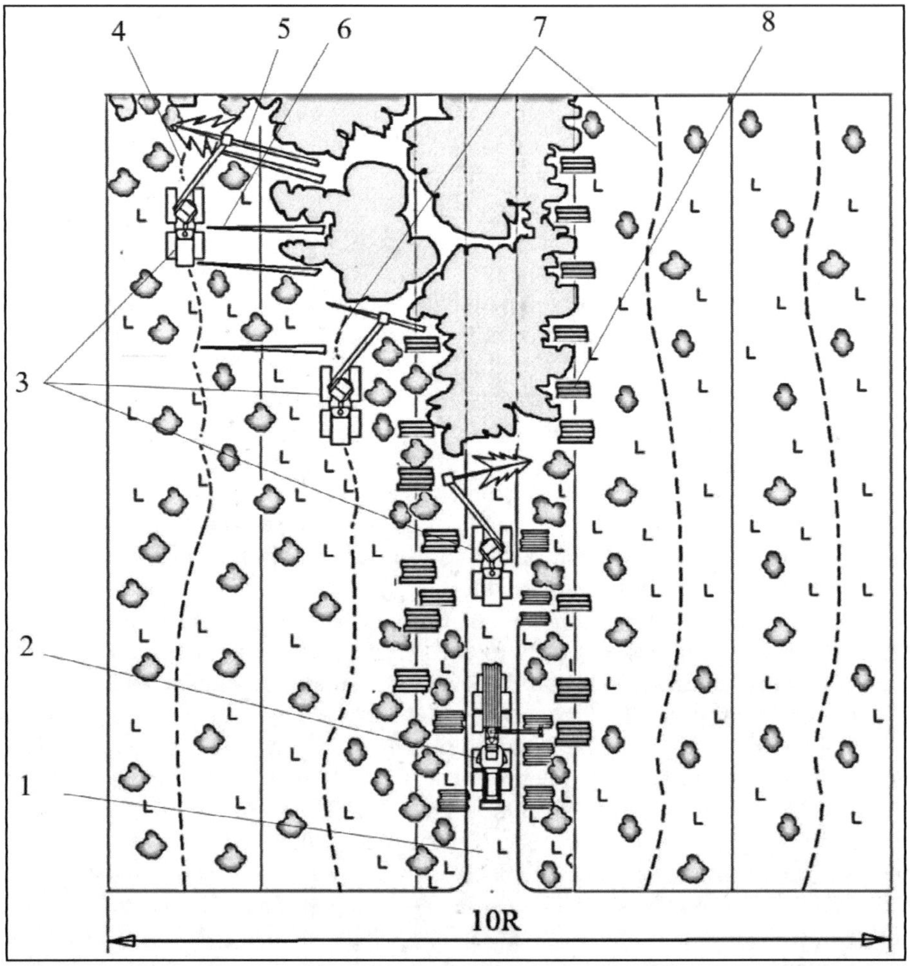

Figure 1. The method of using the harvester for logging in cutting areas with soft terrain.

When the harvester moves along the intermediate trails 4 of the forest swath, which are farthest away from the skidding trail, the harvester performs felling trees 5, limbing and lays sticks 6 in the direction of the skidding trail of this forest swath perpendicular to it. The bases of trees are directed towards the skidding trail.

When the harvester moves along intermediate trails 7, adjacent to the skidding trail, the harvester performs felling, limbing, cutting into lengths, and the harvester also cuts to length of the sticks which lying on the ground. These sticks were harvested earlier at a time when the harvester moves along the intermediate trails of the forest swath, which are farthest away from the skidding trail. The resulting logs are stacked in the direction of the skidding trail of the forest swath perpendicular to it. The harvester stacks the bundles of logs so that the logs can be reached by the manipulator of the forwarder when it moving along the skidding trails.

When the harvester moves along the skidding trail, the harvester performs felling, limbing, cutting into lengths. Bundles of logs 8 are stacked near the skidding trail. The machines operate at a safe distance from each other.

The forwarder begins work at the forest swath after the harvester moves to another forest swath. The forwarder collects logs and skidding them, moving only along the skidding trail. The forwarder does not move along the intermediate trails.

Skidding trails in cutting sites with soft terrain, intended for the movement of the forwarder are strengthened to ensure the movement of the forwarder with logs along them. The bearing power of soil can be enhanced by adding additional elements or materials to the soil of skidding trails and logging tracks. As a result, the pressure caused by the wheels of the forest machines is distributed over a much larger surface. When the area under the contact surface increases, the pressure of the forest machine on the ground decreases. Traditionally, skidding was reinforced with logging residues and low-quality woods. Sometimes, for example, in the summer, this is not enough if work is carried out on peatlands. Serious soil damage can occur even after several passes of harvester or forwarder. To strengthen the skidding of the cutting areas with soft terrain, any currently known technical means can be used. This can be steel landing mats, mats made of wood boards, rubber mats, portable bridges, mats made of vulcanized rubber, etc.

4. Conclusion

Thus, according to the proposed method of cutting operations on cutting areas, the forwarder works only on skidding trails that are located at distances from each other equal to ten distances of movement of harvester manipulator. It allows you to reduce the number of skidding trails for moving the forwarder on logging sites with soft terrain, which are able to provide a one-time movement of a harvester, but require additional costs for strengthening of skidding trails for multiple movements of a forwarder. The proposed technology can be used for clear cutting in cutting areas with soft terrain.

Acknowledgement

The research was conducted with the financial support from the OOO Metropol, (a limited liability company under the laws of Russian Federation) in the frames of the scientific and research project №06.542/19.

References

[1] Laptev A V 2013. Parameters of a working line item of the multioperational machine of handling type. *Forestry Bulletin* 1(93) pp 85–91

[2] Laptev A V and Matrosov A V 2018. Rationale for configuration and geometric dimensions of wheeled harvester working zone. *Forestry Bulletin* **5(22)** pp 77-85

[3] Petelina O A and Maslov O G 2008 Flow sheet of apiaries development by the system of machines "harvester-forwarder" *Forestry Bulletin* **3(60)** pp 128-131

[4] Azarenok V A, Gerc E F, Zalesov S V and Lugansky N A 2012 Cut-to-length technology of logging operations when performing shelterwood felling. [In Russian: Sortimentnaya tekhnologiya lesosechnyh rabot pri ravnomerno-postepennyh rubkah]/ *Agrarian Bulletin of the Urals* **8(100)** p 51-54

[5] Kontinen K 2017 Possibilities for increasing the bearing capacity of soils and roads [In Russian: Vozmozhnosti povysheniya nesushchej sposobnosti pochv i dorog] *Lesprominform* **8(130)** pp 92-94, available at: https://lesprominform.ru/media/_protected/journals_pdf/1428/lesprominform_130.pdf

[6] Rukomojnikov K P 2015 Justification of the methods and parameters of rational technologies for development of forest compartment in forest areas [In Russian: Obosnovanie sposobov i parametrov racional'nyh tekhnologij pokvartal'nogo osvoeniya lesnyh uchastkov]. PhD thesis, Volga State University of Technology

[7] Rukomojnikov K P, Tsarev E M and Anisimov S.E. 2017 Substantiation of average distance of skid logging during integrated development of forest divisions. *Russian Forestry Journal* [In Russian: Lesnoy Zhurnal]. **4(358)** pp 95-105

[8] Rukomojnikov K P 2015 Structuring of Loading Points and Main Skid Road in Conditions of Existing Road Network in Forest Compartment. Journal of Applied Engineering Science. **13(3).** p 167-174. doi:10.5937/jaes13-8866

IOP Conf. Series: Materials Science and Engineering **817** (2020) 012029 doi:10.1088/1757-899X/817/1/012029

[9] Laptev A V 2014 Technological development scheme in cutting areas when performing selective logging using multioperational machines manipulator type. *Forestry Bulletin* 2-S pp 62–69

[10] Rozhkov L N, Eroshkina I F, Klysh A S, Malashevich D G and Protas P A 2018 Technology of partial cutting and natural regeneration of the forest [In Russian: Tekhnologiya nesploshnyh rubok i estestvennogo vozobnovleniya lesa]: teaching aid (Minsk: Belarusian State Technological University) p 180

IOP Conf. Series: Materials Science and Engineering **817** (2020) 012030 doi:10.1088/1757-899X/817/1/012030

The analysis of the interaction of the abrasive wheel working fluid with the treated surface when grinding wood and wood materials

A Sergeevichev[1*], V Sokolova[2], E Vlasov[1], V Sergeevichev[1], A Fedyaev[3]

[1]Department of Applied Mechanics and Engineering Graphics, *Saint-Petersburg State Forest Technical University*, 5 Institutskiy Lane, Saint-Petersburg, 194021, Russian Federation

[2]Department of Mathematical Methods in Management, *Saint-Petersburg State Forest Technical University*, 5 Institutskiy Lane, Saint-Petersburg, 194021, Russian Federation

[3]Department of Technology of Materials, Constructions and Structures from Wood, *Saint-Petersburg State Forest Technical University*, 5 Institutskiy Lane, Saint-Petersburg, 194021, Russian Federation

[*] Corresponding e-mail: 910sav@gmail.com

Abstract. In the mechanical technology of wood, a large amount of work falls on grinding operations, which are final in the manufacture of parts before finishing or gluing. The flow of these technological operations, the appearance and quality of products depends on the roughness of the polished surface. The use of abrasive wheels made of spherecorundum increases the productivity of the process of abrasive machining of parts. The field of application of hard abrasive tools for grinding wood and wood materials is expanding due to the reduction of tool salting and the elimination of burn-in of the treated surface. In the manufacture of an abrasive tool, the characteristics of the internal volumetric structure are regulated - granularity, grain content and ligaments. When grinding wood and wood materials, only the peripheral surface of the circle, its relief, takes a direct part in the work. The paper describes the relief of a circle made of spherecorundum and establishes a relationship between volumetric characteristics and its surface geometric parameters, which are necessary to determine the main indicators of the grinding process.

1. Introduction

The work of many scientists is devoted to the study of the grinding process of wood and wood materials with abrasive tools [1-4]. According to the results of the studies, the use of abrasive wheels for grinding wood and wood materials has certain advantages over the grinding sand, namely: high machining accuracy, good quality of the polished surface, high durability, and long service life of the abrasive tool, relatively low cost. In [5-8], it is noted that, depending on the grinding mode, the abrasive wheel can work with self-sharpening or with blunting.

The main characteristics of an abrasive tool that determine its working properties are graininess, hardness and structure. To obtain different hardness and structure, it is necessary to provide a certain ratio of abrasive grains, ligaments and pores per unit volume of the tool [9-11].

2. Methods of Materials

Two simultaneously proceeding processes accompany grinding wood and wood materials with abrasive wheels: blunting the wheel and self-sharpening it. The predominance of a process depends on the physic-mechanical properties of the grain, binder, characteristics of the wheel, the properties of the material being processed, the cutting mode, and so on.

In the process of grinding with conventional abrasive material, the grain wear sites increase, the radius of curvature and the rake angle increase, the different height of the grains decreases, and their density on the surface of the wheel increases. When the rake angle reaches its maximum value, abrasive grains cease to perform useful cutting work, and only elastic and plastic displacement of the material of the processed surface is performed. As a result, work on overcoming friction, plastic deformation increases, heat dissipation increases, forces acting on the grain's side from the material increase. The further course of the process depends on the type of grain wear and bond strength. If only the abrasion of the cutting elements of the grinding grains occurs with the formation of sites on them and the forces of holding the grain in the bundle are large, then the wheel works in a dull mode and the abrasive cutting tool turns into a friction tool, cauterizing the surface to be treated. In this case, the cutting ability of the abrasive tool is restored by forcibly removing the dull grains during the dressing process. Loss of cutting ability also occurs as a result of filling intergranular space of an abrasive tool with chips.

In the case when, with an increase in the load on each grain, its micro-destruction occurs with the separation of small particles from it or macro-destruction with the separation of large particles commensurate with the grain size, or the complete breaking of the grains from the binder leads to an increase in the distance between the grains and into interaction with the processed material new downstream, dull grains come in. Thus, the working surface of the wheel is constantly updated, and it works in self-sharpening mode.

Self-sharpening is a hallmark of an abrasive tool. In a blade tool, destruction is a critical event leading to breakage and failure. In an abrasive tool, the destruction of cutting grains is a normal working condition, since the acting and permissible loads are quite close in magnitude. In this case, the tool operates in self-sharpening mode.

For abrasive wheels made of spherecorundum, the processes of wear and self-sharpening have their own characteristics. Self-grinding of abrasive wheels of spherecorundum is carried out not only by tearing and removing grains from the surface of the wheel, but mainly by exposing the sharp cutting walls of hollow spherical abrasive grains. Initially, when machining abrasive wheels by means of dressing, the sharp edges of the grains are exposed. However, it is important to determine the conditions under which the exposure of the walls during grinding occurs.

The spherical shape of the grain leads to the predominance of elastic and plastic deformations at most ratios of the penetration depth of abrasive grains into the processed material to the radius of abrasive grains arising during grinding. Destruction of abrasive grain can occur when the load on the part of the material to be processed P_m reaches the ultimate corresponding mechanical strength of the abrasive grain of spherecorundum:

$$P_{\mathrm{m}} = P_{\mathrm{d}} \tag{1}$$

where, P_m – abrasive grain breaking load, N; P_d – load acting from the processed material on the implanting grain, N.

For plastic contact:

$$P_{\mathrm{m}} = S_{\mathrm{c}} \cdot H \tag{2}$$

where, S_c – actual contact area, m^2; H – material hardness, N/m^2.

The actual contact area for a spherical abrasive grain is expressed through its diameter d and through the penetration depth, which in this case is the chip thickness a:

$$S_{\mathrm{c}} = \pi \cdot d \cdot a \tag{3}$$

IOP Conf. Series: Materials Science and Engineering **817** (2020) 012030 doi:10.1088/1757-899X/817/1/012030

Then:

$$P_d = \pi \cdot d \cdot a \cdot H \tag{4}$$

Therefore, the critical chip thickness corresponding to the destruction of abrasive grain:

$$a_{cr} = \frac{P_d}{\pi \cdot d \cdot H} \tag{5}$$

Knowing the mechanical strength of the abrasive grain of spherecorundum, its granularity and hardness of the processed material, we can determine the depth of penetration of the abrasive grain into the material at which the abrasive grain breaks and its cutting edges are exposed.

Calculations made according to formula (5) for different grit numbers of abrasive and wood species showed that the penetration depth necessary for breaking abrasive grain significantly exceeds the maximum possible chip thickness cut by one grain, therefore, in the grinding process, a single abrasive grain of spherecorundum cannot penetrate to such a depth at which the load from the side of the processed material would reach destructive. Therefore, the exposure of the cutting walls of the abrasive grains of spherecorundum during grinding occurs only due to their wear.

The equation for the maximum thickness of chips removed by one grain for the front wall of the grain of spherecorundum has the form [12]:

$$a_{max} = 2\frac{V_s}{V} \cdot \left(\frac{h}{D}\right)^{0.5} \cdot \delta \tag{6}$$

where, V_s – feed rate, m/s; V – cutting speed, m/s; h – grinding depth, mm; D – wheel diameter, mm; δ – grain spacing, mm.

In the general case, substituting its expression (6) instead of a, we obtain a relation between the destructive force P_d and the main parameters of the grinding process and the characteristics of the wheel:

$$P_d = 2\pi \cdot d \cdot H \cdot \frac{V_s}{V} \cdot \left(\frac{h}{D}\right)^{0.5} \cdot \delta \tag{7}$$

The load that destroys the completely abrasive grain of spherecorundum in the grinding process depends on the grinding conditions, the hardness of the processed material and the characteristics of the wheel. Depending on the hardness of the material to be processed, it is necessary to use an abrasive tool of the appropriate characteristics: granularity, hardness, structure, while using spherecorundum having optimal physical and mechanical properties, in particular, the grain wall thickness, which largely determines its breaking load.

The impact of the petals of the working fluid of the grinding wheel from a bulk non-woven material bearing an abrasive, fixed in it with a binder, is expressed as the resultant of friction forces. When installing the petals in the working fluid at an angle to the axis of the wheel, the resultant of the friction forces is the sum of the friction forces in the plane of the petal T_1 and the force T_2 normal to the plane of the petal. The friction force is determined from the expression: $T = f \cdot N$. The normal pressure N will be determined by the direction of the force from the part relative to the petals, the number of petals in this direction and the degree of deformation of the working fluid of the wheel in the radial direction. The stiffness of a single petal in the normal plane is many times less than in the plane of the petal, so the friction coefficients in these planes will be different.

Consider the interaction process of a separate petal mounted at an angle α to the axis of the wheel in its working fluid. We assume that grinding occurs with the same deformation in the radial direction of the wheel and the same forces from the side of the part with tangential and axial directions relative to the wheel. The front edge of the petal first interacts with the part in the direction of rotation (figure 1). The force P_τ, acting on the part side with the tangential direction of feed, is expanded in the plane of the blade and in the plane normal to its side surface: $P_\tau^y = P_\tau \cos\alpha$, $P_\tau^x = P_\tau \sin\alpha$. Based on these formulas, for equal pressure of each petal with tangential and axial directions of feed, the angle α should be equal to 45°. Nevertheless, this is without taking into account the nature of the preliminary deformation of the petal from the action of forces from the part and the interaction of the petal with other petals in the working body of the wheel.

The deviation of the petal from the action of the force P_τ will occur mainly in the direction of least stiffness, that is, along the axis *X-X* in the direction perpendicular to the side surface of the petal. If in this way we sequentially consider the position of the peripheral zone of the petal from the action of the force P_τ, then it will occupy a new position 2 (figure 1). The front edge of the petal deviates by a large amount, since in the direction the stiffness is less due to the small number of petals. Then, the deviation of the petals as the force acts along the peripheral zone of the petal decreases and has a minimum deviation at the end of the petal, since the stiffness of the working medium of the wheel increases due to an increase in the number of petals in the direction of the petal deformation.

Thus, the peripheral zone of the petal initially deforms under the action of the force P_τ, a preliminary bias wave is ahead of the contact zone, and then deformed to the maximum permissible position 2, it exerts an abrasive effect on the part - the friction of the volumetric grinding cloth on the part occurs. If we compare the positions 1 and 2, then the petal is bent in the direction of supply of the part and additionally rotated relative to its original position. Therefore, the petal enters the grinding process rotated by a dynamic angle α'.

With the axial direction of supply of the part, the deformation of an individual petal also occurs mainly in the direction perpendicular to the side surface of the petal (figure 2). At the beginning of the petal, the deformation is small, since the maximum number of petals is in the direction of the P_{OC} force. Then the deformation increases and reaches a maximum at the end of the petal. The peripheral zone of the petal produces grinding being rotated by a dynamic angle α''.

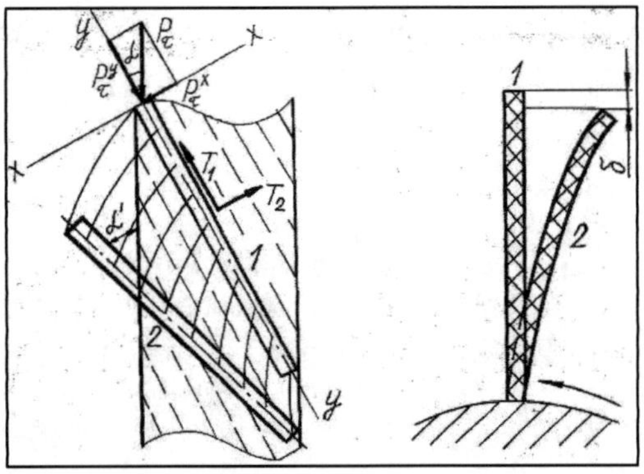

Figure 1. The deformation of the petal with the tangential direction of supply of the part relative to the circle.

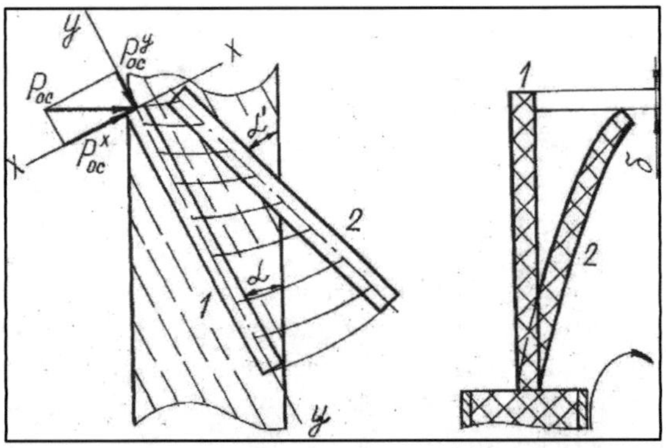

Figure 2. The deformation of the petal with the axial direction of supply of the part relative to the circle.

3. Results and Discussion

In order for the wheel to have equal elasticity in the axial and tangential directions, provide the same working conditions in these directions, it is necessary that the following condition is met: α'= α''= 45°. For abrasive interaction of the petal with the workpiece at an angle of 45°, the following conditions must be met (figure 3):

$$x_1 - x_2 = l(\sin 45^\circ - \cos\alpha) \quad (8)$$

where, x_2 – more deformation of the edge of the petal in the feed direction, mm; x_1 – less deformation of the edge of the petal in the feed direction, mm; α – static angle of rotation of the petals in the working fluid of the wheel with respect to its axis, l – the length of the chord of the petal segment, mm, is determined by the formula:

$$l = 2(R + L)\sin\frac{\beta}{2} \quad (9)$$

IOP Conf. Series: Materials Science and Engineering **817** (2020) 012030 doi:10.1088/1757-899X/817/1/012030

where, R – sanding pad radius, mm; L – the length of the petal free departure, mm; β – center angle of the petal segment,

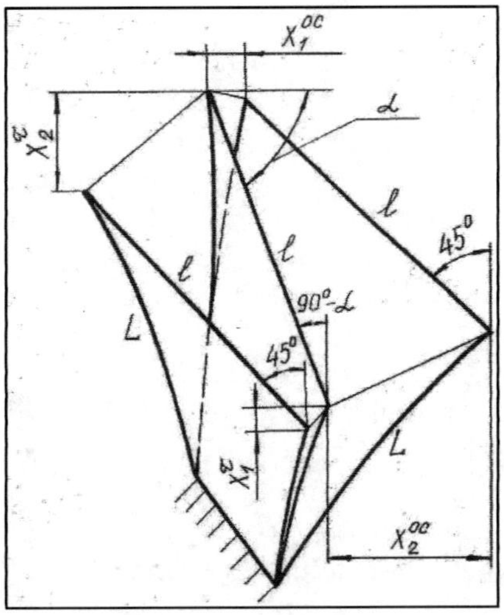

Figure 3. Scheme for determining the angle of rotation of the petals for equal elasticity in the axial and tangential directions of the wheel.

We assume that the deformation of the petals x is inversely proportional to the number of petals n in the first degree:

$$x_2 = \frac{P}{j}, \qquad x_1 = \frac{P}{j \cdot n} \tag{10}$$

where, P – force exerted by the part on the petal, N; j – stiffness of the petal in the direction of the force, N/m.

Substituting (9) and (10) into (8) and transforming, we obtain an expression for determining the static angle α:

$$\alpha = \arccos\left(\frac{1}{\sqrt{2}} - \frac{\frac{P}{j}\left(1-\frac{1}{n}\right)}{2(R+l)\sin\frac{\beta}{2}}\right) \tag{11}$$

This expression establishes the relationship between the angle of rotation of the petals in the working fluid of the wheel relative to its axis, which ensures equal elasticity of the wheel in the axial and tangential directions, with the size and number of elements of the wheel and forces arising from grinding.

4. Conclusions

During grinding, a single abrasive grain of spherecorundum cannot penetrate to such a depth that the load from the material being processed would reach a destructive value. Exposure of the cutting walls of the abrasive grains of spherecorundum during grinding occurs due to wear. The load that destroys the completely abrasive grain of spherecorundum during grinding depends on the grinding conditions, the hardness of the processed material and the characteristics of the wheel.

When the working medium of the wheel interacts from the volumetric sanding belt, the peripheral zone of the lobe initially deforms under the action of forces from the workpiece, a pre-displacement wave ahead of the contact zone, and then the lobe that is deformed to the maximum permissible state has an abrasive effect on the part.

IOP Conf. Series: Materials Science and Engineering **817** (2020) 012030 doi:10.1088/1757-899X/817/1/012030

The angle of rotation of the petals in the working body of the wheel from the volumetric grinding cloth relative to its axis, which provides equal elasticity in the axial and tangential directions of the wheel, is determined by the size, number of petals and forces arising from grinding.

References

[1] Sergeevichev A 2014 Main characteristics of the process of grinding wood and wood materials. Izvestiya Sankt-Peterburgskoj lesotekhnicheskoj akademii] iss 206 pp 124-138

[2] Brinksmeier E et al 2006 Advances in Modeling and Simulation of Grinding Processes *CIRP Annals - Manufacturing Technology* vol 55 chapter 2 pp 667-696

[3] Novoselov Yu 2012 *Dynamics of shaping of surfaces at abrasive processing* [In Russian – Dinamika formoobrazovaniya poverhnostej pri abrazivnoj obrabotke] (Sevastopol: SevNTU) p 304

[4] Rybin B, Sanaev V and Kirillov D 2014 To the question of the choice of parameters of roughness for assessment of roughness of wood *Bulletin of Moscow State Forest University: Forest bulletin* [In Russian – Vestnik Moskovskogo Gosudarstvennogo universiteta lesa: Lesnoj vestnik] vol 4 pp 131-137

[5] Hromchak I 1996 *Abrasive processing of plate materials on mineral binders* [In Russian – Abrazivnaya obrabotka plitnyh materialov na mineral'nyh vyazhushchih] (L'vov) p 47

[6] Gdalevich A 1994 *Finishing with petal circles* [In Russian – Final'naya obrabotka lepestkovymi krugami] (Moscow) p 112

[7] Silin S and Leonov B 1992 *Depth grinding technology optimization* [in Russian – Optimizaciya tekhnologii glubinnogo shlifovaniya] (Moscow) p 120

[8] Sergeevichev A 2015 *Surface formation of wood and wood materials* [in Russian – formirovanie poverhnosti drevesiny i drevesnyh materialov] (St. Petersburg: SPbFTU) p 136

[9] Vinogradov V and Sorokin G 1995 *Abrasive wear* [In Russian – Abrazivnoe iznashivanie] (Moscow) p 224

[10] Carranoa A and James B 2005 Geometric Modeling of Engineered Abrasive Processes *Taylora Journal of Manufacturing Processes* vol 7 chapter 1 pp 17-27

[11] Sanev V, Kamenev B and Sergeevichev A 2018 *Wood cutting* [In Russian – Rezanie drevesiny] (St. Petersburg: Lan') p 456

[12] Ostrovskij V 1981 *Theoretical basis of the grinding process* [In Russian – Teoreticheskie osnovy processa shlifovaniya] (Leningrad) p 144

IOP Conf. Series: Materials Science and Engineering **817** (2020) 012031 doi:10.1088/1757-899X/817/1/012031

Mathematical model for defining peak loads in hydraulic manipulator's joints

N Sidorenkov, B Martynov, A Andronov, M Taraban

[1]St. Petersburg State Forest Technical University, 5 Institutskiy Lane, St. Petersburg 194021, Russian Federation
[2]JSC «Podyemnye Mashiny», 8 Korniyenko street, Velikiye Luki 182112, Pskov region, Russian Federation

Corresponding email: lesbisnes@mail.ru

Abstract. The main component which plays the key role in any cargo lifting machine is a hydraulic manipulator (HM). At the moment our factories manufacture a whole variety of HM, the main disadvantage of which is a short life span (3 to 5 years at full capacity). In order to increase the life span of a HM it is compulsory to continuously control peak load indicators during its utilization, therefore defining the optimal maintenance periods for the machinery. For controlling the technical condition of assembles and connections, non-dismantle methods of diagnostic are being used. The most perspective of them is the method of vibration measurement. The mathematical model of calculating peak shock loads, appearing if HM connections at the initial time and their relation to diagnosis parameter (vibration amplitude) is presented in the following article, frequency range of shock load is defined.

1. Introduction

One of the main tasks facing the forest complex is the creation of new innovative technologies of logging and wood processing production, manufacturing of modern domestic high-performance and reliable machines.

The main mechanism, on which the reliability and productivity of the load-lifting machine depends, is the hydraulic manipulator (HM). At present, our factories produce entire model lines of HM, the main drawback of which is a short service life ranging from three to five years with their full utilization.

Large manufacturers of modern equipment pay great attention to the development of a service network of machines. In order to ensure high performance of equipment, a proactive maintenance and repair strategy (M and R) has become increasingly popular in recent years.

The analysis conducted in [1] allows us to determine the proactive strategy of M and R as the most effective and purposeful for implementation in today's economic conditions. Pro-active strategy combines the advantages of preventive repair effects of the undertaken preventive system and information support of the decision-making processes typical for M and R regarding the technical condition of equipment.

The essence of M's and R's proactive strategy is to systematically eliminate the sources of defects, leading to premature failure of equipment. At the beginning the analysis and causes of occurrence is carried out to find the most frequently occurring defects and their influence on the inter-repair resource. Thus, all possible factors are considered that lead to development of various kinds of malfunc-

IOP Conf. Series: Materials Science and Engineering **817** (2020) 012031 doi:10.1088/1757-899X/817/1/012031

tions. These factors are structural and technological impact on the environment, operating conditions, technical operation of equipment, qualification of personnel, degree of equipment wear, carrying out of M and R etc.

The proactive strategy of M and R is based on an estimation of a technical condition of the equipment that increases mechanism's service life due to decrease in speed of development or elimination of nascent malfunctions at an initial stage of their occurrence; reduces expenses for carrying out of maintenance and R; reduces probability of emergency failures, raises factor of technical use of technics.

The company "Lifting Machines" that is a manufacturer of hydraulic manipulators for various purposes provides the following maintenance intervals:

- maintenance of the M-1 every 50 hours;
- maintenance of the M-2 every 250 hours;
- maintenance of the M-3 every 500 hours;
- maintenance of the M-4 every 1000 hours.

The first failure time is set at 120 hours. The installed technical service life of the HM is determined at the rate of 6000 moto-hours or 3-5 years of operation depending on the intensity of HM use. In addition, according to the manufacturer's data the presented technical resource of HM can be spent for 8-10 years at corresponding system of preventive maintenance.

At the same time, the manufacturer's regulations do not provide a major reduction of service intervals in case of a significant change in loads on the elements of the power structure of the HM or hydraulic system. Shock loads on the elements of the HM power structure at starting-brake or transition modes, gross violations of operating rules, overheating of liquid in the hydraulic system - all these factors directly affect the residual life of the M and should be a sufficient reason to reduce the maintenance intervals between services.

Moreover, the warranty policy of the manufacturer, as a rule, involves the analysis of failed elements and in each specific case, a specialist organizes a quite labor-intensive work to determine the possible factors that have influenced the premature failure in order to recognize the case warranty or non-warranty obligation.

In the presence of system, that monitors continuously the basic parameters influencing operational loadings, the decision-making process could be essentially simplified on recognition of case as the guarantee or not guarantee obligation.

For the end user, such monitoring system could also be useful allowing growth of service intervals in the absence of significant peak loads or overheating during the operation of the HM.

In the automotive industry, such systems have been used for a long time and have already gained the trust of both the manufacturer and the consumer. In particular, the vast majority of service intervals for internal combustion engines are evaluated by the dielectric permeability of the engine oil with a corresponding sensor, which detects the readings at each start of the engine. When oil is contaminated, oxidized and its additives are produced, the dielectric constant changes that leads to recalculation of the oil change interval.

Development of a full-fledged system of continuous monitoring of parameters is impossible without defining the criteria for evaluating the technical condition of the mechanism, so it is necessary to develop the theoretical and practical foundations of service organization, the optimal periods of service and HM prevention. For this purpose, modern methods of diagnosing machines are used. However, practically on all load-lifting machines, all mechanisms, except for HM, are subject to control.

Therefore, in order to increase the life cycle of the HM and the entire machine as a whole, it is necessary to apply the control of the technical condition of the HM in the process of operation and, thus, to determine the optimal periods of prevention for the mechanism, and thus find new methods of managing the technical condition of the mechanism. All listed above is directed on increase of reliability of HM, and increase of their service life that is an actual task.

The aim of the study is to improve the reliability of HM operations. This paper describes the scientific approach of the development of a mathematical model for the calculation of peak loads in the closed circuit on the example of HM VM-10L, used in the forest industry.

2. Methods and Materials

Currently there are conducted a lot of researches of loading a feller-buncherson different modes of their operations such as continuous and selective felling, dismantling of forest after windfall [2].

However, during hydraulic manipulator operation a short-term shock load arises at the initial moment, which renders the big influence on a technical condition of knots and co-stresses of HM. Then comes the established mode of operation of the HM, which is considered in detail in this paper [1].

In this article, the question of determining the impact load in terms of its influence on the technical condition of components and interfaces of the HM is considered.

In work [2] as the diagnostic signal defining a technical condition of the basic compounds of HM the amplitude of vibrations depending on shock loading in knots is defined.

To determine this load, let's consider a simplified scheme of stationary HM, presented in figure 1.

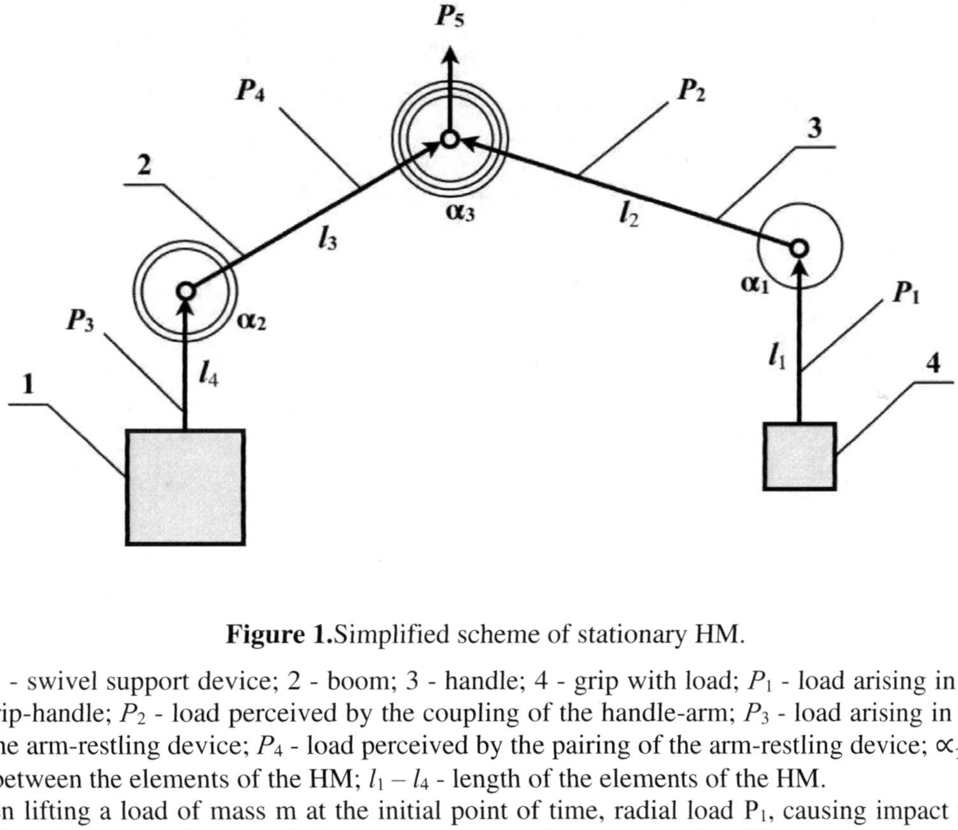

Figure 1. Simplified scheme of stationary HM.

where,1 - swivel support device; 2 - boom; 3 - handle; 4 - grip with load; P_1 - load arising in the coupling grip-handle; P_2 - load perceived by the coupling of the handle-arm; P_3 - load arising in the pairing of the arm-restling device; P_4 - load perceived by the pairing of the arm-restling device; $\propto_1, \propto_2, \propto_3$- angles between the elements of the HM; $l_1 - l_4$ - length of the elements of the HM.

When lifting a load of mass m at the initial point of time, radial load P_1, causing impact pulse g_1, affects the grip support. In general, the impact pulse can be represented as [3].

$$g = Sf\left(\frac{F}{2}\right)^{0.5} \quad (1)$$

where,g is the shock pulse; S is the conjugation gap; f is the recovery factor; F is the conjugation load. Considering the HM's operation at the beginning of time, it should be noted that the main impact is on the interfacing of the handle and the arrow.

Figure 1 shows that:

$$P_5 = [(P_4 + P_2)^2]^{0.5} = (P_4^2 + 2P_4P_2 cos \propto_3 + P_2^2)^{0.5} \quad (2)$$

P_2's load will be equal to

$$P_2 = P_1(l_1 + l_2)sin \propto_1 \quad (3)$$

P_4's load will be equal to

$$P_4 = P_3(l_4 + l_3)\, sin \propto_2 \quad (4)$$

Then, from expression (1), we obtain the formula for the percussive impulse in the pairing of the boom-arm

$$g_5 = S_5 f \left[\frac{(P_4+P_2)^2}{2}\right]^{0.5} \quad (5)$$

To determine the relationship between the amplitude of the vibration signal and the shock load, we use the expressions used in [3].

During the operation of the HM due to the changes in the gap values in the interfaces of the HM alters the value of the disturbing force. This happens because of the presence of amplitude modulation of the signal. Then:

$$g_5 = S_5 f[1 + \varepsilon \sin(\omega t + \varphi)] \left[\frac{(P_4+P_2)^2}{2}\right]^{0.5} \quad (6)$$

where, ε - modulation factor; ω - modulation frequency; φ - modulation phase; t - time.

The amplitude of the vibration signal will be equal to

$$A = \frac{1}{m} S_5 f[1 + \varepsilon \sin(\omega t + \varphi)] \left[\frac{(P_4+P_2)^2}{2}\right]^{0.5} \quad (7)$$

where, m is the weight of the cargo.

3. Results and Discussion

Knowing the amplitude, mass and geometric dimensions of the HM can determine the frequency range in which the shock pulse occurs

Initially, for the stationary variant the optimal position of the HM elements is fixed, the angles between the elements, their length for lifting the load of a given mass are determined, and the nominal frequency at which the impact occurs is calculated. Then experimentally establish deviations of angles and lengths of elements that can appear in conditions of work HM. In this way, the frequency range of the impact impulse generated is determined.

According to the results of studies conducted on the HM VM-10L, the frequency range is 0.5 - 3 kHz.

Thus, knowing the frequency and determining the amplitude of the vibration signal, it is possible to control the technical state of the HM couplings.

References

[1] Bobrovitskiy V I and Sidorov A V 2011 Improvement of the system of maintenance and repair of equipment in the conditions of centralization of the repair service of the enterprise(in Russian – Soverchenstvovanie sistemy TO I R oborudovaniya v usloviyax centralizacii remontnoii slugbi predprieatiea) *Vibration of machines: measurement, reduction, protection.* (Donetsk: Don-NTU), **1(24)** pp 23–28

[2] Gasimov G Sh and Alexandrov V A 2014 Dynamics of the felling and bundling machines (in Russian – Dinamika valohno-paketirujchih machin) (St. Petersburg: Polytechnic Units) p 244

[3] Martynov B G 2012 Fundamentals of diagnostics of the LZM (in Russian– Osnovy diagnostiki mexanizmov i agregatov LZM) (St. Petersburg: St. Petersburg State Technical University), p 100

IOP Conf. Series: Materials Science and Engineering **817** (2020) 012032 doi:10.1088/1757-899X/817/1/012032

Optimization of forest road network layout problem

A Simonenkova[1*], M Simonenkov[2], I Bacherikov[2]

[1]Department of Economic cybernetics Saint-Petersburg State University, 7/9 University Embankment , St. Petersburg 199034, Russian Federation
[2]Institute of Technological Machines and Forest Transportation, St. Petersburg State Forest Technical University, 5 Institutskiy Lane, St. Petersburg 194021, Russian Federation

*Corresponding email: sukhodolskaya.anastasia@gmail.com

Abstract. The article describes a three-stage hierarchical method of solving the optimization of forest road network layout problem taking into account road construction and transportation costs using a mathematical model based on the multi-commodity flow model. Optimization of the forest road network layout problem is formulated as an alternative of the Steiner tree problem in graphs. The first stage involves optimization of the forest road network layout within clusters. Many harvesting areas have already been divided into clusters using a clustering optimization model. The second stage involves solution of the similar problem between clusters. At the third stage, the final optimization is performed based on the results obtained at the first two stages. The developed methodology of heuristic solution of optimization of forest road network layout problem allows to improve solving time by 3-6 times with a deviation from the optimal result in the range of 0.5-8%.

1. Introduction

Forestry companies face a complex problem of planning the layout, construction and use of forest roads when making tactical-level management decisions [1]. This task requires significant resources and is one of the main efficiency drivers of the forestry companies in the future. Therefore, a lot of studies have been dedicated to the forest roads network layout optimization problem. Some authors solve this problem together with the optimization of harvesting area selection [2, 3], some authors solve these problems separately [4, 5]. Several approaches may be used for optimization of the forest roads network layout. The problem can be solved using iterative algorithms based on the sequential application of a modified algorithm for searching the shortest path [2, 5], the evolutionary algorithm [6], and other algorithms; using MIP models [7], or dynamic programming [8]. Comparative analysis of heuristic algorithms for finding a problem solution is given in the work of Shirasawa [9]. A similar task of searching for the shortest route in networks based on deep learning method of artificial neural networks is a case of great interest [10].

Complexity of the considered problem is driven by the need for detailed planning of road network layout in large forest areas. The existing approaches applicable to large areas are based on the algorithms of searching for the shortest path and they are not able to design the optimal forest road network and assess the optimality of the provided solution. An optimal or near-optimal solution can be obtained by formulating the problem as an alternative of NP-hard Steiner tree problem in graphs. Stückelberger [4] used the same approach and formulated the problem as a Steiner tree problem alternative. However, his heuristic solution method is not applicable or is limited in application to large forest areas. Solving time for an area of 3 500 hectares ranged from 0.5 to 12 hours. One of the best practical solutions of a Steiner tree problem in graphs with mathematical programming and preprocessing of a graph is given in [11], which considers graphs up to 300 000 vertices and 400 000 edges. Solution of the optimization problem for the forest road network layout requires creation of a graph with a far greater number of vertices and edges. It should be noted that the authors are not aware of any studies involving assessment of impact of timber transportation costs on the optimal layout of the forest road network.

The developed methodology for solving a forest road network layout optimization problem for large areas with a detailed graph is aimed at finding a near-optimal solution which is based on application of linear programming, is hierarchical, and takes into account wood transportation costs. Within the framework of the methodology, an optimization problem of the forest road network layout is formulated as an alternative of the Steiner tree problem in graphs. Basic block diagram of the developed methodology is given in figure 1.

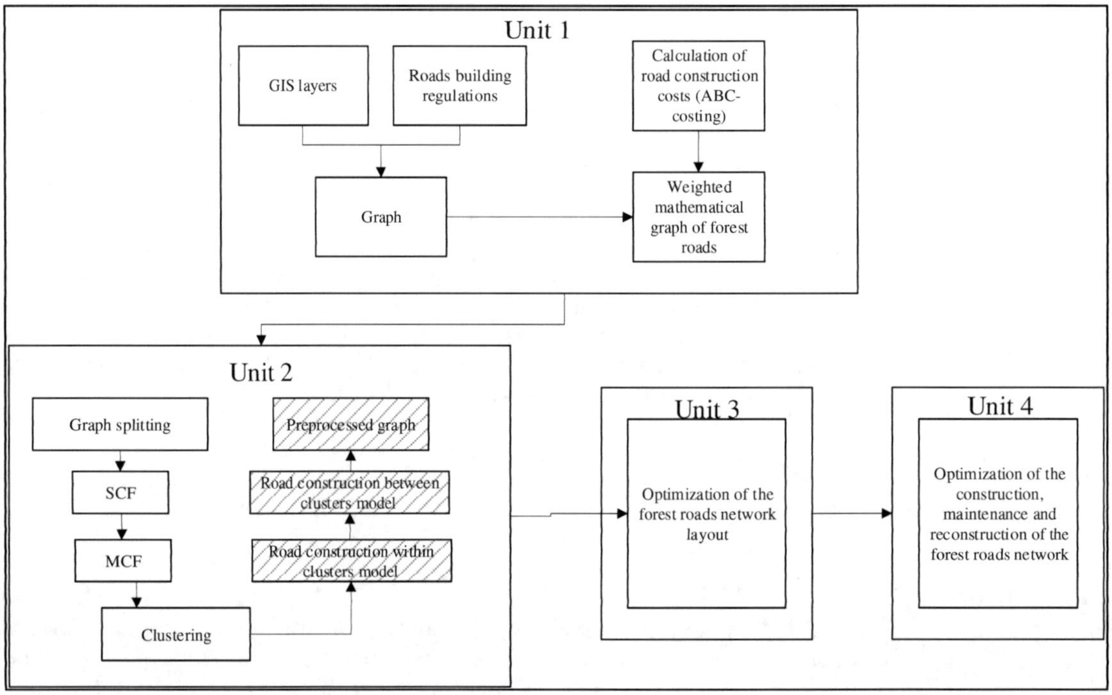

Figure 1. Framework of the optimization of forest road network layout problem solution methodology.

The method of searching minimum cost paths based on road construction costs between harvest blocks using network flow models (SCF and MCF in figure 1) is described in the paper [12]. To solve the forest road network layout problem for large areas and improve the solving time, it is proposed to use a hierarchical approach based on clustering of the entire set of harvesting areas. The developed method of clustering is a mixed integer linear programming optimization model and is described in the paper [13]. The purpose of the model is to divide the entire set of harvesting areas into groups with the lowest costs of road construction from harvesting areas inside the cluster to its center, which is also a harvesting area, given the restrictions on the minimum and maximum number of harvesting areas in the cluster. The result of clustering is presented as a set of harvesting areas with the lowest road construction costs within the cluster to its center divided into clusters/groups. Below are the examples of clustering with division of the harvesting area set into 4 clusters (with the minimum number of supply-nodes in the cluster equaling 3) (figure 2) and division of the harvesting area set into 3 clusters (with the minimum number of supply-nodes in the cluster equaling 5) (figure 3).

The article focuses on the hierarchical method of searching for an optimal forest road network layout within and between clusters based on a mathematical model of linear programming (highlighted by hatching in figure 1).

2. Methodology

A weighted, directed mathematical graph with extraverted edges was created to test the methodology. The vertices of the graph are located in the centroids of a squared grid with a 100 m × 100 m resolution. Valence of the graph vertices is limited to 16. The weight of the graph edge represents construction cost of a potential road. Cost calculation of constructing the edges is based on the penalty functions method similar to that described in [2, 5]. The basic cost of construction of one kilometer of forest road is multiplied by penalties depending on the grade, intersection with talwegs, and watersheds. Vertices, grade, talwegs, and watersheds were calculated based on digital elevation model data of the Shuttle Radar Topography Mission (SRTM) [14]. The graph with 11 919 vertices and 183 843 edges is used in this study for consideration. 18 graph vertices are accepted as harvesting areas. Each harvesting area was randomly assigned an inventory from 5 000 to 15 000 m^3.

We propose a process of solving an optimization problem of forest road network layout within the clusters and between them hierarchically in 3 stages.

Stage 1: Solve the linear programming optimization model of forest road network layout within each cluster based on the mathematical model of multi-commodity flow.

Stage 2: Solve the linear programming optimization model of forest road network layout between cluster's centers using the model applied in Stage 1and results of Stage 1.

Stage 3: Solve the linear programming optimization model of forest road network layout for all harvesting areas using the model applied in Stages 1 and 2 where results of the two previous stages will serve as input data.

2.1. Stage 1

Mathematical model of forest road network layout based on multi-commodity flow model:

Objective function:
$$\sum_{\substack{v\_i\in sV,\\ v\_j\in sV}} vR_{v\_i,v\_j} pC_{v\_i,v\_j} + \sum_{\substack{v\_i\in sV,\\ v\_j\in sV,\\ p\in sP}} vF_{v\_i,v\_j,p} pD_{v\_i,v\_j} pT pI_p \rightarrow \min \quad (1)$$

Table 1. Indices and sets.

Set	Index	Description
sV	v, v_i, v_j	Vertices of the graph
sP	p	Products, $\lvert sP\rvert=\lvert sV\rvert-1$

Table 2. Parameters.

Parameter	Description
pX_v	Coordinate X of the vertex v
pY_v	Coordinate Y of the vertex v
pD_{v_i,v_j}	Distance from the vertex v_i to the vertex v_j, km
pU_{vp}	1, if the vertex v is a harvesting area (supply-node) and the source of the product p, 0 – otherwise
pM_{vp}	1, if the vertex v is a demand-node of the product p, 0 – otherwise
pB_{vp}	$1\lvert pU_{vp}=1;\ 0\lvert pU_{vp}+pM_{vp}=0;-1\lvert pM_{vp}=1;$
pT	Roundwood transportation rate, thousand rubles / (m^3 * km)
pI_p	Inventory at the harvesting area (supply-node) corresponding to the product p
pC_{v_i,v_j}	Construction cost of the edge from the vertex v_i to the vertex v_j, thousand rubles

Table 3. Variables.

Variable	Description
$vF_{v\_i,v\_j,p}$	Product (p) quantity, transported from the vertex v_i to the vertex v_j, units
vR_{v_i,v_j}	1, if the graph's edge from the vertex v_i to the vertex v_j is built, 0 – otherwise

Table 4. Variables properties.

Variable	Type	Lower bound	Upper bound	Index domain
$vF_{v\_i,v\_j,p}$	Continuous	*0*	*1*	$v\_i, v\_j \in sV \mid pC_{v\_i,v\_j}>0$
vR_{v_i,v_j}	Continuous	*0*	*1*	$v\_i, v\_j \in sV \mid pC_{v\_i,v\_j}>0$

Constraints:

$$\sum_{v\_j \in sV} vF_{v\_i,v\_j,p} + pB_{v\_i,p} = \sum_{v\_j \in sV} vF_{v\_j,v\_i,p} \quad \forall v\_i \in sV,\ p \in sP \tag{2}$$

$$vR_{v\_i,v\_j} \geq vF_{v\_i,v\_j,p} \quad \forall v\_i, v\_j \in sV \mid pC_{v\_i,v\_j}>0,\ p \in sP \tag{3}$$

$$vR_{v\_i,v\_j} + vR_{v\_j,v\_i} \leq 1 \quad \forall v\_i, v\_j \in sV \mid pC_{v\_i,v\_j}>0 \tag{4}$$

Sets and indices are given in Table 1. Parameters of the model and their description are given in Table 2. Variables of the model and their properties are given in Table 3 and Table 4.

The objective function (1) of the model is to minimize road network construction and wood transportation costs.

The equation (2) is a transshipment constraint where for each vertex the sum of incoming and outgoing flows of each product is equal to 0. Herewith, in order to model the supply and take into account the corresponding costs the following parameters are used: 1 for the vertex representing the harvesting area (supply node) of the corresponding product, -1 for the demand-node of each product and 0 for all other vertices. At Stage 1, we consider the harvesting areas – cluster centers – as demand-nodes. The constraint (3) prohibits transportation of products using unbuilt roads (edges). The constraint (4) partially eliminates symmetry in allowable range of model solutions. Logically, they only allow construction of roads in one direction. The constraints (4) are similar to those (12) in Naderializadeh [2]. As a result of the mathematical model solution for each cluster the optimal network of forest roads was developed, an example of the model solution for four clusters (with the minimum number of supply-nodes in the cluster equaling 3) (figure 4) and three clusters (with the minimum number of supply-nodes in the cluster equaling 5) (figure 5).

2.2. Stage 2

At the second step, the following changes are introduced into the mathematical model described above:

Objective function:

$$\sum_{\substack{v\_i \in sV,\\ v\_j \in sV}} vR_{v\_i,v\_j} pC^{stage2}_{v\_i,v\_j} + \sum_{\substack{v\_i \in sV,\\ v\_j \in sV,\\ p \in sP}} vF_{v\_i,v\_j,p} pD_{v\_i,v\_j} pTpI_p \rightarrow min \tag{5}$$

Table 5. Parameters.

Parameter	Description
$pC^{stage2}_{v\_i,v\_j}$	0.001 thousand rubles for the edges that were included in the solution of the task of Stage 1; cost of road construction from vertex v_i to vertex v_j, thousand rubles – for all other edges

A new parameter for Stage 2 model and its description is given in Table 5.

The results of the first stage were used in the second stage as input data, i.e. the edges included in the solution in the first stage were assigned a minimum weight of one ruble. This was done in order to ensure that the road network between the cluster centers would be partially routed along the edges which were included in the optimal solution in the first stage. A new parameter for road construction cost is introduced. The value of this parameter is equal to 1 ruble for all the edges included in the solution of stage 1, for all other edges the value of the parameter does not change. At Stage 2, a harvesting area (V11688) is considered to be a demand-node of all products appearing in the cluster centers.

As a result of the solution of the mathematical model between the cluster centers, the optimal network of forest roads was developed, an example of the model solution for four clusters (with the minimum number of supply-nodes in the cluster equaling 3) (figure 6) and three clusters (with the minimum number of supply-nodes in the cluster equaling 5) (figure 7).

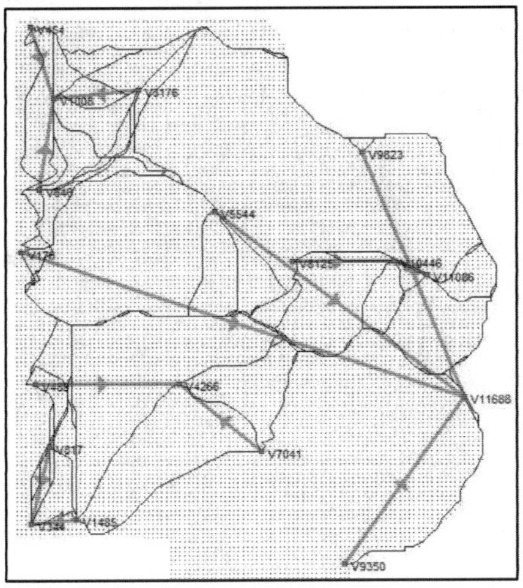

Figure 2. Example of dividing the harvesting area aggregation into 4 clusters (with the minimum number of supply-nodes in a cluster equaling 3).

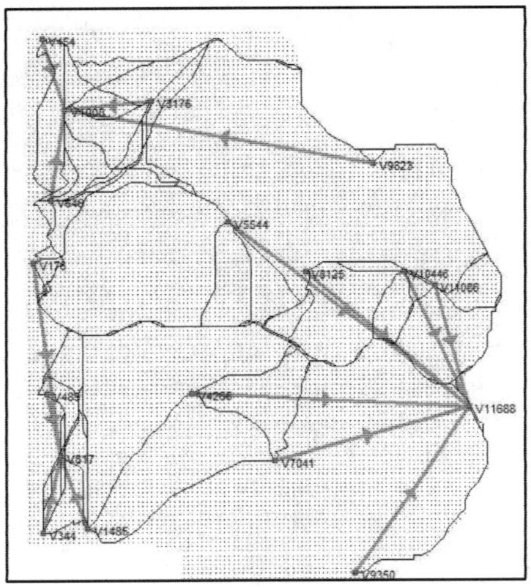

Figure 3. Example of dividing the harvesting area aggregation into 3 clusters (with the minimum number of supply-nodes in a cluster equaling 5).

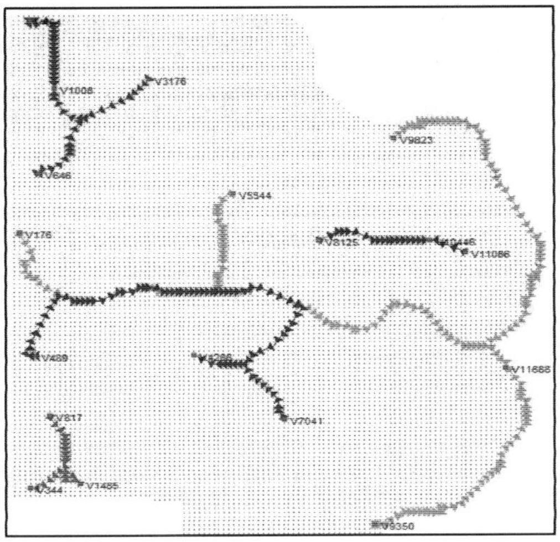

Figure 4. An example of Stage 1 model solution for a set of 4 clusters, where the cluster centers are the vertices (with the minimum number of supply-nodes in a cluster equaling 3).

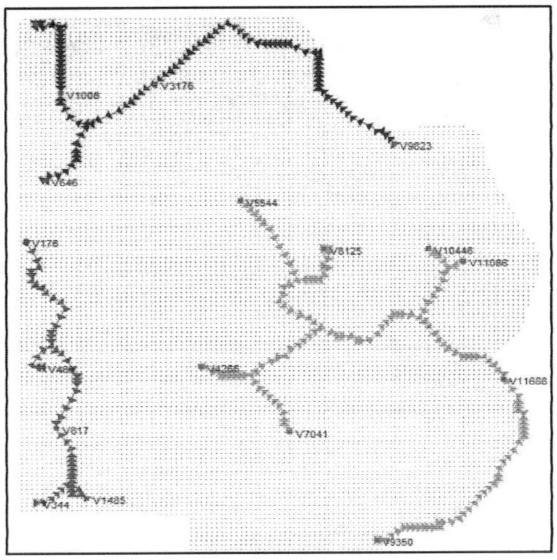

Figure 5. An example of Stage 1 model solution for a set of 3 clusters, where the cluster centers are the vertices (with the minimum number of supply-nodes in a cluster equaling 5).

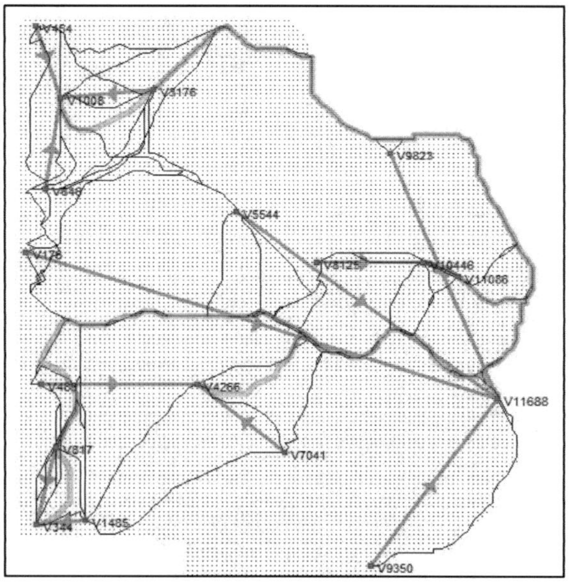

Figure 6. An example of Stage 2 model solution for a set of 4 clusters, where the cluster centers are the vertices (with the minimum number of supply-nodes in a cluster equaling 3).

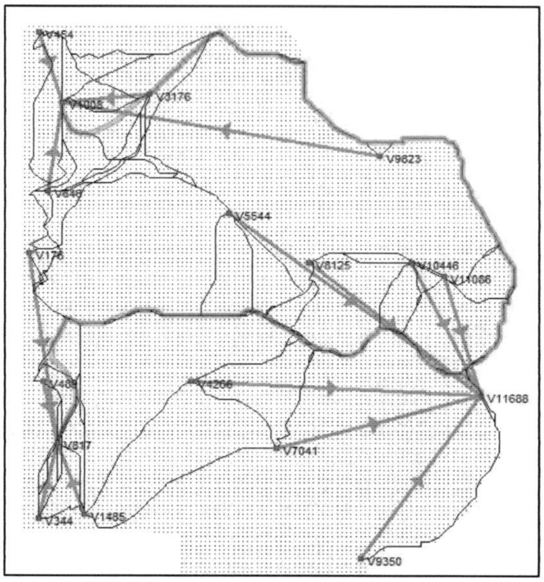

Figure 7. An example of Stage 2 model solution for a set of 3 clusters, where the cluster centers are the vertices (with the minimum number of supply-nodes in a cluster equaling 5).

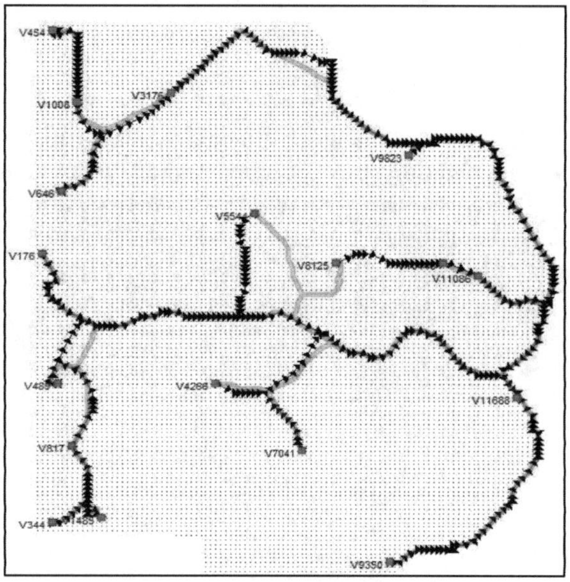

Figure 8. Comparison of the optimal solution with the solution obtained by the hierarchical approach (minimum number of supply-nodes in the cluster is 3).

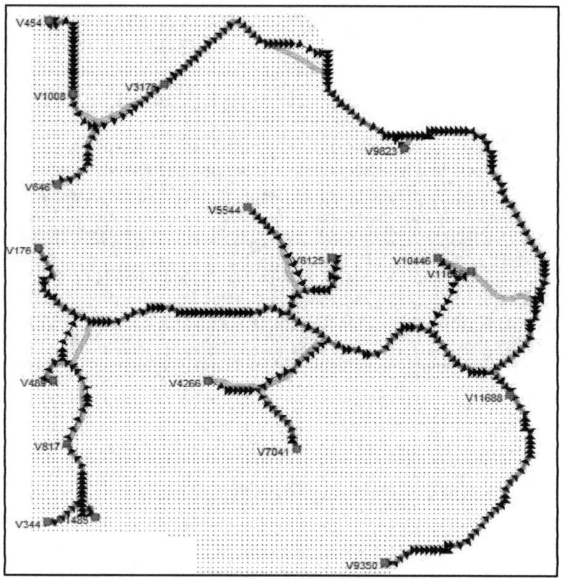

Figure 9. Comparison of the optimal solution with the solution obtained by the hierarchical approach (minimum number of supply-nodes in the cluster is 5).

2.3. Stage 3. Final optimization

At the third stage, the harvesting area (V11688) – is considered as a demand-node of all products appearing in each supply-node. Here, optimization is carried out only among those edges which were delivered as models solution results at the first two stages (Stage 1 and Stage 2). The following changes are introduced into the mathematical model described in Stage 1.

Table 6. Parameters.

Parameter	Description
$pE^{stage1}_{v\_i,v\_j}$	1 – if the edge from the vertex v_i to the vertex v_j is included in the optimal solution at Stage 1, 0- otherwise
$pE^{stage2}_{v\_i,v\_j}$	1 – if the edge from the vertex v_i to the vertex v_j is included in the optimal solution at Stage 2, 0 - otherwise
pE_{v_i,v_j}	1 – if $pE^{stage1}_{v\_i,v\_j}+pE^{stage2}_{v\_i,v\_j}\geq 1$, 0- otherwise

Table 7. Variables properties.

Variable	Type	Lower bound	Upper bound	Index domain
$vF_{v\_i,v\_j,p}$	Continuous	0	1	$v\_i, v\_j \in s V \mid pC_{v\_i,v\_j}>0 \ \& \ pE_{v\_i,v\_j}=1$
vR_{v_i,v_j}	Continuous	0	1	$v\_i, v\_j \in s V \mid pC_{v\_i,v\_j}>0 \ \& \ pE_{v\_i,v\_j}=1$

Constraints:

$$vR_{v\_i,v\_j}\geq vF_{v\_i,v\_j,p} \ \forall v\_i, v\_j \in s V \mid pC_{v\_i,v\_j}>0 \ \& \ pE_{v\_i,v\_j}=1, \ p \in s P \tag{6}$$

$$vR_{v\_i,v\_j}+vR_{v\_j,v\_i}\leq 1 \ \forall v\_i, v\_j \in s V \mid pC_{v\_i,v\_j}>0 \ \& \ pE_{v\_i,v\_j}=1 \tag{7}$$

New parameters for Stage 3 model and their description are given in Table 6, new variables and their properties are given in Table 7.

Having solved the model described in Stage 1, provided that the demand-node of each product located in the harvesting area V11688, it is possible to find an optimal solution to the forest road network layout problem.

An example of a model solution for Stage 3 for four clusters (with the minimum number of supply-nodes in a cluster equaling 3) (figure 8) and three clusters (with the minimum number of supply-nodes in a cluster equaling 5) (figure 9). Figures in green show the optimal forest road network layout, black color highlights the network layout obtained by the hierarchical approach.

3. Results and discussion

The calculations were performed using a MSI GT72VR 7RE laptop, mathematical models were written in the AIMMS v4.64.1.0 mathematical modeling system, the solutions were found directly using commercial solver CPLEX v12.6 with the standard parameters except for the algorithm of searching for a solution to the linear programming problem where Simplex method was replaced by the method of internal point (LP Solve – Barrier) [15]. The results delivered during the computational experiment are given in Table 8.

The values of the objective function and the time required to find a solution were compared with those of the optimal solution developed based on the computational experiment. The following conclusions can be made on the basis of the analysis:

- in 5 out of 7 cases, the deviation of the objective function value from the optimal value was less than 1%;
- in all cases, it was possible to significantly reduce the solving time by a factor of approximately 3-6.

It should be noted that the value of the objective function obtained at the first stage of the solution, which is the sum of the values of the objective functions of the model solution for each cluster, increases as the minimum number of supply-nodes in a cluster grows, which can be explained by the fact that the smaller the number of harvesting areas in the cluster, the less the length and cost of road network construction in the cluster. On the other hand, the smaller number of supply-nodes in the cluster and the higher number of clusters result in expansion of the road network between the clusters and, consequently, to its cost increase.

It is impossible to determine the optimal value of the minimum number of supply-nodes in a cluster based on one computational experiment, which makes it necessary to take further steps:

- develop an approach to determine the optimal minimum number of supply-nodes in one cluster;
- conduct a large number of computational experiments using different data sets to determine the optimal minimum number of supply-nodes in one cluster;
- conduct computational experiments using other clustering methods in order to select the best method;
- test the methodology on a large data set after determining the optimal minimum number of supply-nodes in a cluster and selecting the best harvesting area clustering method.

Table 8. Results.

Minimum number of supply-nodes in a cluster	3	4	5	6	7	8	9
Number of clusters	4	4	3	3	2	2	2
Stage 1. Value of the objective function (thousand rubles)	60181	64285	69405	75063	75685	77040	80431
Stage 1. Solving time (sec.)	151.15	156.76	220.74	196.46	400.95	377.36	312.61
Stage 2. Value of the objective function (thousand rubles)	31826	27256	23653	18971	0[a]	0[a]	4649
Stage 2. Solving time (sec.)	50.6	29.52	14.88	14.33	0[a]	0[a]	6.95
Stage 3. Value of the objective function (thousand rubles)	85020	86168.29	84400.58	84770.5	84648.8	84493	90792
Stage 3. Road construction costs	74532	75834	73958	74305	74244	74031	80833
Stage 3. Transportation costs	10488	10334	10443	10466	10405	10449	9959
Stage 3. Solving time (sec.)	202.02	186.34	235.7	210.85	332.56	377.36	319.62
Optimal value of the objective function (thousand rubles)				84002.98			
Optimal road construction costs				73784			
Optimal transportation costs				10219			
Solving time (sec.)				746.41			
Deviation from the optimum objective function value (%)	1.21	2.58	0.47	0.91	0.77	0.58	8.08
Time deviation from the optimal solution solving time (%)	146.69	300.56	216.68	254.00	124.44	97.80	133.53

[a] the Stage 2 model was not solved because the Stage 1 model solution had already provided roads within the clusters connecting the clusters to each other

Acknowledgments
The research was partially supported by FASIE. Grant № 14592ГУ/2019 by July 23, 2019. We would like to acknowledge and thank AIMMS B.V. for Academic License and Zinchuk Marina for helping with translation.

References
[1] Heinimann H R 2017 Forest Road Network and Transportation Engineering – State and Perspectives *Croatian Journal of Forest Engineering* **38(2)** pp 155-173
[2] Naderializadeh N and Crowe K A 2018 Formulating the integrated forest harvest-scheduling model to reduce the cost of the road-networks *Operational Research* pp 1-24 DOI: 10.1007/s12351-018-0410-5
[3] Epstein R, Weintraub A, Sessions J, Sessions B, Sapunar P, Nieto E, Bustamante F and Musante H. 2001 *PLANEX: A system to identify landing locations and access* Proceedings of the International Mountain Logging and 11th Pacific Northwest Skyline Symposium pp 190-193
[4] Stückelberger J A 2007 *A weighted-graph optimization approach for automatic location of forest road networks* Eidgenössische Technische Hochschule ETH Zürich, Zürich p 127 DOI: 10.3929/ethz-a-005552657
[5] Anderson A E and Nelson J 2004 Projecting vector-based road networks with a shortest path algorithm *Canadian Journal of Forest Research* 34 pp 1444-1457. DOI: 10.1139/x04-030
[6] Woodam C and Sessions J 2010 *Designing a Forest Road Network using Heuristic Optimization Techniques* Council on Forest Engineering (COFE) Conference Proceedings: "Appalachian Hardwoods: Managing Change
[7] Akay A E, Aruga K, Bettinger P and Sessions J 2013 Using optimization techniques in designing forest roads and road networks *Journal of the Bartin Faculty of Forestry* 15(1) pp 49-62
[8] Sokolov A P 2016 *Feasibility of technologies and parameters of processes of integrated development of forest resources bases on the basis of logistic approach* DSc Thesis. (Petrozavodsk: Petrozavodsk State University) p 323
[9] Shirasawa H and Hasegawa H 2014 A comparative study of heuristic algorithms for the multiple target access problem *Journal of Forest Research* **19(5)** pp 437-499 DOI: 10.1007/s10310-013-0432-z
[10] Zhou T 2018 *Deep Learning Models for Route Planning in Road Networks* Master's thesis School of Electrical Engineering and Computer Science p 67
[11] Leitner M, Ljubic I, Luipersbeck M, Prossegger M and Resch M 2014 *New Real-world Instances for the Steiner Tree Problem in Graphs* available from https://www.academia.edu/21293263
[12] Simonenkova A, Bacherikov I, Simonenkov M 2019 Optimization of forest road network layout problem. Method of searching minimum road construction cost paths between harvest blocks. *The current issues in forestry: Conference of Young Scientists*, November 06-08, 2019, St.Petersburg State Forest Technical University. (St.Petersburg: SPbSFTU) pp 119-126
[13] Simonenkova A, Bacherikov I and Simonenkov M 2019 *Optimization of forest road network layout problem. Clustering of logging sites. Mathematical model for minimizing the total cost of road construction to cluster centers* In print
[14] Jarvis A, Reuter H I, Nelson A and Guevara E 2008 *Hole-filled seamless SRTM data V4, International Centre for Tropical Agriculture (CIAT)* available from http://srtm.csi.cgiar.org
[15] Press W H, Teukolsky S A, Vetterling W T, Flannery B P *Numerical Recipes: The Art of Scientific Computing* (Cambridge: Cambridge University Press) pp 1262

IOP Conf. Series: Materials Science and Engineering **817** (2020) 012033 doi:10.1088/1757-899X/817/1/012033

Perspectives of bark dump recycling at wood processing enterprises

F Svoykin*, A Birman, I Bacherikov, O Mater, V Bozhbov

[1]*St. Petersburg State Forest Technical University*, 5 Institutskiy Lane, St. Petersburg 194021, Russian Federation

*Corresponding email: svoykin_fv@mail.ru

Abstracts. Woodworking and pulp-and-paper mill areas accumulate significant volumes of wood debarking waste, which is stored in bark dumps with a volume of up to 1 000 000 m^3. Methods for recycling bark, both fresh and stagnant, are proposed through the production of fuel briquettes and pellets, composting, building materials, and the use of combustible gases that accumulate in the depths of the old bark layers. Characteristics of extractive substances obtained from bark are given and their fields of application are considered. Possible ways of recultivation of old bark dumps, ways of waste water treatment are offered. The recommendations on reduction of technogenic hazard of bark dumps are given and the necessity of constant monitoring of such objects is grounded. The technological scheme of complex use of bark dump materials on the basis of developments of Saint-Petersburg State Forest Technical University is proposed.

1. Introduction

Almost all timber materials are subject to debarking at woodworking and pulp and paper mills (PPM) [1, 2]. Manufacturers need to choose rational methods of bark recycling, which accounts for 10-15% of the volume of stem wood. Recycling is necessary because bark contains substances extracted by water, which have a harmful effect on the flora and fauna of water and soils.

Wastewater flows containing bark are a consequence of almost all operations related to the processes of processing and storage of bark. Thus, the bark-containing effluents of the pulp and paper mill make up 10-12% of the total amount of wastewater. Table 1 shows the characteristics of only the press equipment of the pulp and paper mill from the position of formation of the bark-containing sewage. Practically the presses are not able to dehydrate the bark to a relative humidity of 55-60% [3].

Table 1. Parameters for pressing the bark with moisture 84%.

Type and model of press	Humidity after the press, %	Wastewater volume, m^3/h	Bark concentration in wastewater, mg/l
Roller press from «Fulton»	64	1.1	1 200
Screw press with screw diameter; mm; 400; 550; 750	62	0.1; 0.2; 0.3	1 500
Chain press: DO-318; FMP-39; FMP-51	64	1.0; 1.4; 1.8	2 100; 2 000; 1 900
Drum press from «Enso»	55	1.5	—
Piston press from «Kovan»	55	1.75	2 100

It should be noted that various physico-chemical agents are widely used to clean bark dump infiltration from suspended and dissolved particles. They allow to remove almost all suspended and biochemical non-oxidizing substances, as well as to perform their recovery. The most likely combination of cleaning methods is the use of membrane and anaerobic methods.

Membrane methods of wastewater treatment are widely used at the enterprises of chemical wood processing [4]. The main membrane separation methods include reverse osmosis, ultrafiltration, electrolysis, evaporation through the membrane, electro-osmofiltration [5].

At power inputs of about 3 kW/h per 1 m^3 of water, at a temperature of 25°C, pressure drop of 3.6 MPa on 0.46 m^2 of membranes, at the initial content of dissolved substances of 3.5 g/l it is possible to reduce the filtrate contamination in biochemical oxygen demand and chemical oxygen demand by 93%, in dissolved substances – by 95% [6].

In most cases, millions of cubic meters of bark are taken to waste dumps with the area of tens of hectares. These areas are completely removed from economic use. Rotting bark becomes a breeding ground for plant and insect disease carriers. Runoffs from rainfall and snowmelt worsen the ecological situation in the surrounding areas and water areas. Due to self-oxidation, fires occur, causing fires and polluting the air basin [7]. Modern world practice of bark utilization is primarily designed to work with fresh bark. Here a number of technically and economically justified directions of processing are opened. Below are the main directions and approaches to the recycling of bark, both fresh and rotting [8-11].

2. Methods and Materials

2.1. Burning of bark to produce heat and/or electricity

Bark is a low-grade fuel. The fractional composition of the bark formed on debarkers is a mass with pieces from 5 to 200 mm (sometimes larger) with uneven humidity. Absolutely dry bark weighs 1 solid m^3, e.g. 740 kg for spruce and 530 kg for pine, which does not differ much from wood. Compared to fossil fuels (e.g. coal), bark ashes contain less silica, aluminum and iron oxide and lime [12]. On average, the composition of combustible bark and wood is the same, table 2.

Table 2. Composition of the combustible mass.

Fuel type	Composition of the combustible mass, %			
	C	H	O	N
Wood	51.0	6.1	42.3	0.6
Bark	51-52	5.9-6.3	40.8-42.5	0.6-0.9

The humidity of bark produced by debarking is usually higher than critical. Before such bark enters the furnace it is necessary to reduce it to 50-60%. This is achieved by means of bark squeezing presses or by drying the bark with boiler exhaust gases, ventilation emissions and other secondary thermal energy resources.

The higher heat of bark combustion is not much different from the higher heat of wood combustion, and sometimes exceeds it, because the bark contains more lignin and resins than wood. Thus, the highest heat of dry mass combustion of pine wood is 4 186.8 J/kg, pine bark – 4 850.8 J/kg [13]. Analysis of numerous experimental data on the dependence of the lower working heat of combustion Q_H on humidity shows that it can be expressed by the formula:

$$Q_H = (4660 - 55 \times W) \times 4186.8\ J/kg, \quad (1)$$

where, W is the humidity of the bark, %.

The theoretical burning temperature of the bark at a humidity of 50% varies between 1 473–1 673°K. Thus, the composition of the combustible mass, the higher heat of combustion, the volatile matter yield and the combustion temperature of dry bark do not differ much from that of wood [14]. Poor bark permeability is caused by its jamming or compaction in bunkers and transport systems [15].

IOP Conf. Series: Materials Science and Engineering **817** (2020) 012033 doi:10.1088/1757-899X/817/1/012033

There are ways to reduce the humidity of the bark without the formation of waste water. One of them is bark drying, which is appropriate if the initial bark humidity does not exceed 75%. Modern technology offers a range of machines and mechanisms with different degrees of efficiency allowing to clean the bark from foreign inclusions, grinding and dehydration of the bark [15-17].

The leader in the production of bark-oriented boilers is the Finnish company «Värtsila» (boilers of «Bio Power» series). The cost of a complete installation, depending on the capacity and type of energy produced (hot water, steam, electricity) varies between 10 000 000 – 15 000 000 euros. It is known from the scientific sources that capital investments in bark combustion plants are 3-4 times higher than for fossil fuel plants. However, calculations show that these costs pay off in less than 1.5 years compared to light fuel and in 3.5 years compared to heavy fuel [8].

2.2. Obtaining extractive substances from the bark

The bark of a number of tree species (oak, willow, spruce, fir, larch) contain tannides – substances used for tanning leathers. The coniferous bark is subsequently extracted to produce tanning extract and pectin. Resinous substances derived from the bark of coniferous trees are a complex of higher fatty acids, aliphatic alcohols, sterols and esters. From these compounds the form of the food additive for animals is received.

The solid residue of the bark after extraction is a porous carbon material with high absorbability. This property is used for the application of nitrogen and phosphorus-containing additives necessary for the production of quality fertilizers. In the course of complex processing of coniferous bark it is possible to receive fir oil, coniferous balm, tannins, dyes and carbon sorbents (active coal). Aspen bark is used to produce vitamin and fodder additives and fertilizers. Birch bark is used to produce betulin and suberine substances to treat wounds and burns, antioxidants and sorbents. There is a lot of experience in processing birch bark with tar with a yield of 25% of the mass of dry birch bark. The forehead, the content of which in the birch bark is 60-80% of its mass, can be processed into flour suitable as a filler of synthetic adhesives. Tannins, after extraction of tannides, are partially or fully used as fuel or raw material for the production of cardboard, furfural, ethanol, fodder yeast, wood-based panels and other materials [1, 2].

2.3. Bark fertilizer

Work on the use of bark of different species and the degree of reclining on agricultural fertilizers has been actively carried out in various countries over the past sixty years. Bark waste is a natural organic fertilizer and has great humus potential. Phloem and near-cambial parts of the bark are rich in nutrients that are essential for the development of microorganisms. Possessing natural porosity and high moisture capacity, the bark accumulates and retains moisture well, promotes soil aeration and stimulates soil bacteria activity. The bark contains organically bound nitrogen in the amount of up to 3.83%, which becomes available for plant nutrition with its slow, lasting 5-7 years, decomposition in the soil. Along with nitrogen, the bark contains the following chemical elements: calcium up to 2 500, magnesium up to 600, potassium up to 300, phosphorus up to 200, manganese up to 40, boron up to 3 mg per 100 g of dry matter. Fresh bark of bark dumps is stored in pits up to 5–10 m high on hard-coated sites and periodically turned over for more effective aeration. In some cases compost is mixed with manure, bird droppings and mineral fertilizers containing nitrogen and phosphorus [2]. The complexity of this direction of processing is mainly in the economic plane, in the absence of guaranteed sales of bark fertilizer at prices acceptable to the producer and potential consumer.

2.4. Bark utilization for fuel briquettes and raw materials for pyrolysis

Debarking waste is used to produce fuel briquettes. The density of briquettes is 1 000-1 100 kg/m^3, the bending strength is 1.5 MPa, and the combustion heat is 16-17 J/kg. [18, 19] The bark is ground in a hammer crusher or rotor crusher, then dried in drum or pneumatic dryers [20, 21]. Dry matter (up to 15% moisture content) is pressed in matrix, screw or roller presses. The quality of fuel cells (briquettes or pellets) obtained from bark is relatively low due to high ash content and sulphur content. They belong to the «black» or industrial class and, accordingly, their cost is low.

The St. Petersburg State Forest Technical University (SPbFTU) has developed a project for obtaining liquid products – fuel briquettes and raw materials for pyrolysis. The essence of the development is shown in the diagram, figure 1. This technology is implemented using both Russian and imported machinery and equipment. At the same time, it is economically reasonable to have own pyrolysis production, and, most importantly, to obtain the finished product – charcoal – at the end of the process [22].

2.5. Building and structural materials made of bark

Experimental studies conducted in Russia and abroad have allowed to develop a number of construction and structural materials on the basis of bark and various binders: «Korolit», «Korobeton», bark slabs and bark-plastic materials.

Korolit is an analogue of arbolit, in which bark crushed into fractions of 2 to 10 mm in size in combination with cement is used as a filler. The density of the obtained boards is 750...800 kg/m^3, the compressive strength limit is 2.5...3.5 MPa. This material is fire and frost resistant. The use of korolit along with arbolite and instead will reduce construction costs, save wood used as a filler in the arbolite.

The most interesting are bark slabs without binder, whose function is performed by natural phenolic compounds, which are in the bark in greater quantities compared to wood. As a source material spruce bark is used, which is crushed to particles with sizes of 10...12 mm, dried to a humidity of 10...12%, sent for secondary crushing in a hammer mill and sieved through a sieve with cells of 3 mm. The pressing is carried out in two stages: firstly, cold pressing, then hot pressing at a pressure of 5 MPa with a holding time of 1 minute per each millimeter of the plate thickness at a temperature of 150°C. The resulting boards are used as an insulating structural material. Development of industrial technology for production of bark plates without is a rational direction of plate production from this type of raw material.

Wood-bark boards with binder can be single-layer (100% debarking waste) and three-layer with special chip outer layers and an inner layer of bark. The optimal size of bark particles for single-layer boards and the inner layer is a fraction of 2...7 mm. The density of the boards is 800 kg/m^3. Urea-formaldehyde resins can be used as a binder in barkwood boards. Three-layer bark wood-based panels with outer layers of wood chipboard have high physical and mechanical properties, do not differ in appearance from chipboard, and in terms of thermal conductivity, bio- and fire resistance exceed them [23].

Bark can be used in the inner layer of three- and five-layer chipboards. Three-layer boards are 1.2-1.5 times stronger than single-layer boards when containing bark in the inner layer. The inner layer can consist of 40-50% shredded birch bark. Lower results are obtained by using aspen bark – it is necessary to increase the binder content in the inner layer from 9-9.5 to 10.5-11% (from the weight of absolutely dry wood).

2.6. Recommendations for possible processing and reduction of the technogenic hazard of old bark dumps

Old bark dumps in the forest complex of the Russian Federation are not atypical objects. Their age is more than 40 years, the volume is 500 000-1 000 000 000 m^3. Such bark dumps are typical for enterprises that have no possibility to process waste and have to get rid of bark in order not to clutter production areas and carry the so-called "burdensome waste". Such enterprises should be provided with a reasonable set of long-term measures to dispose of or conserve the bark dump.

Layer-by-layer analysis of the bark dump material allows us to speak about significant differences in the physical and chemical properties of its material, which differs in density, water saturation, degree of biological activity, fractional composition and a number of other indicators. Correspondingly, today there is no single technological scheme, which would allow to dispose of all layers of the bark dump with equal efficiency on a single type of liquid products.

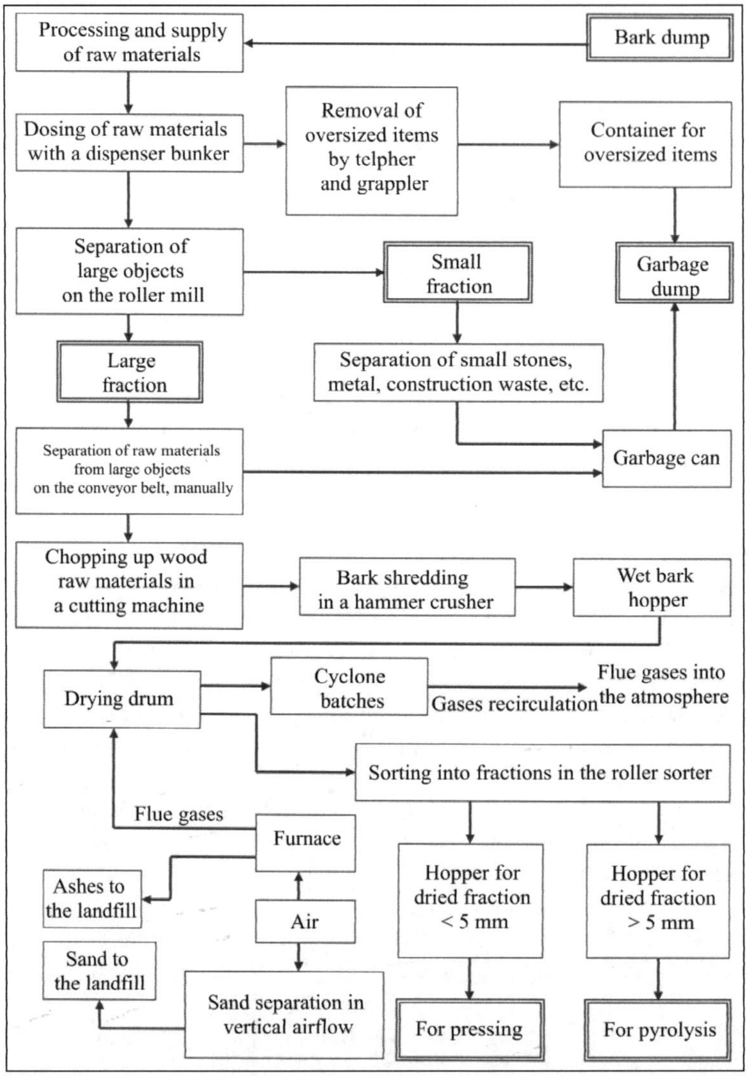

Figure 1. Technological scheme of complex use of bark dumps.

Using the experience gained at SPbFTU, a number of scenarios for the disposal (conservation, elimination) of the old bark dump can be proposed:

1. Complete removal of the bark dump with subsequent soil recultivation and landscape works in its place.
2. Preservation of the bark dump within the existing boundaries, ensuring environmental safety standards, with the possibility of collecting methane emissions for the purpose of its utilization.
3. Gradual processing of bark dump filler into liquid products with subsequent soil recultivation.

The above options have their advantages and disadvantages that are solely in the financial and economic plane. Any of the above scenarios are subject to constant monitoring and environmental protection measures related to effective treatment of infiltration and rainwater.

2.7. Bark dump conservation with use of methane emissions

This option is the least costly and includes:

– Design and organization of a unified drainage system around the bark dump to prevent breakthrough of sewage.

– Organization of collection, transportation, purification and combustion of emission methane and its conversion into electric energy.

The power generating unit is powered by a system of perforated pipes embedded in an array of gas generating material and consists of two modules – gas purification and gas turbine. The use of the modules makes it possible to solve a number of tasks in a complex: to exclude methane accumulation in the bark mass, to improve the ecological condition of water and air basins, to obtain liquid products in the form of electric power of industrial parameters. According to the data of the researches carried out at the bark dump of Mondi Syktyvkar JSC, the content of contaminants in the bark dump at different depths makes up Table 3, it is established that the maximum one-time methane concentration in the bark dump material is 2 times higher than the maximum permissible concentration at the depth of 5 meters, and at the depth of 10 meters – 6 times.

Table 3. Content of pollutants in the soil air at different depths of the bark dump of Mondi Syktyvkar JSC.

Name of the substance	Concentration of the substance, mg/m³		Maximum allowable concentrations [a], mg/m³
	5 m	10 m	
Sulfur dioxide, SO_2	<0.05	<0.05	0.5
Hydrogen sulfide, H_2S	<0.003	<0.003	0.008
Phenol	<0.004	<0.004	0.01
Weighted substances	<0.26	<0.26	0.5
Carbon oxide, CO	1.5	1.8	5.0
Methane, CH_4	103.0	300.0	50.0
Alkane	35.0	220.0	—[b]

[a] according to GN 2.1.6.3492-17
[b] isn't regulated.

According to SP 11-102-97 «Engineering environmental site investigations for construction» (para. 4.63), soils with methane content >0.1% and CO_2 >0.5% are considered to be potentially dangerous in gas and geochemical terms; in dangerous soils, methane and CO_2 content >1% and CO_2 content is up to 10%; fire-hazardous soils contain methane in soil air > 5% and CO_2 –> n-10%.

Let us transfer the obtained values of methane content in the soil air of the examined bark dump from mg/m³ to the volume percent [24]. According to the Avogadro law, one mole of any gas at the same temperature and pressure occupies the same volume at standard temperature and pressure equal to 22.41383 l or 0.02241383 m³. One mol of methane (CH_4) has a mass of 12 + 4 × 1 = 16 g. The concentration of methane at a depth of 5 m is equal:

$$C = \left(103\,mg \div 1\,m^3 \times 100\%\right) / \left(16000\,mg \div 0.02241383 m^3\right) = 0.0144\% \quad (2)$$

The concentration of methane at a depth of 10 m is equal:

$$C = \left(300\,mg \div 1\,m^3 \times 100\%\right) / \left(16000\,mg \div 0.02241383 m^3\right) = 0.0420\% \quad (3)$$

Thus, according to SP 11-102-97 "Engineering and environmental studies for construction" (p.4.6.), the bark dump material should be considered potentially non-hazardous in gas and geochemical terms, as the methane content in its soil air is less than 0.1%. However, it is higher than 0.01%. Therefore, the bark dump material should be considered a gas-generating soil and, according to SP 11-1-2-97 "Environmental engineering research for construction" (para. 5.39), the bark dump should be identified as a gas-geochemical anomaly associated with gas-generating soils [25].

Based on the data of experimental studies, it can be concluded that it is expedient to organize the collection of methane emissions from the bark dump for further processing into thermal or electrical energy. At the same time, it is recommended to recultivate the upper layer and to plan the surfaces of the bark dump in order to ensure effective drainage, as well as systematic planting of the bark dump surface with fast-growing, zoned tree species with developed root system. In the case of landscape-growing works on the surface of the bark dump will require the delivery of fertile soil on its upper part, installation of anti-slip meshes on inclined surfaces. As the preferred species of tree plantations can be recommended specimens, which are already found on the surface of the dump.

2.8. Complete elimination of the bark dump with the following soil recultivation

This option refers to the hypothetical possibility, as it requires high financial costs. The essence of it is to fully move the contents of the bark dump (after removal of inorganic components) to a prepared landfill. Requirements for the landfill: remoteness from water systems and residential areas, bunding along the perimeter, waterproofing of the base. The contents of the bark dump should be distributed over the landfill in a layer not exceeding 2 m, pressed. Natural soil is used as the top layer. Landscaping and landscaping measures are required in the future.

2.9. Gradual processing of bark beetle filler for liquid products with subsequent soil recultivation

If there is a task of stage-by-stage processing of the whole volume of bark and products of its decay into liquid products, it is necessary to split the task into two independent directions: full utilization of fresh debarking products (new bark dump) without their further accumulation; stage-by-stage utilization of the old bark dump for the products of various degrees of profitability with the further recultivation of the soil and recreation of the natural landscape. In any case, it is necessary to be aware of the fact that these areas of work are costly activities and are designed for a long period of implementation.

3. Results and Discussion

It should be noted that bark dumps, as technogenic objects, require constant environmental and fire safety monitoring. Concentration of long-term storage bark is a serious threat to the ecosystems of water bodies and adjacent territories. The bark dump produces infiltrate of VII degree of contamination and, due to the release of methane, is a potentially flammable object.

In conclusion, the coronary dump disposal should be comprehensive. Bark from onerous recycled materials can become an economically viable resource. For the majority of wood-based materials (lumber, plywood, slabs) the share of raw materials in the cost of production is about 50%. Bark is a free raw material, the money for which the enterprise has already paid.

References

[1] Nikishov V 1985 *Complex use of wood: Handbook* [In Russian – Kompleksnoe ispol`zovanie drevesiny`] (Moscow: Forest industries) p 264

[2] Mikhailov G and Serov N 1988 *Ways to Improve the Use of Recycled Wood Raw Materials* [In Russian – Puti uluchsheniya ispol`zovaniya vtorichnogo drevesnogo sy`r`ya] (Moscow: Forest industries») p 224 ISBN 5-7120-0050-4

[3] Chibirev O, Katsadze V, Vlasov Yu and Kucher S 2017 *Modern press equipment for wood waste briquetting* [In Russian – Sovremennoe pressovoe oborudovanie dlya briketirovaniya drevesny`x otxodov] Forests of Russia: politics, industry, science, education Proc. II Int. Sci. Tech. Conf. (Saint-Petersburg: SPbFTU) pp 48-51

[4] Yudkevich Y, Vasil`ev S and Yagodin V 2002 *Obtaining chemical products from wood wastes* [In Russian – Poluchenie ximicheskix produktov iz drevesny`x otxodov] (Saint-Petersburg: SPbFTU) p 84

[5] Sokolova V, Gryazkin A, Beliaeva N, Petrik V and Smirnov A 2018 *Bipolar Electrokinetic Dehydration of Wood by Electro-Osmosis of Various Breeds* Thermal Science **22(1A)** pp 285-294 DOI: 10.2298/TSCI160418024S

[6] Bastrikov D, Vlasov Yu, Ilyushenko D and Kucher S 2017 *Experimental studies of energy consumption of pine debarking wastes shredding* [In Russian – E`ksperimental`ny`e issledovaniya e`nergoemkosti drobleniya otxodov okorki sosny`] Izvestia Sankt-Peterburgskoj Lesotehniceskoj Akademii (SPb: SPbFTU) **219** pp 193-205

DOI: 10.21266/2079-4304.2017.219.193-205
[7] Khramov Y 2000 *Physical and chemical methods of industrial sewage treatment: Handbook* [In Russian – Fiziko-ximicheskie metody` ochistki proizvodstvenny`x stochny`x vod: Uchebnoe posobie] (Saint-Petersburg: SPbFTU) p 76
[8] Murashkin N, Gazizov A, Gumerova O and Kupriyanova E 2009 *Economic issues related to the use of debarking waste in market relations at the enterprises of the timber industry* [In Russian – E`konomicheskie voprosy` ispol`zovaniya otxodov okorki v usloviyax ry`nochny`x otnosheniyax na predpriyatiyax LPK] Technology and Equipment of the Timber Industry Complex Collection of scientific papers **4** (Saint-Petersburg: SPbFTU) pp 46-53
[9] Krivonogova A 2015 *Methodology of improvement of technology preparation of soft hardwood as raw material for production of charcoal* [In Russian – Metodika sovershenstvovaniya texnologii podgotovki drevesiny` myagkix listvenny`x porod kak sy`r`ya dlya proizvodstva drevesnogo uglya] Advanced technology in the forest sector Proc. Sci. Tech. Conf. (SPb: SPbFTU) pp 6-18
[10] Chernykh A 2012 *Residential building envelopes with wall panels on a wooden frame* [In Russian – Ograzhdayushhie konstrukcii zhily`x zdanij s primeneniem stenovy`x panelej na derevyannom karkase] Federal construction market **1** pp 38-39
[11] Pásztory Z, Mohácsiné I, Gorbacheva G and Börcsök Z 2016 *Utilization of Tree Bark* BioResources **11(3)** pp 7859-7888 DOI: 10.15376/biores.11.3.Pasztory
[12] Chibirev O, Katsadze V, Vlasov Yu and Kucher S 2017 *Fuel briquettes from woodworking waste* [In Russian – Toplivny`e brikety` iz otxodov lesopererabotki] Forests of Russia: politics, industry, science, education Proc. II Int. Sci. Tech. Conf. (Saint-Petersburg: SPbFTU) pp 46-48
[13] Bastrikov D, Vlasov Yu and Kucher S 2018 Investigation of the energy consumption of barking waste grinding by installation with a knife working body [In Russian – Issledovanie e`nergoemkosti izmel`cheniya otxodov okorki ustanovkoj s nozhevy`m rabochim organom] Lesotehničeskij žurnal [Forestry engineering journal] **1(29)** pp 120-128
[14] Safonov A and Zotova E 2014 *Specifics of estimating ash content in wood pellets* [In Russian – Osobennosti opredeleniya soderzhaniya zoly` v drevesny`x pelletax] Aktual'nye napravleniâ naučnyh issledovanij XXI veka: teoriâ i praktika Proc. Int. Sci. Tech. Conf. «I European Forestry Youth Forum» (Voronezh: VSFTU) **3(2)** pp 347-351
[15] Bacherikov I 2017 *Improvement of the functioning of bins and silos for bulk wood materials* [In Russian – Sovershensatovavnie funkcionirovaniya bunkerov i silosov dlya sy`puchix drevesny`x materialov] PhD Thesis (SPb: SPbFTU) p 138 https://dlib.rsl.ru/01008953662
[16] Bastrikov D, Katsadze V, Vlasov Yu and Kucher S 2017 *Equipment for wood waste shredding* [In Russian – Oborudovanie dlya izmel`cheniya drevesny`x otxodov] Forests of Russia: politics, industry, science, education Proc. II Int. Sci. Tech. Conf. (Saint-Petersburg: SPbFTU) pp 11-13
[17] Birman A, Korkka A, Khitrov E, Patyakin V, Korshak and Varakosov M A 2010 *The device for manufacturing of wood briquettes* [In Russian – Ustrojstvo dlya izgotovleniya drevesny`x briketov] Patent 96350
[18] Ilyushenko D and Costin I 2011 *Some problems with the use of bark in industry* [In Russian – Nekotory`e problemy` ispol`zovaniya drevesnoj kory` v promy`shlennosti] Systems. Methods. Technologies (Bratsk: BrSTU) 2011 pp 121-124
[19] Tsyvin M 1973 *Use of wood bark* [In Russian – Ispol`zovanie drevesnoj kory`] (Moscow: Forest industries) p 96
[20] Bacherikov I, Lokshtanov B and Rzhavtsev A 2013 *Waste Treatment Technology for Incineration in CHPPs at Pulp and Paper Mills* [In Russian – Texnologiya pererabotki otxodov dlya szhiganiya na TE`Cz cellyulozno-bumazhny`x kombinatov] Actual problems of forest complex development: conf. proc. (Vologda: VSTU) pp 104-108

IOP Conf. Series: Materials Science and Engineering **817** (2020) 012033 doi:10.1088/1757-899X/817/1/012033

[21] Gryazkin A, Beliaev V, Beliaeva N, Petrik V, Kuznetsov E and Shakhov A 2017 *The Logging Waste as Inexhaustible Resource for Alternative Energy* Thermal Science **21(2)** pp 1135-1142 DOI: 10.2298/TSCI150306047G

[22] Tuncev D, Safin R, Kasimov A, Kitaev S and Shakirov A *2018 Investigation of the process of rapid thermoconductive pyrolysis of bark and peat* [In Russian – Issledovanie processa by`strogo termokonduktivnogo piroliza drevesnoj kory` i torfa] («Derevoobrabativaushaya promishlennost'» (Woodworking industry)) **3** pp 34-39

[23] Chernykh A 2012 *Application of wall panels on a wooden frame for residential and public buildings* [In Russian – Primenenie stenovy`x panelej na derevyannom karkase dlya zhily`x i obshhestvenny`x zdanij] Light-transparent constructions **1** pp 4-8

[24] Bastrikov D, Katsadze V, Vlasov Yu and Kucher S 2017 *Mathematical expressions for evaluation of bark shredding* [In Russian – Matematicheskie vy`razheniya dlya ocenki raboty` izmel`cheniya kory`] Forests of Russia: politics, industry, science, education. Proc. II Intern. Sci. Tech. Conf. (Saint-Petersburg: SPbFTU) pp 13-16

[25] Tuncev D, Safin A, Kasimov A, Antipova E and Shakirov A 2017 et al. *Improvement of technology and equipment for thermochemical processing of wood waste into high-calorie generator gas* [In Russian – Sovershenstvovanie texnologii i oborudovaniya termoximicheskoj pererabotki drevesny`x otxodov v vy`sokokalorijny`j generatorny`j gaz] (Kazan: KNRTU Press) p 132 ISBN 978-5-7882-2321-6

IOP Conf. Series: Materials Science and Engineering **817** (2020) 012034 doi:10.1088/1757-899X/817/1/012034

International projects experience for the development of interaction between education and logistics business in forest sector

T Tereshkina[1*], S Tereshchenko[2]

[1]Marketing and Logistics Department, *Saint Petersburg State University of Industrial Technologies and Design,* 4 Ivan Chernykh street, Saint Petersburg 198095, Russian Federation
[2]Directorate of International cooperation, *Saint-Petersburg State Forest Technical University*, 5 Institutskiy Lane, Saint-Petersburg, 194021, Russian Federation

*Corresponding email: ttp_big@mail.ru

Abstract. This article will analyze the impact of international projects on cooperation between business and universities. Multifaceted aspects of interaction between universities and logistics business in the forest sector are considered, the influence of this interaction on the quality of the educational process is determined. The main forms of cooperation of organizations of the forest sector with Russian higher educational institutions engaged in training in the field of logistics and supply chain management are identified. It will be shown that international projects that create new networks between universities and business in the forest sector increase the effectiveness of such cooperation and create new forms of cooperation. The experience of implementation of international projects in Russian universities is analysed and the possibilities of its use for the development of interaction between business and education are determined. The results of an empirical study of cooperation between educational institutions and organizations of the forest sector are presented. Conclusions are drawn about the impact of international projects on the intensity and effectiveness of interaction between higher education institutions and enterprises of the forest sector, which ensures the growth of professionalism of teachers, competence of students and focus on the real needs of the labor market.

1. Introduction

In the conditions of modern development, characterized by rapid obsolescence of knowledge, increasing the level of innovation of products and services, globalization of the economy, is increasingly important the success of universities and business cooperation. A state that promotes the creation of links between business, universities and government structures, gaining a competitive advantage through rapid transfer of new knowledge created during implementation of innovative products, and companies and research organizations increase their chances of developing more quickly and effectively, taking advantage of new opportunities [1, 2]. In addition, there are a number of universities that have become drivers of regional strategic development because of their role as key suppliers of innovation in the knowledge economy, and because of their cooperation with business. There are different investigations about university-business cooperation [3-5].

However, there are no studies assessing the impact of international projects on the intensity and quality of such cooperation. International projects make it possible to intensify this cooperation, to use the best practices and forms of cooperation of partner universities.

The consideration of these issues is particularly relevant for forest sector enterprises and universities implementing educational programs in the field of logistics and supply chain management. Over the past decade, the use of innovative logistics technologies has enabled forest sector enterprises to improve customer service, optimize logistics costs, which are second in importance after material costs, and thereby increase their competitiveness [6].

Implementation of international projects will allow to develop new forms of cooperation between business and education, to assess the effectiveness of existing ones, allowing to accelerate the transfer of knowledge and improve the quality of the educational process, including in the field of logistics of forest sector enterprises [7-9].

2. Methods and Materials

The research method used is mainly qualitative [10]. Several international projects, connected with university-business interaction were analysed from the point of view of effectiveness and influence of the international projects on such cooperation. The questionnaire was also used to receive the opinion as of the business and universities representatives about cooperation. There were several questions included into the questionnaire.

There were two types of questionnaires: one for enterprises and one for universities. The questionnaire for the enterprises and for universities consisted of two parts. In the first part of the questionnaire for enterprises data connected with the background information about the enterprise were gathered. The information about the name of the enterprise, location, field of business and size was collected. In the second part of the questionnaire the questions concerning the cooperation of enterprises and business before the participation in the international project and after were asked. Also there were asked some questions concerning the identification of the interviewed enterprises' development needs in the frame of cooperation with universities and concrete details of such collaboration. The first part of the questionnaire for universities was devoted to the identification of the university and the second was connected with the identification of the needs of collaboration with business from the point of view of university and the situation about university - business interaction before and after international projects realization. Since the level and nature of enterprise - university collaboration differs in different cooperation network it was decided to make a qualitative analysis of the interview results.

In the frame of the research, in-depth interviews were conducted among different enterprises of logistic business in forest sector, which participated jointly with universities in the international projects. The aim was to find out the best practices of the university-business partnership, as well as getting their opinion about the prospects and problems of cooperation with universities. The purpose of the surveys was to identify the influence of international projects on the development of university-business partnership. The existing forms of cooperation between the universities and logistic business, in forest sector as well as the identification of factors contributing to the development of interaction between the education system and business, and the barriers preventing this process were examined. In addition, aspects related to expectations were studied.

3. Results and Discussion

Cooperation between logistic business in forest sector and universities influence on both sides and may raise the effectiveness of their work. But it is proved that benefits from such cooperation may be received not only by universities and logistic business in forest sector, but also by other stakeholders, which receive benefits. The content of the benefits from university-business cooperation for different groups of stakeholders is shown in the table 1.

Table 1. Stakeholders and their benefits from university and logistic business in forest sector cooperation [11, 12].

Stakeholders	Benefits received
Students	Access to companies for internship programs. Access to experiential and entrepreneurial education as well as opportunity to have 'real' professional life exercises.
Teachers	Access to industrial research fund. Opportunities for organizing consulting. Opportunity to formulate the students competences in accordance with industry's needs .
University	Raising the quality of education. Flexibility in accordance with the needs of labor market.
Business	Opportunity to receive complementary expertise that may be needed in the business. The possibility to hire skilled employees.
Community	Contribution to the economic development through events sponsored by the businesses, run jointly between universities and businesses. Job creation, a more thriving entrepreneurial local environment and increased tax revenues are also found to be important benefits to the community.

The survey conducted within the framework of international projects by two Russian universities: Saint-Petersburg State University of Industrial Technologies, Design, and Saint-Petersburg State Forest Technical University revealed a number of regularities in the development of forms of university logistic business in forest sector interaction [9, 11].

Based on the respondents' opinion may be concluded that there are many forms of university logistic business cooperation in forest sector. Among most frequently used are: visits of student groups, training, student thesis writing based on the data, received from the enterprise, giving work-related cases, participation of business representatives in the presentations of the Bachelors' and Masters' thesis, joint scientific research, lectures of business life representatives in universities etc. The forms of cooperation between universities and logistic business in forest sector in Russia are shown in table 2.

Table 2. The most useful ways of cooperation between universities and logistic companies in forest sector by business representatives' opinion in Russia (developed by the authors).

Types of cooperation	Share in the amount of the enterprises participated in the investigation, %
Visits of student groups	22.5
Internships	85.0
Training	97.5
Representatives of the companies participate in the presentations of Bachelors' and Masters' thesis	34.8
Joint projects	47.5
Joint scientific research	12.5

Lectures of the representatives of the companies in the universities	57.5
Working life cases from the companies are used in educational process	12.5

The most effective forms of cooperation were identified as: training (97,5 %), lectures of business life representatives in universities (57,5 %), student thesis writing based on the data, received from the enterprise (53,8 %), participation of business representatives in the presentations of the Bachelors' and Masters' thesis (34,8 %), internships (85,0 %).

The companies of forest sector stressed that the collaboration with universities should be on the win-win basis. The most effective forms of collaboration from their point of view are: joint projects and study based on work-related cases. Another critical success factor is accessibility of human resources and its continuous development. Very important driver of better collaboration is studying in a workplace when students are offered multiple opportunities to develop their learned skills in practice. In order to develop such collaboration, the international projects may be used. These facts were stressed by the companies, which participated in the international projects.

The companies that participated in the international projects jointly with universities showed that there are several outcomes from such collaboration. First, international projects allow creating new networks between universities and business, which is helping to organize more closer and effective cooperation. Second, some international projects that involve business representatives as participants, may help companies develop the human capital by means of education of their employers with up-to-date knowledge. So the companies are greatly interested in the participation in such projects and cooperation with universities in the frame of them. Third, the international projects allow receiving some answers for the urgent problems that companies have by including students in such projects. Very good example of international project, in which the students are involved, is INNOEVENT project, which is realized in a number of foreign universities in several countries and in two Russian universities: Saint-Petersburg State Forest Technical University and Saint-Petersburg State University of Industrial Technologies and Design [4]. In the frame of such project companies are giving tasks to the groups of students and would like to receive decisions to their problems that they have. The INNOEVEN project is realized for four years from 2015. The companies are satisfied by the results of such project. Fourth, the university-business cooperation in the frame of international projects may allows to get concrete results, which, as companies believe, are the basis for effective cooperation between universities and business.

From the point of view of universities, international projects, in which companies from forest sector were involved, were very useful for the development of educational process. The positive results of participating in the international projects with companies for the universities maybe found in different directions. First, it allows to establish relations with new companies and to develop relations with the company networks that already existed. Second, participation in such projects will allow raising the quality of education in the universities, because it will be connected with practical knowledge based on the real situations in business. Forest related high education should reflect changing needs in business environment to ensure graduates to be competitive in real business world. Third, the participation in the international projects may allow raising the quality of human capital of the universities. Teachers may receive new information about the real situation in forest sector companies and will be able to teach the students based on the examples from real life. They also will feel themselves able to teach the representatives from the business life. Fourth, the universities participation in the international projects with forest related companies will give the students better competencies within forest sector and students will get the opportunity to not only establish a stronger network with the companies and organizations they work with, but also help solve some of the companies' or organizations' challenges, while they may develop a specific "know-how" with the other partners involved [13]. Fifth, the international projects with forest related companies allow to intensify the practice-oriented research. The real results, obtained during such research will be the win-win situation as for the university and for companies also. It also helps for the universities to be more competitive in the educational markets.

4. Conclusion

The following conclusions can be drawn from the analysis of the situation in cooperation of logistic business in forest sector and universities in Russian Federation and influence of international projects on it:

1. Collaboration of higher education institutions with the logistic business in forest sector in Russia was studied empirically in the present article.
2. The qualitative research method and the major principles of collecting data by means of interviewing sets of enterprises, working in logistic business in forest sector were given in the article.
3. As for the general outcome of the conducted research, cases of effective cooperation of logistic business in forest sector with the universities are presented in all international projects. The most used forms of cooperation are: training, internships, participation in joint projects, lectures of the representatives of the companies in the universities.
4. The representatives from universities and the logistic business in forest sector in Russia stressed that they are interested in organizing cooperation based on win-win approach. The international projects from their point of view are helping in establishment new networks between universities and the logistic business in forest sector, in finding modern effective forms of cooperation.

However, the conducted research showed that the needs for the development of cooperation between universities and the logistic business in forest sector in Russia should be examined more precisely by the educational sector and forest sector in order to intensify truly effective collaboration to improve the quality of the educational process in logistics. The practical value of the article is in its analysis of how international projects influence on collaboration between universities and the logistic business in forest sector in Russia.

References

[1] Higher education governance between democratic culture, academic aspirations and market forces 2006. ed J Huber and J Kohler (Strasbourg Council of Europe Publishing)
[2] MacGregor K 2015 Six key elements of an entrepreneurial university. http://www.universityworldnews.com/article.php?story=20151106141848199. Referred: 21.10.2016. Updated: 7.11.2015
[3] Barsony J 2003 Towards the Entrepreneurial University. SEFI 2003 Conference–Global Engineer pp.229-239
[4] Christiansen P K 2014 INNOEVENT – An approach to World of Work, In, ed Kakkonnen Innovative Teaching and Learning Methods in Multicultural Environments, (Mikkeli: Mikkeli University of Applied Sciences) pp 78-89
[5] Hofer A R and Potter J 2010. Universities, innovation and entrepreneurship: criteria and examples of good practice. OECD Local Economic and Employment Development (LEED) Working Papers pp 431-433
[6] Tereshkina T, Tereshchenko S, Bezrukova T, Drevaleva E and Pecherskaya O 2017 Improvement of Methods of Analysis of Effectiveness of Stock Management at Industrial Enterprises. Integration and Clustering for Sustainable Economic Growth pp 461-468
[7] Itkonen K, Tereshchenko S and Tereshkina T 2013. Experiences of intensive weeks as a way of internazionalization. Bridging entrepreneurship education between Russia and Finland. Conference Proceedings. Mikkeli University of Applied Sciences. Research Reports pp 93-103
[8] OECD 2012. A Guiding Framework for Entrepreneurial Universities. WWW-document. https://www.oecd.org/site/cfecpr/EC-OECD%20Entrepreneurial%20Universities%20Framework.pdf. Referred: 21.10.2016. Updated: 18.11.2019

IOP Conf. Series: Materials Science and Engineering **817** (2020) 012034 doi:10.1088/1757-899X/817/1/012034

[9] Tereshkina T R, Tereschenko S V, Kyarnya M 2014 Connecting learning and working life in higher education. Economic and management technologies of the XXI century: theory and practice, training of specialists: materials of the International methodological and scientific-practical conference. St. Petersburg, November 20, 2014 pp 72-75

[10] Saunders M, Lewis P and Thornhill A 2009 Research Methods for Business Students (5th Edition). Prentice Hall.

[11] Nemelentsev M, Christiansen P, Peter Storm Henningsen, Lund S, Tereshchenko S and Tereshkina T 2016 Collaboration of higher institutions with enterprises// Development and facilitation of innovative pedagogy in the Nordic-Russian, Mikkeli (Mikkeli University of Applied Sciences) pp.76-91

[12] Study of University Business cooperation in the US 2013. Final report. www.lse.ac.uk/enterprise. Referred: 21.10.2019. No updating information

[13] Doutriaux J and Barker M. 1996 University and Industry in Canada. A Changing Relationship. Industry and Higher Education 10 (1) pp 88–103

IOP Conf. Series: Materials Science and Engineering **817** (2020) 012035 doi:10.1088/1757-899X/817/1/012035

Method of increasing percentage of raw wood materials in plywood production

S A Ugryumov[1*], N A Vokhmyanin[1], S I Zatenko[2], S V Spiridonov[1], M B Taraban[2]

[1] Institute of technological machines and forest transport, *Saint-Petersburg State Forest Technical University,* 5 Institutsky lane, Saint-Petersburg 194021, Russian Federation

[2] Institute of forest business and innovation, *Saint-Petersburg State Forest Technical University,* 5 Institutsky lane, Saint-Petersburg 194021, Russian Federation

[*] Corresponding email: ugr-s@yandex.ru

Abstract. The article deals with the issue of increasing efficiency of plywood production by recycling waste from format plywood trimming. The study proposes manufacturing a composite material with outer layers of full-format peeled veneer and an inner layer based on slats from format trimming of plywood installed on the edge with a different gap. Results of physical and mechanical properties testing of proposed composite material show that using of the slats from plywood trimming allows forming lightweight material holding sufficient strength. Due to return use of plywood production waste, manufacturing proposed composite material reduces total material consumption, which provides cheaper products with effective waste disposal.

1. Introduction

Plywood and plywood products made on the basis of wood shelled veneer and synthetic adhesive compositions are one of the main structural materials for construction, auto, wagon, container production, furniture production and other industries [1]. Plywood production is an integral part in the system of integrated wood processing, and this is typical not only for Russian industry, but also worldwide. Production and consumption of plywood increase annually, which requires increasing wood harvesting. Deficit of wood raw materials noted at the end of the twentieth century is still in present. The main direction of covering this deficit lays in deep processing of wood, including processing of secondary wood resources and waste products. Development of methods to reduce the resource consumption in plywood production is a part of a valuable multifaceted work to create new types of materials with controlled properties depending on intended areas of further usage, with improved utilization and processing of wood raw material and return using of the woodworking waste.

At plywood enterprises, in the process of preparation of wood raw materials, peeling and processing of veneer, gluing and processing of finished products, woodworking waste in form of scraps of various sizes or small particles is formed inevitably. The amount of the waste depends on the level of organization of production and its equipment, but the share of the waste in any case is quite large and is up to 50% or more relative to initial volume of timber received by an enterprise [2]. In modern technological processes of plywood production, a significant part of resulting waste is processed in the main production cycle or sent to processing particle boards and other pressed materials production for structural purposes. Often, waste from debarking of raw wood materials, veneer-flaw, substandard lump veneer and other waste are sent to combustion in production furnaces to obtain heat energy used in the main production cycle, e.g. for heating the pools during hydrothermal wood processing, drying the veneer, heating heat carriers of the presses other purposes [3]. However, a more efficient use of the waste is its processing into materials with a certain added value. So, in recent years, many plywood enterprises organize workshops for fuel pellets and briquettes production, which allows processing significant share of the waste and receiving additional profits.

When cutting plywood in a format required, lump waste in form of slats, which length and thickness corresponds to the size of the cut sheet of plywood, and the width ranges from 2.5 to 4.0 cm, is formed. The volume of such waste is from 7 to 11% of the volume of unedged plywood, depending on the format of the plywood produced and specified allowance for trimming. Recycling and disposal of such waste is difficult. Burning is not allowed due to emission of toxic substances from the decay of the cured adhesives during combustion. Therefore, the waste is usually grinded into wood chips used in production of particle boards. It should be borne in mind that presence of adhesive layers and high density of format scraps negatively affect processes of wetting, spreading and absorption of glue, as well as compaction and contact of the particles in piezothermic treatment process, which leads to deterioration of particle boards strength. Therefore, the chips obtained from the format scraps of plywood are used only as an additive to the main raw material. In the absence of plate production in transport accessibility from plywood production, the problem of processing and disposal of these wastes is an urgent task.

There are various technological methods of using wood waste, including waste plywood production and composite wooden materials. For example, production of flat lattice panels based on woodworking processing industries waste with facing layers of cardboard or other sheet materials can significantly reduce material consumption while reducing density of the material [4]. However, the strength of the material is insufficient due to the low density and the presence of voids in the structure.

There has been developed a method of manufacturing a laminated wooden plate material, consisting in forming the inner layer and strips from lumpy wood waste. The veneer or scraps are used laying edge-to-edge or with an offset in width. [5]. This design allows the waste to be recycled while reducing the density and material capacity in the products, but its performance properties are low due to heterogeneity of the inner layer structure, which results in dissection and cracking of the material, especially in alternating temperature and humidity conditions.

In recent years, lightweight materials, such as particle boards "DendroLight", with an inner layer of woodworking waste, connected to a rack or lattice structure, are produced and used in construction and furniture production [6]. Such boards can be made of almost any thickness, have a low density, but their strength and ability to withstand shock loads are insufficient.

A prospective direction for return use of format plywood trimming waste is production of composite materials, in the inner layer of which plywood trimming slats are used, and the outer layers consist of one or more layers of veneer [7]. In the framework of this study, the structural features of the formation of such materials are analysed and their physical and mechanical properties are experimentally evaluated.

2. Methods and materials

Slats from plywood trimming should be re-cut to form a uniform width. In this study, slats with nominal widths of 10 and 20 mm were used. From the structural point of view, the composite material is formed with outer layers of mutually perpendicular sheets of format peeled veneer and inner layers formed by installing the slats obtained from the trimming on their edge, laying with an offset.

Urea-formaldehyde glue based on KF-G resin and acid curing catalyst was used as the main binder in this material, taking into account adhesive properties of the glue [8, 9].

To form the outer layers, shelled birch veneer of grade *I* was used with a nominal thickness of 1.5 mm, slats with a thickness and width of 10 mm. Nominal thickness of the finished composite material was 14 mm.

The assembly of plywood packages consisted in selection and laying the veneer sheets and slats of the inner layer with different distances between them in accordance with a given design. The glue was applied on both sides of the inner (even) layers of veneer and on the end surface of the rails. In addition to these samples, plywood samples were made from format veneer as a standard.

Hot pressing of the collected packages was carried out with a laboratory hydraulic press at the modes recommended by technical literature [10]:

- pressing temperature 125°C,
- specific pressing pressure (based on the total area of contact of the surfaces of the rails with the veneer) 2 MPa,
- moulding time under pressure 7 min.

Cutting on the samples for testing was carried out in a day after unloading.

The bending strength was estimated according to GOST 9625-2013, the tensile strength when chipping on the adhesive layer was estimated according to GOST 9624-2009 [11, 12].

To assess the bending strength along the fibers of the outer layers, samples were made with a width of 50 to 60 mm in such a way that in the cross section there were at least two rails in the inner layer. The bending strength across the fibers of the outer layers was evaluated on samples with a thickness of 250 mm and a width of 50 mm.

3. Results and Discussion

At the first stage of research, a composite material with an inner layer of vertically mounted rails with a displacement was manufactured and tested, with the displacement varying from 0 to 20 mm. The assembly scheme of the material is presented in figure 1. Figure 2 shows appearance of a composite material with an inner layer of vertically mounted offset rails.

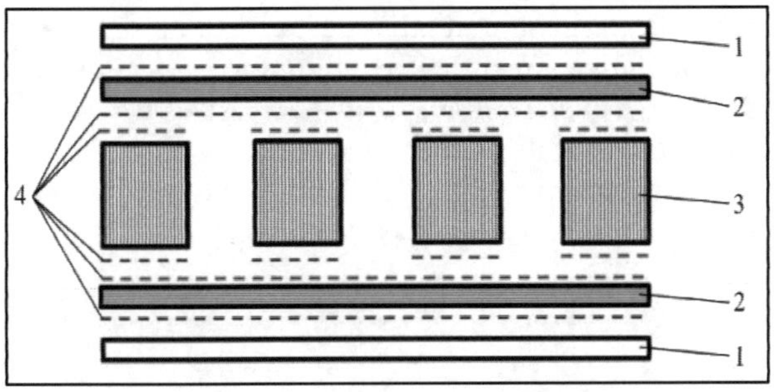

Figure 1. Composite material construction: 1 - veneer layer with longitudinal direction of fibers; 2 - veneer layer with transverse direction of fibers; 3 - plywood trim strips; 4 - adhesive layers.

Figure 2. Appearance of plywood panel with an inner layer of vertically mounted slats.

Table 1 presents summary physical and mechanical characteristics of the material.

Table 1. Physical and mechanical properties of the composite material.

Material	Distance between rails, mm	Density, kg/m^3	Tensile strength at bending, MPa		Breaking strength, MPa	
			along the fibers	across the fibers	dry	after soaking
General purpose plywood	-	723	110.6	69.2	4.16	2.21
Composite material with inner layer of slats	0	678	102.1	37.9	2.74	1.71
	5	521	62.5	25.8	2.52	1.69
	10	458	43.2	18.6	2.51	1.70
	15	417	31.9	13.8	2.51	1.65
	20	388	27.8	11.2	2.41	1.61

Analysis of the obtained results showed that composite plywood with inner layers of slats from format plywood trimming has lower strength compared to general purpose plywood, which is explained by the presence of voids in the inner layer and generally lower density of the material. However, the bending strength along the fibers of all of the samples exceeds its normalized value according to GOST 3916.1-96. The peculiarity of this material is its low density and possibility of recycling plywood production waste.

When forming the strips along the fibers direction in the outer layers of plywood, the greatest strength is achieved in the longitudinal direction of the composite material, but the strength in the perpendicular direction is significantly less. Reaching the equal strength in the format of the composite material is possible by laying the rails in the inner layer in a mutually perpendicular direction. Technologically, this can be achieved by assembling the lattice structure, installing the slats in pre-cut adjacent grooves (figure 3). Slats can be assembled into a grid at intervals of different sizes depending on the required strength of the material and the intended areas of its further use.

Outside the lattice layer, several layers of full-length peeled veneer are glued, forming a composite material. Figure 4 shows appearance of the composite material produced by this method.

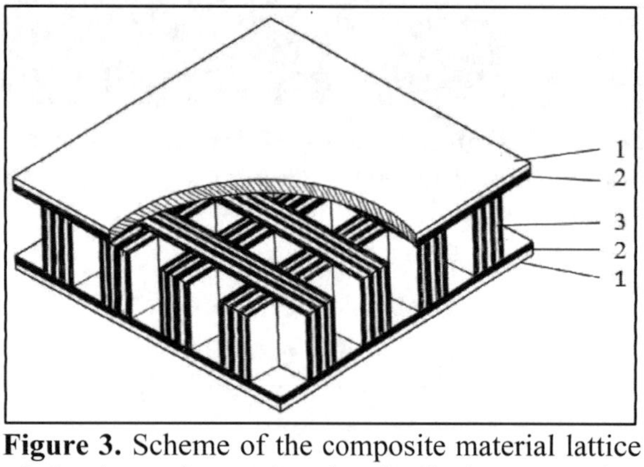

Figure 3. Scheme of the composite material lattice of the inner layer: 1 - longitudinal veneer; 2 - transverse veneer; 3 - lattice of slats

Figure 4. Appearance of the composite material lattice of the inner layer

Table 2 presents summary physical and mechanical properties of the composite material with the inner layer of the lattice structure.

Table 2. Physical and mechanical properties of the composite materialwith the inner layer of the lattice structure.

Material	Distance between rails, mm	Density, kg/m^3	Tensile strength at bending, MPa		Breaking strength, MPa	
			along the fibers	across the fibers	dry	after soaking
Composite material with ribs and inserts	15	442	52.7	5.9	2.51	1.65
Composite material with lattice construction	15	508	40	29	2.62	1.68

The analysis of the results showed that the bending strength along the fibers of the composite material of proposed design exceeds the normalized value according to GOST 3916.1-96, an is equal to 25 MPa. The use of the lattice in the inner layer allows to increase the strength of the composite material in the transverse direction.

4. Conclusion

The proposed designs allow to form lightweight composite materials with sufficiently high physical and mechanical properties. The materials can be successfully used as flooring and cladding materials, as well as slab materials in various structures manufacture. Recycling the waste of plywood production reduces material consumption and provides more efficient way to use the waste, while decreasing the number of sheets of peeled veneer with the adhesive, results in reducing adhesive consumption per unit of production.

References

[1] Lukash A 2014 The new technology of laminated materials [in Russian – Tekhnologiya novyh kleenyh materialov] (Saint-Petersburg : Lany) p 304

[2] Volynsky V 2003 Technology of glued materials [in Russian – Tekhnologiya kleenyh materialov] (Arkhangelsk : Arkhangelsk State Technical University) p 280

[3] Lyubov V, Popov A, Ivut A, Kondakov S and Sedleetsky N 2016 Increasing efficiency of energy use of plywood production waste *Bulletin of Cherepovets state University* [Vestnik Cherepoveckogo gosudarstvennogo universiteta – in Russian] 4(73) pp 28-32

[4] Lucash A, Shitikova A and Cherenkov M 2014 Patent 143493 RU Honeycomb wall panel from wood-based materials p 4

[5] Malicki A 2016 Patent 159337 RU Structural panel p 12

[6] Nikolaev A 2010 Dendrolight Plates are environmentally friendly, durable, easy and profitable *Equipment and tools for professionals* [Oborudovanie i instrument dlya professionalov – in Russian] 1 pp 20-22

[7] Mitrofanov V 2018 Rational use of waste from format trimming of plywood *Russia and the world: national security, challenges and prospects* [Rossiya i mir: nacional'naya bezopasnost', vyzovy i perspektivy: materialy mezhdunarodnoj mezhdisciplinarnoj nauchnoj konferencii – in Russian] 2 pp 135-136

[8] Malysheva G 2005 Physical chemistry of adhesive materials *Materials Science* [Materialovedenie – in Russian] 6 pp 38-40

[9] Petrova A and Malysheva G 2017 Adhesives, adhesive binders and adhesive prepregs [in Russian – Klei, kleevye svyazuyushchie i kleevye prepregi] (Moscow : Russian Research institute of Aviation Materials) p 472

[10] Volkov A 2010 Handbook of plywood [in Russian – Spravochnik fanershchika] (Saint-Petersburg : Publishing house of Polytechnic university) p 486

[11] GOST RF 9625-2013 *Laminated glued wood. Methods for determination of ultimate and modulus*

of elasticity in static bending. 2013 (Standartinform, Moscow)
[12] GOST RF 9624-2009 *Laminated glued wood. Method for determination of shear strength.* 2009 (Standartinform, Moscow)

IOP Conf. Series: Materials Science and Engineering **817** (2020) 012036 doi:10.1088/1757-899X/817/1/012036

Classification of factors influencing the reliability of the driver-vehicle-road-environment (DVRE) system in the conditions of mountain quarries

A M Umirzokov[1], U M Mallaboev[2], S S Saidullozoda[3], Kh Kh Khabibullozoda[3]*

[1]*Tajik National University*, 17, Rudaki Avenue, Dushanbe 734025, The Republic of Tajikistan

[2]*Tyumen Industrial University*, Tobolsk Industrial Institute (branch), 5, zone of Universities, Tobolsk 626158, Russian Federation

[3]*Polytechnic Institute, South Ural State University*, 76, Lenin Avenue, Chelyabinsk 454080, Russian Federation

*Corresponding email: saivali.saidullo@mail.ru

Abstract. The article provides a classification of factors affecting the efficiency of the DVRE system in operating conditions in mountain quarries. A detailed classification of factors affecting the reliability of mining heavy-duty dump trucks operating in mining quarries for such groups of features as design, technological, operational and road-climatic is provided. In addition, the article provides a classification of factors based on controllability. To assess the influence of the stochastic nature of external influences that determine the dynamics of the interaction of the vehicle with the roadway, rational methods based on probability theory are substantiated. Three of them are recommended for an adequate assessment of the influence of external influences on the output indicators of a career heavy-duty dump truck, in particular for evaluating the process of interaction of a vehicle with a roadbed, which is probabilistic in nature. Among the recommended estimation methods are included such as the method of approximating the external load with a periodic function such as a sinusoid with certain values of the amplitude and period of oscillation, the method of expanding the periodic function in Fourier series and the probabilistic-statistical method, which is based on taking into account the action of many random factors characterized by a stable frequency.

1. Introduction

The vehicle is a fundamental and most vulnerable element of the DVRE system and acts as a dominant factor in the formation of the system efficiency.

The DVRE system is both large and complex. The DVRE system is complex because it contains ambiguous relationships and regularities of the interconnection between its elements, the system is multi-criteria, and the processes taking place in it are stochastic, it is difficult to model the DVRE system. The DVRE system can be classified as complex because of its multidimensionality, diversity of the nature of elements, connections, heterogeneity of the structure, as well as because the system operates in the conditions of a significant environmental impact uncertainty. Its complexity is exacerbated by the random nature of changes in its indicators [1-3].

Almost all input and output parameters of the DVRE system are stochastic. The stochastic behavior of the DVRE system reliability is formed from the probabilistic nature of the reliability of each of its components. The probabilistic nature of the DVRE system reliability is especially clearly manifested when heavy duty vehicles are operated in high-mountain regions on gravel roads. Among the elements of the DVRE system, the "vehicle" subsystem is the dominant element determining the complex stochastic nature of the system in general. The main assessment parameter of the stochasticity level is the oscillations of the vehicle's tractive resistance, which is caused by such factors as: uneven torque and braking moments, as well as the vehicle speed, complexity of the ground profile, road profile and the condition of the road carpet, random distribution of the normal load on the vehicle axles and wheels, as well as frequency of maneuvers, etc.

2. Methods and Materials

Depending on the design-engineering features, operational modes and road-climatic conditions, the dynamics of the vehicle interaction with the road may differ. At the same time, depending on the number of external loads acting on the vehicle, their weightage, nature of action and dynamism, the interaction between the vehicle and the road can be conditionally classified as simple, medium and complex. As for the nature of the interaction of a heavy duty truck with a haul road in high mountain conditions, it can be definitely classified as complex, depending on many different factors and varying in rather wide ranges. At the same time, we should take into account that when heavy duty haul trucks are operated during the construction of high mountain hydraulic facilities (or when working in a high mountain quarry), there appear new factors which significantly differ from those met when the vehicle is operated on a plain [4-8]. These factors include the probabilistic nature of external actions on the vehicle, characteristics of the road carpet, altitude above the seal level, peculiar road profile, the condition of the yard, etc. The variety of the factors affecting the life and reliability of heavy duty haul trucks operated in the conditions of mountain quarries can be conditionally divided into three groups: design-engineering, operational and road-climatic (figure 1).

The actions of design-engineering factors by the composition and structure of the influence on the reliability of haul trucks are aggravated in severe extreme conditions of their operation in mountain quarries [9, 10].

The actions of road-climatic factors on the reliability of heavy duty haul trucks in mountain conditions are characterized by their probabilistic nature and are preconditioned by the altitude above the sea level, unevenness of the roadway, plan and profile complex road geometry, frequency of maneuvers, driving skills, etc.

Based on controllability, it will be expedient to divide the factors affecting the life of heavy duty haul trucks operated in the conditions of mountain quarries into three categories: fully controlled, partially controlled and uncontrolled (figure 2). The classification based on the controllability feature significantly differs from the classification based on the cause-and-effect feature. Despite the differences, these two classifications are of unrivaled practical importance for establishing ways to improve the reliability of the DVRE system.

The most significant factors determining the life of heavy duty vehicles operated in mountain quarries include the categories of operational and road-climatic factors by the cause-and-effect feature, and partially controlled and uncontrolled factors by the controllability feature.

The climatic factors, which are characteristic of mountain quarries and significantly reduce the life of heavy duty haul trucks, include such factors as the ground profile, road profile, type and condition of the road carpet, ambient temperature and altitude above the sea level [2, 3, 11].

The longitudinal profile of mountain haul roads is an alternation of rises and falls of different sizes and lengths, the complexity of which is preconditioned by the following circumstances [3]:

- a sufficiently large difference in the altitude above the sea level in individual road sections, where the longitudinal road grades reach 10-12% and even higher values in some short sections (both in the cross-section and longitudinal directions);
- construction of a road on mountain slopes, when the choice of the route of the constructed road is determined by the complex mountain terrain and landscape;
- random arrangement of the yard, etc.

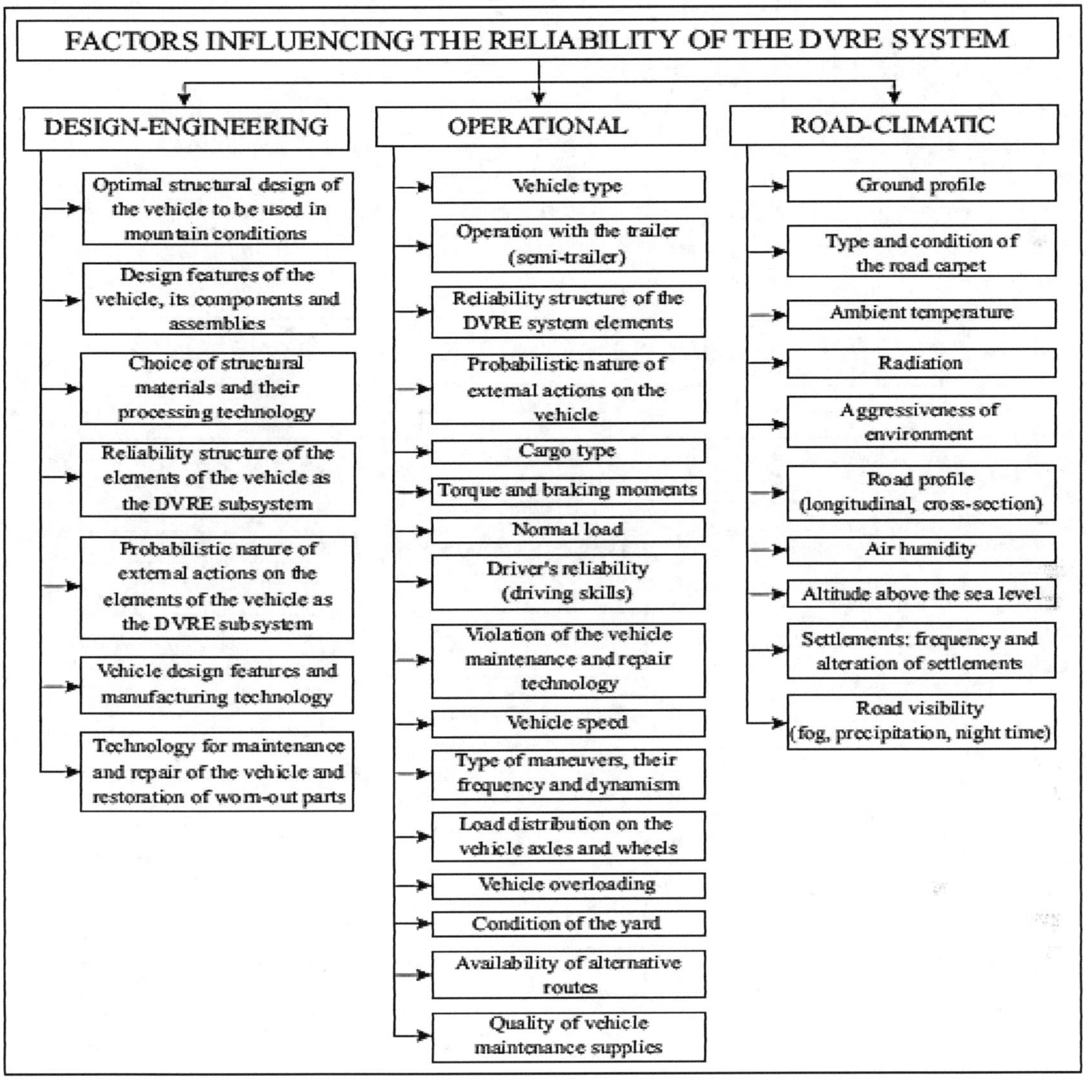

Figure 1. Classification of factors affecting the life of heavy haul trucks operated in the conditions of mountain quarries.

Out of the many factors determining the reliability of heavy duty trucks operated in severe road-climatic conditions, the probabilistic nature of external actions on the vehicle is an important factor. Strictly speaking, all the factors determining the reliability of heavy duty trucks operated in severe road-climatic conditions are probabilistic in one or another way or fluctuate randomly. Despite this, the main factor forming the complex stochastic nature of external actions on the vehicle used for transport operations in the conditions of constructing high-altitude hydraulic facilities is the oscillation in the vehicle's tractive resistance. In turn, oscillations of the vehicle's tractive resistance are caused by the following factors: uneven torque and braking moments, as well as vehicle speed, complexity of the ground profile, road profile and road carpet condition, random distribution of the normal load on the vehicle axles and wheels, as well as frequency of maneuvers, etc.

As proposed by academician V N Boltinsky, oscillations of resistance forces of a heavy duty vehicle used for transport operations in the conditions of mountain quarries can be divided into two categories [4-6]:

• high-frequency (short-term periodic oscillations with a period of $T \leq 1...2$ s and an insignificant amplitude), which can be overcome by the mass inertia of the heavy duty truck and almost do not affect the oscillations of its tractive resistance.

• low-frequency (oscillations with a period of $T \geq 1...2$ s), mainly preconditioned by the joint manifestation of the above factors, which significantly affect the oscillations of the vehicle's tractive resistance. They include: uneven moments on the wheels, vehicle speed, complex ground profile, road profile and extreme road carpet condition, random distribution of the normal load on the vehicle axles and wheels, as well as frequency of maneuvers, etc.

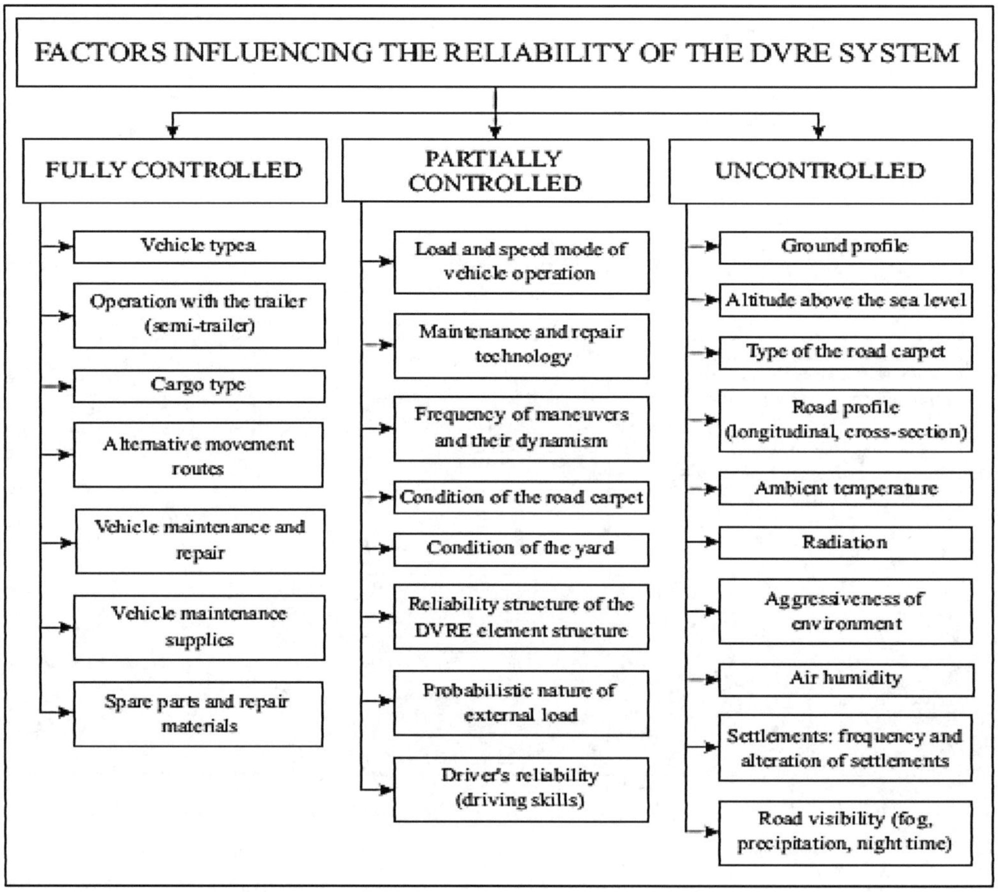

Figure 2. Classification of factors affecting the life of heavy haul trucks operated in the conditions of mountain quarries by the controllability feature.

Moving along such a road, a heavy duty haul truck is subject to the influence of the torque and braking moment which are probabilistic in nature.

Taking into account the probabilistic nature of external actions on the vehicle, the vehicle contacts the roadway in an unstable and dynamic mode, which significantly contributes to the development and acceleration of the decrease in energy, technical and economic indicators, as well as its safety margin and durability. The decrease depends on the value of the variation coefficient of the wheel torque and the tractive force of the vehicle. We can assume that the vehicle's variation coefficient can be in the range of 10-15% depending on the complexity of the operating conditions (it increases with an increased load), while the energy, technical and economic indicators, as well as the safety margin and durability of heavy duty haul trucks, will decrease to 10%.

The influence of the oscillations of external actions determining the dynamics of the vehicle's interaction with the roadway, depending on the sample size, the nature of the process, the condition of the object under study and other factors, is assessed by several methods based on probability theory. Three of them can be recommended for an adequate assessment of the influence of external actions on the output indicators of a heavy duty haul truck, in particular for assessing the process of the vehicle's interaction with the roadway, which has a probabilistic nature:

- the method of approximating the external load with a periodic sinusoid-type function with certain values of the amplitude and period of oscillations [4];
- the method of expanding a periodic function in a Fourier series [5];
- the probabilistic-statistical method which based on taking into account the action of sets of random factors characterized by a stable frequency [6].

The first method proposed by academician V N Boltinsky to simulate the effects of the unsteady external load on an agricultural unit using a periodic function at fixed values of the amplitude A_M and period T_M of oscillations, can be used to simulate the effects of the unsteady external load (torque, power) on the wheels of a heavy duty vehicle, i.e.:

$$M_t(t) = \bar{M}_k + A_M \sin mt = \bar{M}_k(1 + 0.5\delta_M \sin mt), \quad (1)$$

where, $\bar{M}_k = 0.5(M_{max} + M_{min})$ is the average value of the modulus of resistance on the vehicle wheels; $A_M = 0.5\bar{M}_k\delta_M$ is the amplitude of oscillations; $\delta_M = 2A_M / \bar{M}_k$ is the irregularity degree of the value M_k; $m = 2\pi / T_M$ is the frequency of a periodic oscillation; T_M is the oscillation period of the value M_k; $M_{max} = \bar{M}_k(1 + 0.5\delta_M)$ and $M_{min} = \bar{M}_k(1 - 0.5\delta_M)$ are the maximum and minimum values of the modulus of resistance on the vehicle wheels.

The second method proposes to assess the influence of external actions by the method of expanding a periodic function in a Fourier series, the essence of which is as follows: if the periodic function *f(t)* with the period T_o satisfies the Dirichlet conditions, it can be approximated by the expression:

$$f(t) = A_o + \sum_{k=1}^{\infty} (A_k \cos k\omega_o t + B_k \sin k\omega_o t), \quad (2)$$

where, A_o is the constant term (average value of the function); k is the external harmonic order (k = 0; 1; 2; ...; ∞); A_k and B_k are the coefficients of a trigonometric Fourier series; $\omega_o = 2\pi / T_o$ is the circular frequency of the first harmonic; t is the period of the first harmonic equal to the period of the function.

The coefficients A_o, A_k and B_k are determined from the expressions:

$$\left.\begin{aligned} A_o &= \frac{1}{T_o}\int_{-T_o/2}^{T_o/2} f(t)dt; \\ A_k &= \frac{2}{T_o}\int_{-T_o/2}^{T_o/2} f(t)\cos\omega_o t dt; \\ B_k &= \frac{2}{T_o}\int_{-T_o/2}^{T_o/2} f(t)\sin\omega_o t dt. \end{aligned}\right\} \quad (3)$$

To achieve the required accuracy and simplify the assessment of the influence of external actions on a heavy duty haul truck using the second method, in many cases it is sufficient to consider the first 10 ... 12 harmonics of the periodic function *f(t)*.

The vehicle operation in the conditions of mountain and high mountain quarries can be assessed by a complex probabilistic-statistical method using analytical dependencies according to the vehicle's deterministic (installation) features.

To determine the vehicle power or the vehicle wheel torque in the operating conditions, we can apply the formula proposed by Professor S I Iofinov [11-14].

$$N^{\mathrm{o}} = N_{\mathrm{o}} \cdot \lambda_{\mathrm{d}} \cdot \lambda_{\mathrm{t}} \cdot \lambda_{\mathrm{v}}, \quad (4)$$

where, N_{o} is the operating vehicle power according to the installation (deterministic) feature; λ_{d} is the dynamic coefficient (proposed by academician V N Boltinsky), which takes into account the decrease in the vehicle power under unsteady loads; λ_{t} is the time coefficient, which takes into account the decrease in the vehicle power caused by wear, adjustments and aging of the vehicle; λ_{v} is the probabilistic coefficient, which takes into account the change in the vehicle power caused by the probabilistic nature of the factors determining the life of heavy duty vehicles in mountain conditions.

3. Results and Discussion

To determine the main probability and statistical estimates of the influence of factors determining the vehicle life in severe road-climatic conditions of operation, it is advisable to use the method of random argument functions, in which the input x and output y variables are determined by the deterministic functional dependence $y_{\mathrm{i}} = f(x)$, which is established during the approximation of the vehicle's typical (installation) feature [15, 16].

Taking into account the probabilistic nature of the factors affecting the operation of the vehicle x_{i}, the output variables y_{i} are random values, the quantitative characteristics of which are generally calculated by the following formulas:

- mean value

$$\bar{y} = \int_{-\infty}^{\infty} f(x)\varphi(x)d(x); \quad (5)$$

- variance

$$D(y) = \int_{-\infty}^{\infty} [f(x) - \bar{y}]^2 \varphi(y)d(x); \quad (6)$$

- mean-square deviation (standard)

$$\sigma_{\mathrm{y}} = \sqrt{D(y)}; \quad (7)$$

- variation coefficient

$$\upsilon_{\mathrm{y}} = \sigma_{\mathrm{y}} / \bar{y}. \quad (8)$$

Notably, the low-frequency components of operational and road-climatic factors change according to the normal distribution law $y = f(x)$, which probability density function is calculated from the expression

$$\phi(x) = (\sigma_{\mathrm{x}}\sqrt{2\pi})^{-1} \exp[-\frac{(x-\bar{x})^2}{2\sigma_{\mathrm{x}}^2}], \quad (9)$$

where, $\bar{x}$, σ_{x} and σ_{x}^2 are the mean value, standard, and variance of the random value x, respectively.

Thus, to obtain basic estimates of the reliability and durability of heavy duty vehicles taking into account the probability and statistical nature of the factors affecting their reliability, we should establish the laws of distribution of the input *φ(x)* and output *φ(y)* variables, as well as the connecting functions *f(x)* included in expressions (5) ... (9).

4. Conclusions

1. The factors related to operational and road climatic groups and determining the life of heavy duty vehicles in the conditions of mountain quarries are almost never manifested separately, and in most cases there is a complex combination of many factors. To this end, to determine the life of heavy duty vehicles in severe operational and road-climatic mountain conditions, it is necessary to use the system approach to assess the joint action of the factors.

2. To simulate the influence of a complex combination of many factors on the reliability of the vehicle as a dominant subsystem in the DVRE system, we can use the torque on the vehicle axles (vehicle power in serviceable conditions) or the vehicle's tractive force as a combined indicator.

3. The vehicle power in serviceable conditions connected with the probabilistic nature of the factors determining the mileage changes in inverse proportion to the growth of the variation coefficient.

4. The power loss connected with an increase in the probabilistic nature of the factors is largely spent to reduce the reliability of the DVRE system in general, and to reduce the reliability of the vehicle and the road in particular.

References

[1] Umirzokov A M, Mambetalin K T, Saydullozoda S S, Saibov A A, Abaev A Kh and Berdiev A L 2019 *Classification of the DVRE system.* [in Russian-Klassifikatsiya sistemy VADS] (Polytechnic Bulletin. Series "Engineering Research) **2019** 1 (45)

[2] Umirzokov A M, Saibov A A, Mazhitov B Zh, Berdiev A A and Tursunov F A 2016 *Evaluation of the efficiency of operating vehicles in mountain conditions of the Republic of Tajikistan.* [in Russian - Otsenka ehffektivnosti ehkspluatatsii avtomobilej v usloviyakh vysokogor'ya Respubliki Tadzhikistan] (Vehicles proceedings of the XVIIIth International Scientific and Practical) Conference of Nov p 110 – 117

[3] Umirzokov A M, Mambetalin K T, Saydullozoda S S and Saibov A A 2019 *A conceptual model for evaluating the effectiveness of the "Driver - Vehicle - Road – Environment" system.* [in Russian- Kontseptual'naya model' otsenki ehffektivnosti sistemy «Voditel–Avtomobil–Doroga–Sreda»] Bulletin of SUSU. Series "Machine Building". **2019** DOI: 10.14529/engin190104

[4] Boltinsky V N 1962 Theory, design and calculation of tractor and automobile engines (Moscow: Selkhozgiz) p 392

[5] Korn G and Korn T 1977 Mathematical Handbook for Scientists and Engineers (Moscow: Nauka Publishing house) p 832

[6] Ageev L Ye and Bakhriev S Kh 1991 Operation of energy-saturated tractors (Moscow: Agropromizdat) p 271

[7] Kolmogorov A N 1986 Probability theory and mathematical statistics (Moscow: Nauka) p 535

[8] Bazhenov Yu V 2006 Basics of the theory of machine reliability: textbook (Vladimir: Publishing house of Vladimir State University) p 157

[9] Boldin A P and Sarbaev V I 2010 Reliability and technical diagnostics of road transport vehicles: textbook. Moscow: Publishing House of the Moscow Aviation Institute p 206

[10] Krutov V I, Glushko I M, Popov V V, Savelyev A Ya, Sumarokov L N, Venikov V A, Kogdov N M, Timofeeva O V, Chus A Vand Momot A I 1989 Fundamentals of scientific research (Moscow: Vysshaya Shkola) p 400

[11] Tursunov A A and Umirzokov A M 2009 *The influence of the metrological conditions of a mountain region on the efficiency of the ATS operation.* [in Russian -Vliyanie metrologicheskikh uslovij gornogo regiona na ehffektivnost' raboty] Scientific Bulletin of

Moscow State Technical University of Civil Aviation (MSTU of CA). **2009** 147(10)

[12] Tursunov A A, Abdulloev M A, Umirzokov A M and Majidov B J 2010 *Evaluation of the influence of the mountain environment parameters on the energy indicators of vehicles' power plants.* [in Russian - Otsenka vliyaniya parametrov gornoj sredy na ehnergeticheskie pokazateli ehnergoustanovok transportnykh mashin] Proceedings of the International Scientific and Technical Conference. – Tyumen. **2010** p 330-334

[13] Rotenberg R V 1986 *Basics of reliability of the driver-vehicle-road-environment system* (Moscow: Transport) p 215

[14] Karunin A L, Buznik E N and Dashchenko O A 2005 *Automotive manufacturing technology*: university textbook Moscow p 624

[15] Yahyaev N Ya and Korablin A V 2009 *Basics of reliability theory and diagnostics*: university textbook (Moscow: Academy Publishing Center) p 256

[16] Shashkin V V and Karzov G P 1992 Reference book. *Reliability in machine building.* Politekhnika p 719

IOP Conf. Series: Materials Science and Engineering **817** (2020) 012037 doi:10.1088/1757-899X/817/1/012037

Logistic approach to a company's performance assessment based on a KPI system

D Voronova[1*], **L Berezhnaya**[2]

[1]Department of Marketing and Logistics, *St. Petersburg State University of Industrial Technologies and Design,* 4 Ivana Chernykh street, St. Petersburg 198095, Russian Federation
[2] Department of Management, *Orenburg State University*, 13 Victory Avenue, Orenburg 460018, Russian Federation

*Corresponding email: djusha_k@list.ru

Abstract. The aim of the article is to develop guidelines for the use of the logistic approach to evaluation of a company's performance based on the Key Performance Indicators (KPI) system. The literature review presents the results of scientific research of Russian and foreign scientists who actively use the KPI system to evaluate the logistic activities of companies. As part of the study, the concept of a logistic approach to evaluating a company's activities based on KPIs is presented. The algorithm of development and implementation of the KPI system based on the logistic approach is presented and described. As the key performance indicators the following logistic criteria have been used: Right product, Right consumer, Right quantity, Right quality, Right place, Right time, Right costs, based on which the KPIs have been developed. In the end, conclusions have been drawn and further directions of research presented within the framework of this scientific topic.

1. Introduction

The issues of using management tools based on the logistic approach that allows companies to gain competitive advantage in the market by reducing costs and increasing resistance to market fluctuations, as well as to enter a new qualitative level of development, are always relevant. An important aspect is the availability to managers of complete and reliable information about the results of a company's activities. Obviously, this requires an appropriate tool to conduct a timely and reliable assessment, to identify shortcomings and deviations of the obtained results from the planned ones and to take an adequate managerial decision to eliminate problems that impede the achievement of the goals of the company. One such tool may be a specially designed KPI system.

In the economic literature, various systems of this type are widely represented, which are modern concepts of strategic management based on a targeted approach. The most effective and popular among companies from various spheres of activity is the balanced scorecard (BSC) system, developed by Kaplan and Norton (1996 [1], 2001 [2]); the system uses four key aspects to determine a company's effectiveness: (1) the perspective of "training and growth", (2) the perspective of internal business processes, (3) the perspective of customers, and (4) the financial perspective. This concept allows implementation of a company's strategy through its decomposition to the level of operational management and control based on KPIs. BSC is designed to strike a balance between financial and non-financial performance indicators, between the interests of internal and external stakeholders, as

IOP Conf. Series: Materials Science and Engineering **817** (2020) 012037 doi:10.1088/1757-899X/817/1/012037

well as between presenting past results and forecasting future ones. It allows a company to directly extract long-term (strategic) goals from the overall strategy and associate them with short-term (operational) goals [3].

It is worth noting that recent economic trends and company development processes indicate the need to use logistic criteria to evaluate companies' activities, since the management of a modern company has already stepped over the scope of a single company and is associated with the organization of joint work of all participants in the formed supply chains.

In scientific works of foreign and Russian scientists, theoretical and methodological provisions on the application of the logistic approach to the development of a KPI system as an instrument for assessing and making optimal managerial decisions are insufficiently presented. The use of only classic success factors - "finance", "customers and marketing", "business processes", "personnel", limits companies in applying the concept of logistics and supply chain management to assessing their performance. The use of logistic criteria in the formation of a KPI system also has not received the necessary coverage in the works of scientists.

In general, both foreign and Russian researchers are involved in the application of BSC to the process of carrying out logistic activities. Foreign authors, as a rule, more widely apply this approach in their work. So, the features of using BSC were considered, for example, for organizing humanitarian supply chains [4], developing a multi-criteria approach to selecting suppliers [5], evaluating the effectiveness of marketing channels and distribution of products [6]. The possibilities of using BSC are so wide that they can be used not only for a particular company, but also for new forms of cooperation due to the rapid development of information and communication technologies: virtual enterprises, network companies and professional virtual communities [7].

Russian scientists mainly concentrate their attention on the implementation of the logistics activities of individual companies, by elaborating in more detail the methodological aspects of applying BSC. Tulikova *et al.* [8] proposed a methodological approach to improving logistics systems based on the use of BSC. The fundamental structure of the system for the formation and evaluation of key indicators of the logistics management of an industrial enterprise was studied by Greyz *et al.* [9]. Saplina and Kurlykova [10] proposed a matrix approach to the analysis of four projections in a balanced scorecard. However, as noted by Kuzheva [11], BSC itself will not provide a qualitative leap in logistics management and improvement of performance: for this it is necessary to translate the strategic goals of the company into the operational goals of departments and employees and evaluate the results of their activities in the context of implementation of the strategy with the help of key performance indicators.

2. Methods and Materials

To develop a company's strategy and evaluate its management effectiveness, a set of defined KPIs is needed that measure the progress towards the achievement of the company's goal. Based on the principles of the logistic approach, the achievement of goals is determined by the fulfillment of seven conditions, the so-called "seven rules of logistics": the required products (services), in the required quantity, of the required quality, at the right time, at the right place, with the lowest costs, to the pre-defined consumer.

Based on the abovementioned, we offer the following sequence of development of a KPI system for assessing the effectiveness of a company based on a logistic approach presented in figure 1.

As shown in figure 1, in accordance with a company's strategy based on the logistics concept, areas of activity that are significant for the enterprise are identified, for which a set of KPIs is established that are subject to a regular measurement based on data.

IOP Conf. Series: Materials Science and Engineering **817** (2020) 012037 doi:10.1088/1757-899X/817/1/012037

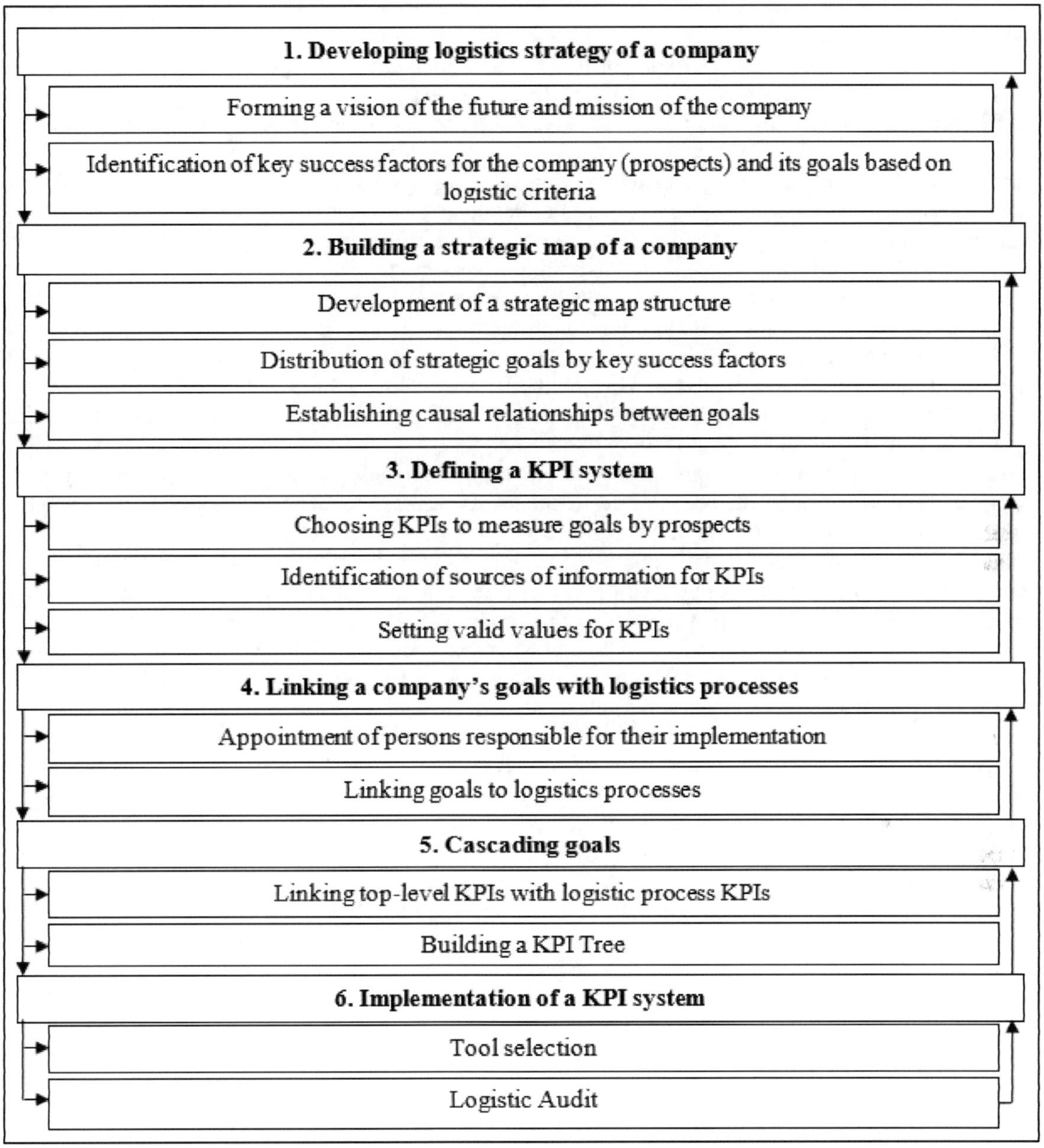

Figure 1. Logistic approach to the development and implementation of a KPI system.

Stage 1. Long-term directions of a company's development are identified on the basis of the principles of logistics and supply chain management, which are based on cross-functional and inter-organizational coordination and integration, and strategic goals are identified. An essential condition for achieving the goals of a company will be their formalization, assessment and adjustment. To do this, it is necessary to highlight the key factors for the success of a company (prospects). Application of a logistic approach to a company's management determines the use of logistic criteria as such factors - the right product or service (Right product), the right consumer (Right consumer), the right amount (Right quantity), the right quality (Right quality), to the right place (Right place), at the right time (Right time), with the right costs (Right costs).

Stage 2. A strategic map is designed, which is a graphical description of the key success factors (projections) that determine their strategic goals and cause-effect relationships between them.

Stage 3. The numerical characteristics of the selected goals are determined - KPIs, allowing them to be measured. KPIs selection is subject to special requirements.

After the selection of KPIs, the next important task is to determine the sources of information for their assessment, the difficulty of which is to develop and implement a mechanism for searching and collecting the necessary and adequate information on a regular basis, which is practically absent in Russian companies. Basically, they use forms to collect data used to assess financial performance, and a limited amount of data to determine the time of loading, unloading, transportation, the degree of loading of vehicles, and the distance from the point of departure to destination.

However, to obtain a more complete and clear picture of the functioning of a company and the effectiveness of its logistics activities, only quantitative data are not enough. It is also necessary to take into account the qualitative components of the assessment: perceptions, feelings, opinions, responsibility and trust, using the following methods: observation, interview, and questionnaire. At this stage, acceptable values for KPIs are also established, so that during the evaluation to further compare the actual values of indicators with them.

Stage 4. The goals of a company will be achieved if responsible persons are appointed for their implementation. This can be the heads of departments, as well as any of their employees who need to be motivated to achieve them through the same KPIs.

The binding of goals to the logistic processes is carried out by distribution over projections (logistic criteria). In the future, the structure of the KPI system may be periodically reviewed. In this case, it is necessary to adhere to the principles of effective interaction of supply chain participants [12]:

- documenting cause-effect relations. The company performance evaluation system is designed in such a way that the relationships between factors, goals and indicators are clear. Each parameter included in a KPI system becomes an element of the chain of cause and effect relationships with which the company receives information about the strategy of a business unit;
- the relationship with financial results (costs). KPIs are aligned with financial goals through a chain of cause-effect relationships. This allows us to more effectively manage the movement of financial flows, taking into account the implementation of the overall strategy of the company and the implementation of individual business processes;
- factors for achieving results. The assessment system based on the logistic approach is a complex of KPIs and activity factors, ensuring a balance of results and factors for their achievement.

Stage 5. To conduct an objective assessment of the effectiveness of a company, as well as to develop adequate solutions on its basis, the integration of many KPIs into one or more global systemic indicators is carried out. As a result, there is a "linking" of the KPIs of the upper level with the KPIs of the process level, forming a "KPI Tree".

Stage 6. The use of information systems that allow monitoring and optimization of logistics processes can help introduce a KPI system based on a logistic approach in a company. The information system supports the rapid exchange of data with performers, intermediaries, suppliers and customers. That will allow us to quickly carry out a logistics audit and make appropriate management decisions based on it.

3. Results and Discussion

In accordance with the above algorithm, a strategic map was developed to evaluate a company's activities, the components of which are formed on the basis of logistic criteria (figure 2). This fundamentally distinguishes it from the classic interpretation of the BSC of Norton-Kaplan.

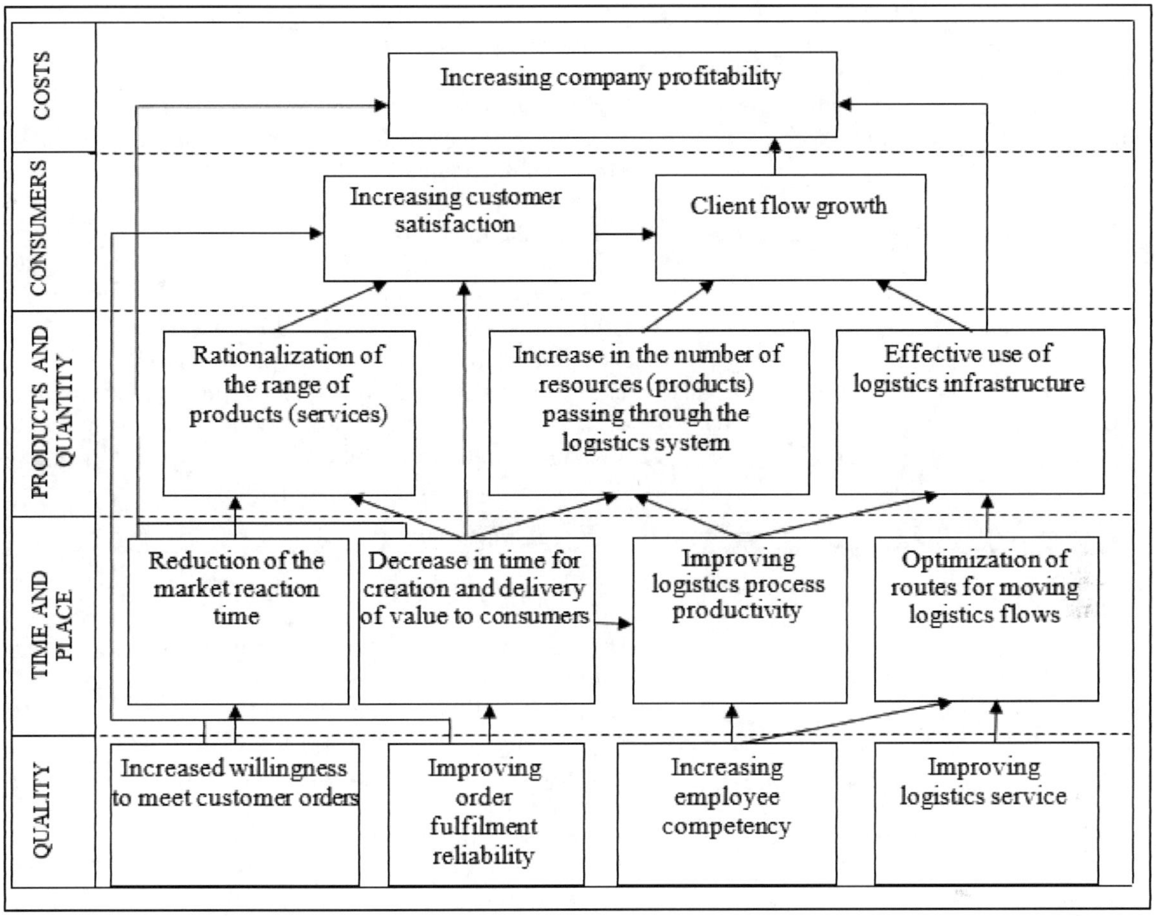

Figure 2. Logistic approach to designing a strategic map of a company.

To measure the goals presented in figure 2 using a special technique considered in the authors' works earlier[13, 14], KPIs were selected and distributed according to the projections as follows (table 1).

It should be noted that the presented KPI system is regularly reviewed in accordance with customer requirements and the company's ability to satisfy them.

Thus, we can conclude that the application of a logistic approach to evaluating a company's activities based on KPIs will allow the management to analyse the status and prospects of current activities, taking into account supply chain development trends, quickly bridge the gaps between the existing strategy and its implementation, and also respond in time to emerging issues.

Table 1. KPI system for a company's assessment based on a logistic approach.

Purpose	KPI
KPI projections "Costs"	
Increasing company profitability from logistics activities	Unit costs for logistics, p. / t (kg, l)
	Losses during delivery, %
	Profitability of logistics activities (general and by function), %
KPI projection "Consumers"	
Increasing customer satisfaction	Customer Satisfaction Level,%
	Customer Loyalty Index, %
Client flow growth	The proportion of repeated applications, %

IOP Conf. Series: Materials Science and Engineering **817** (2020) 012037 doi:10.1088/1757-899X/817/1/012037

Purpose	KPI
	Relative Market Share
KPI projection "Products and quantity"	
Providing the consumer with the necessary resources, products, services	Correspondence of the parameters of the delivered resource (product) to the requirements of a customer,%
	The breadth and depth of the range of products (services)
Effective use of logistics infrastructure	Vehicle idle rate
	Fleet update rate
	Warehouse load factor
	Utilization rate of material handling equipment
	Dynamic utilization of payload per revolution
Increase in the number of resources (products) passing through the logistics system	The volume of delivered resources (products), t (kg, l)
	Fulfillment of logistic plans (for providing resources, for transportation, for sales),%
	Warehouse turnover, t (kg, l) / period
	Warehousing, p.
KPI projection "Time and place"	
Decrease in time for creation and delivery of value to consumers	Coefficients of uniformity and rhythm of order fulfillment
	The coefficient of timely delivery of resources (products)
Improving logistics service Productivity	Labor productivity, t · km / person
	Productivity per one transport-ton-hour of work, t
	The turnover rate of material resources (products), days
Optimization of material flow routes	Route Optimization Factor
	Reliability of adhering to the on-site supply schedule (share of resources (products) delivered to the right place),%
KPI projection "Quality"	
Improving logistics service	The level of quality of service,%
Increasing employee competency	Employee Competency Index
Improving order fulfillment reliability	The coefficient of unprocessed applications
	Downtime of the enterprise in connection with the shortage of necessary resources, pcs.
	The share of delivered goods of the required quality,%
	The share of delivery errors (deliveries with defects, deliveries with erroneously sent products, incomplete deliveries, etc.),%
	The coefficient of discipline of storage (errors in the distribution of goods for storage)
Increased willingness to meet customer orders	Operation Flexibility Ratio
	The coefficient of readiness for the execution of orders
	Vehicle Technical Availability

4. Conclusions

As a result of the study, the following conclusions have been formulated:

- the developed algorithm for the formation of a KPI system for assessing the effectiveness of a company on the basis of the logistic approach is a logical sequence of steps related to the development of the company's logistics strategy, building a strategic map and defining a KPI system, linking the company's goals with the logistics processes, cascading goals and introducing this system into the system monitoring, planning and controlling the company;

- the presented KPI system, which includes six main blocks of strategic goals, distributed according to the criteria of "costs", "consumers", "product and quantity", "time and place" and "quality", and measuring their indicators, will make it possible to comprehensively evaluate the activities of companies focused on logistic management principles, and timely develop and implement adequate management decisions.

Thus, the KPI system, developed on the basis of the logistic approach, is one of the main tools ensuring the stability and prosperity of companies, entering new markets and constantly growing customers, increasing profits even under adverse economic conditions. It is a necessary component of modern strategic management, the effectiveness of which is confirmed by the experience of successful international companies. Further areas of the study of the logistic approach to assessing the activities of companies on the basis of a balanced scorecard are related to the testing of the developed methodological provisions on the example of specific firms in order to verify the presented materials from the point of view of their practical significance.

References

[1] Kaplan R and Norton D 1996 *The balanced scorecard. Translating strategy into action* (Boston: Harvard Business School Press)

[2] Kaplan R and Norton D 2001 *The strategy-focused organization. How balanced scorecard companies thrive in the new business environment* (Boston: Harvard Business School Press)

[3] Van Looy A and Shafagatova A 2016 Business process performance measurement: a structured literature review of indicators, measures and metrics. *SpringerPlus* **5** 1797

[4] de Leeuw S 2010 Towards a Reference Mission Map for Performance Measurement in Humanitarian Supply Chains *Collaborative Networks for a Sustainable World PRO-VE 2010 (IFIP Advances in Information and Communication Technology* vol 336) ed L M Camarinha-Matos, X Boucher and H Afsarmanesh (Berlin, Heidelberg: Springer)

[5] Verdecho M and Alfaro-Saiz J and Rodríguez-Rodríguez R 2010 A Multi-criteria Approach to Select Suppliers Based on Performance *Balanced Automation Systems for Future Manufacturing Networks BASYS 2010 (IFIP Advances in Information and Communication Technology* vol 322) ed A Ortiz, R D Franco and P G Gasquet PG (Berlin, Heidelberg: Springer)

[6] Bazett M, Bowden I, Love J, Street R and Wilson H 2005 Papers: Measuring multichannel effectiveness using the balanced scorecard. *J Direct Data Digit Mark Pract* **6** pp 224–231

[7] Chituc C and Azevedo A 2005 Multi-Perspective Challenges on Collaborative Networks Business Environments *Collaborative Networks and Their Breeding Environments PRO-VE 2005 (IFIP - The International Federation for Information Processing* vol 186) ed L M Camarinha-Matos, H Afsarmanesh and A Ortiz (Boston, MA: Springer) pp 25–32

[8] Tulikova O, Tatarnikov D and Bykova A 2009 Methodological approach to improving logistics systems through the use of a balanced scorecard [in Russian – Metodologicheskiy podkhod k sovershenstvovaniyu logisticheskikh sistem s ispol'zovaniyem sbalansirovannoy sistemy pokazateley] *Vologda Readings* **74** pp 133–135

[9] Greyz G, Katochkov V and Volojanin V 2014 The basic structure of the system for the formation and evaluation of key indicators of the logistics management of an industrial enterprise [in Russian – Bazovaya struktura sistemy formirovaniya i otsenki klyuchevykh pokazateley upravleniya logistikoy promyshlennogo predpriyatiya] *Bulletin of the South Ural State University: Economics and Management* **8** pp 128–135

[10] Saplina T and Kurlykova A 2012 Formation of a system of balanced indicators of the logistics activity of an enterprise [in Russian – Formirovaniye sistemy sbalansirovannykh pokazateley logisticheskoy deyatel'nosti predpriyatiya] *Vestnik of Orenburg State University* **13** pp 314–317

[11] Kuzheva S 2014 Using a balanced scorecard in logistics management [in Russian – Ispol'zovaniye sbalansirovannoy sistemy pokazateley v upravlenii logistikoy] *Bulletin of the Omsk University, series "Economics"* **4** pp 88–94
[12] Berezhnaya L 2015 Principles of effective interaction of participants in supply chains [in Russian – Printsipy effektivnogo vzaimodeystviya uchastnikov v tsepochke postavok] *Risk: resources, information, supply, competition* **4** pp 60–64
[13] Viskova D, Kutsenko E and Lavrenko E 2016 *Management of transport and storage facilities* (Orenburg: Orenburg State University)
[14] Voronova D 2016 Formation of logistics solutions in the management of transport and storage facilities [in Russian – Formirovaniye logisticheskikh resheniy v upravlenii transportno-skladskimi ob"yektami] *Intellect. Innovations. Investments* **5** pp 18–24

AUTHOR INDEX